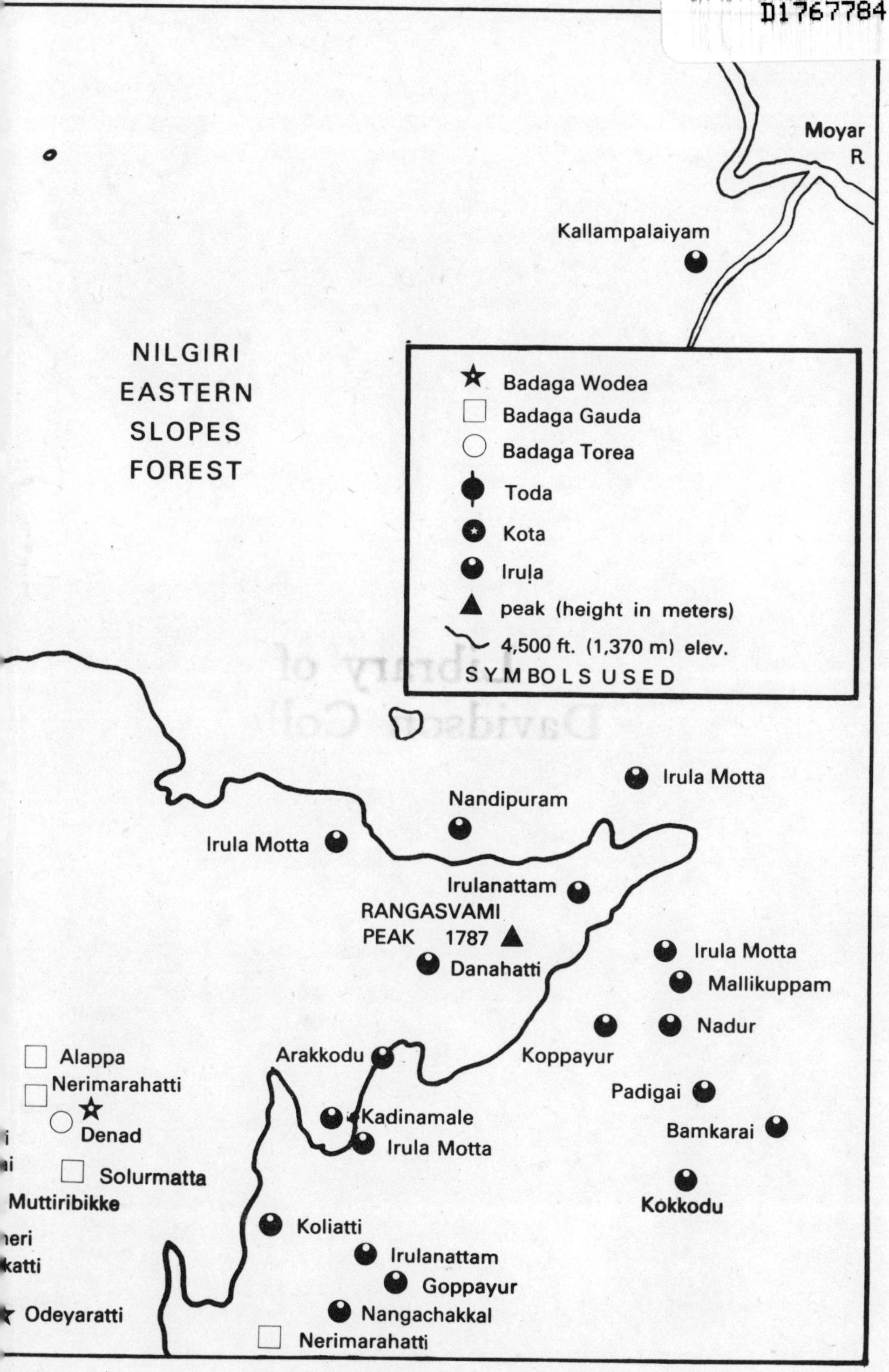
Moyar
R
Kallampalaiyam
NILGIRI
EASTERN
SLOPES
FOREST
Badaga Wodea
Badaga Gauda
Badaga Torea
Toda
Kota
Iruḷa
peak (height in meters)
4,500 ft. (1,370 m) elev.
SYMBOLS USED
Irula Motta
Nandipuram
Irula Motta
Irulanattam
RANGASVAMI
PEAK 1787
Irula Motta
Danahatti
Mallikuppam
Nadur
Alappa
Arakkodu
Koppayur
Nerimarahatti
Padigai
Denad
Kadinamale
Bamkarai
Irula Motta
Solurmatta
Muttiribikke
Kokkodu
Koliatti
Irulanattam
Goppayur
Odeyaratti
Nangachakkal
Nerimarahatti

BLUE MOUNTAINS

BLUE MOUNTAINS

The Ethnography and Biogeography of a South Indian Region

edited by
Paul Hockings

Delhi
OXFORD UNIVERSITY PRESS
Oxford New York
1989

Oxford University Press, Walton Street, Oxford OX2 6DP
New York Toronto
Delhi Bombay Calcutta Madras Karachi
Petaling Jaya Singapore Hong Kong Tokyo
Nairobi Dar es Salaam
Melbourne Auckland
and associates in
Berlin Ibadan

SBN 0 19 562177 8

Phototypeset by CBT (Indraprastha Press), New Delhi 110002
Printed by Rekha Printers Pvt. Ltd., New Delhi 110020
and published by S. K. Mookerjee, Oxford University Press
YMCA Library Building, Jai Singh Road, New Delhi 110001

Preface

The Nilgiris District of South India is a hilly area of 982 sq. miles, bounded on the west by Kerala, on the north by Karnataka, and on the southeast by Coimbatore District. It therefore occupies the highest and westernmost parts of Tamil Nadu State. For administrative purposes the district is divided into three *taluks*, as follows: Gudalur (in the northwest) 289 sq. miles; Ootacamund (central) 462 sq. miles; Coonoor (to the east) 241 sq. miles. (Coonoor was very recently subdivided into Coonoor and Kotagiri *taluks*.) These areas have been stable for over a century, fortuitous circumstances for the person interested in comparing census returns over that period. Grigg (1880: 11, n.2) gives the total area a century ago as 988 sq. miles, but 239 sq. miles had been annexed from an adjoining district in 1877.

The entire district is hilly, a much-worn massif of gneiss and granite, and is divisible into two natural regions (1) The Nilgiri Plateau, 35 miles from east to west and 20 from north to south; a high tableland, deeply indented, with an average elevation of 6500 ft. (2) The southeast Wynaad, a tableland at about 3000 ft. covered with bamboo forests, paddy flats and peat bogs.

The Nilgiri Plateau rises sharply from the surrounding country, and is divided by a range of peaks running in a general north-south direction. The highest of these, lying somewhat to the east, is Doddabetta. This range has the effect of sheltering the eastern part of the plateau from the southwest monsoon, and the western part from the northeast monsoon. The climate on these two parts of the Nilgiri Plateau thus tends to show marked differences.

The district is cut by numerous streams, all draining either into the Moyar, which flows eastwards through a deep gully on the northern border of the district, or into the Bhavani, which flows along the southern border. Both the Bhavani and the Moyar eventually feed into the Kaveri.

The whole Plateau rises abruptly from the plains, its sides being sheer rocky walls on the west. The interior of the plateau consists mainly of undulating grassy hills divided by narrow valleys, each one containing a stream or swamp surrounded by shola thickets. On the higher slopes the soils are shallow but on the lower land there are deep rich loams and some yellow ochrous clays and black peaty soils. Much of the land will not respond easily to fertilizer, and consequently is suitable only for grass and timber. However, a mixture of red and black soils with small stones on the low level land is well suited to the cultivation of crops. Although the higher slopes are not suited to crops, they are in great demand for tea plantations. Wind is important in inhibiting forest growth on these higher parts of the mountains.

The vegetation of the valleys is typically a dense and rather low (i.e. 50–60 ft.) forest with much undergrowth and many epiphytes, mosses and ferns. Both tropical and temperate flora occur, including magnolia, rhododendron and laurel. Occasional peat bogs break the character of an otherwise rich parkland. Today much of the district is forested.

Botanically, as well as zoologically and ethnologically, the Nilgiris forms a distinct ecological realm of its own, but with affinities to the flora of Assam. Fast-growing Australian eucalyptus and acacias have been introduced throughout the district to supply the hill-stations with firewood, and control erosion, as well as to supply the pulpwood, tannin and rayon industries.

There are no irrigation works in Ootacamund Taluk, as the country is hilly, and the population there depends on imported grains from the lowlands. In Coonoor Taluk cultivation relies on rain, apart from a low area which is irrigated by the Moyar. Famine is unusual in that *taluk*. In Gudalur Taluk it would also be difficult to harness irrigation waters, again because of the terrain; but there are patches of paddy flats depending on the rain. No famine has been known in this *taluk* either.

The economy of the district is most atypical for India. Of over 100,000 acres under cultivation, only 20,000 are devoted to foodgrains, including some 5000 acres of rice. Much of the economy revolves around the hill-station of Ootacamund and its two satellites, Coonoor and Wellington. Local industry therefore stresses the usual functions of a resort. But over the past fifty years another equally important utilization of natural resources has led to rather a different pattern of industrialization. The Nilgiris have offered great potentialities for the development of hydro-electric power. Indeed, the government cordite factory at Wellington had one of the India's earliest hydro-electric power installations; but this was overshadowed in 1932 by the opening of the Pykara Falls dam.

Industrialization has continued since then with the development of many more dams and power plants, so that the district today is a major exporter of electricity throughout southern India. Several factories have recently opened, and more of these developments can be expected in the future as a partial answer to the needs of the rapidly expanding Nilgiri population, which today stands at over half a million.

The above paragraphs offer a brief sketch of the environment in which these people find themselves. A detailed analysis of that environment and of at least some of the distinctive ethnic groups here forms the subject matter of the present book.

This book grew out of the realization that there are scattered around the world a dozen scholars who have done serious and original research on the anthropology and geography of the Nilgiris; and that taken together the knowledge of these few people was encyclopaedic in its scope. A review of all available literature on the Nilgiris (Hockings 1978) led to the awareness of an unexpected paradox. The total of some 3,000 books and articles yields a density of over three publications per square mile—one could almost literally paper the district with them—and shows this to have been perhaps the most intensively studied part of rural Asia east of the Holy Land. And yet, in all this welter of material, there has been no general survey of the geography and anthropology since the district gazetter was published eighty years ago (Francis 1908). In view of the great advances in our knowledge since that time, and the vast politico-economic changes which have occurred, it seemed to the editor and his colleagues that it was high time to attempt a new synthesis.

The result is a book that covers most of the diverse ethnic groups in the district and includes information which is relatively new. Frank E. Poirier's chapter is the first published survey of primate ethology for the region. Anthony R. Walker's account of the Todas is the first to point out that polyandry is a thing of the past and that their economy is no longer dependent on buffalo pastoralism, but primarily on land and cash. Dieter B. Kapp's data on the 'Kurumbas' makes it clear that this is not the designation of one Nilgiri tribe, as had previously been supposed, but seven discrete ethnic groups in the district. Murray B. Emeneau, the dean of American scholars of Dravidian, has written the first overview of the Nilgiri languages. Similarly, William A. Noble has produced the first modern overview of Nilgiri prehistory. Ted Adams has written the first sociolinguistic study of Nilgiri town. And each of the other papers in this book brings to the reader's attention original and, we hope, interesting observations about this most diverse small region.

We must record with regret the passing of three valued friends and colleagues during the preparation of this book. H.R.H. Prince Peter of Greece and Denmark died in October of 1980. He was a prominent ethnographer of the Toda tribe, who began work with them in 1939, incorporating his findings in numerous articles and the well-known book, 'A Study of Polyandry' (Peter 1963). Dr D. B. Sastry, long associated with the Anthropological Survey of India, died in June of 1980. We had hoped that his unequalled knowledge of the biological anthropology of Nilgiri communities would be drawn together in a chapter for this book, but it was not to be. David G. Mandelbaum is also no more. He died in 1987.

—Paul Hockings

Contents

Preface—Paul Hockings v

List of Figures xi

List of Tables xii

List of Plates xiii

1. *The Nilgiris as a Region*—David G. Mandelbaum 1
2. *The Nilgiri Environment*—Hans J. von Lengerke and François Blasco 20
3. *The Non-Human Primates of the Nilgiris*—Frank E. Poirier 79
4. *Nilgiri Prehistoric Remains*—William A. Noble 102
5. *The Languages of the Nilgiris*—M. B. Emeneau 133
6. *The Kotas in their Social Setting*—David G. Mandelbaum 144
7. *Toda Society between Tradition and Modernity*—Anthony R. Walker 186
8. *The Badagas*—Paul Hockings 206
9. *The Kuṟumba Tribes*—Dieter B. Kapp and Paul Hockings 232
10. *An Introduction to the Naikens: The People and the Ethnographic Myth*—Nurit Bird-David 249
11. *The Iruḷas*—A. William Jebadhas and William A. Noble 281
12. *The Muḷḷu Kuṟumbas*—Rajalakshmi Misra 304
13. *Gudalur: A Community at the Crossroads*—T. Adams 318
14. *British Society in the Company, Crown and Congress Eras*: 334
15. *The Cultural Ecology of the Nilgiris District*—Paul Hockings 360

Bibliography: (A) Natural Sciences 377
(B) Human Sciences 384

Biographical Notes 399

Index 403

List of Figures

A Portion of the Northeastern Nilgiri Massif.		end papers
2.1	The Nilgiris: Topographical Outline.	22
2.2	Two Geological NNW-SSE Profiles of the Nilgiris.	23
2.3	Seasonal Changes of Wind Direction and Velocity.	25
2.4	Average Annual Course of Six Atmospheric Elements.	26
2.5	Annual Course of Screen Temperatures at Selected Stations.	27
2.6	Frost at Three Selected Stations.	28
2.7	Average Annual Rainfall.	29
2.8	Average Annual Number of Rainy Days.	30
2.9	Average Monthly Rainfall.	31
2.10A	Average Monsoon Rainfall, December-March.	32
2.10B	Average Monsoon Rainfall, April-May.	33
2.10C	Average Monsoon Rainfall, June-September.	34
2.10D	Average Monsoon Rainfall, October-November.	35
2.11	Average SW Monsoon Rainfall.	36
2.12	Bioclimatic Map of the Nilgiris.	37
2.13	Profile of a Shola.	56
2.14	Profile of a Shrub Savanna.	57
3.1	Nilgiris District.	80
4.1	Nilgiris Archaeological Sites.	107
4.2	Selected Megalithic Sites	108
4.3	Further Megalithic Sites.	110
4.4	Iron Artifacts from One Stone Circle.	123
4.5	Carvings on Orthostats.	124
5.1	Stemma for South Dravidian.	134
8.1	The Badaga Week.	216
8.2	The Family of Hette.	226
9.1	The Two Ālu Kuṟumba Phratries and their Clans.	234
10.1	Schematized Map of a Local Community.	264
11.1	Rangaswami Betta Iruḷa Worship Centre.	282
11.2	Koppayur.	288
15.1	Schematic Cross-Section of the Nilgiris.	362
15.2	Nilgiri Ecological Niches.	366

List of Tables

3.1	Vegetational Usage by Nilgiri Langurs.	81
3.2	Faunal Complement of Tract 3.	82
3.3	Comparison of some Major Traits of Tracts 2 and 3.	82
3.4	Comparison of Major Traits of Lion-Tailed and Bonnet Macaques.	90
3.5	Comparison of some Major Traits of Nilgiri and Hanuman Langurs.	96
3.6	Comparison of Colobine Group Structures.	98
8.1	The Proportion of Badagas to Two Other Local Communities.	208
8.2	Marriage Reciprocity.	213
8.3	The Badaga Year.	218
10.1	The Distribution of Naiken Population.	266
11.1	Gathered Products (mostly from Plants).	295
13.1	Population Growth in the Nilgiris, 1901–1981.	321
13.2	Population of Gudalur Town, 1978 (based on the author's census).	325
14.1	The Class System among Nilgiri Europeans, 1850–1950.	345
14.2	Change in the European and Eurasian Population.	351
14.3	Value of rho.	359
15.1	Toda Population over the Past Fifty Years.	364
15.2	Nilgiri Memorial Ceremonies.	367
15.3	Foraging, Hunting, Pastoral and Cultivation Strategies.	372
15.4	Average Daily Food Consumption.	374

List of Plates

(following page 406)

1 The Nilgiris from Space.
2 View from Coonoor Ghat towards Mettuppalaiyam.
3 Patches of Evergreen Montane Forest.
4 Vegetation on Difficult Terrain.
5 In the Kundah Range.
6 Upper Bhavani Reservoir.
7 Tree Fern.
8 Physiognomy of a Shola Tree.
9 Fire in a Shrubbery Savanna.
10 Early Morning Hoar-Frost.
11 The Wenlock Downs.
12 Intensive Agriculture.
13 Features of Megalithic Sites (A-I).
14 Pottery, Terracotta and Bronze Finds (1–38).
15 Carved Orthostats (A–H).
16 Po.*s*: A Traditional Toda Hamlet.
17 Modern Toda Girls in Tamil-style Dress.
18 Toda Woman.
19 Toda Man.
20 A Toda Herdsman become Farmer.
21 Irula Sites (A–H).
22 Irula Cultivation and Worship (A–H).

The Nilgiris as a Region

DAVID G. MANDELBAUM

1

In the south of the Indian peninsula, where the coastal ranges converge, there rises the hilly plateau called the Nilgiris (Blue Mountains) or the Nilgiri Hills. The Nilgiris District, one of the districts of the state of Tamil Nadu, includes the plateau, the jungle-clad slopes of the uplands, and some adjoining lowland tracts. Much has been written about the Nilgiri region; Paul Hockings has commented that it is perhaps the most intensively studied part of rural Asia (1978: 2). This volume brings together studies of salient material and human features of the Nilgiri region. To give some context for understanding these features, we consider first what is meant by a region.

There is a territorial component to all human societies that is a constant, though not always recognized, factor in social relations. It is a universal, underlying influence rather than, as with many animal species, an overriding instinctual command. Human territoriality derives from common residence within a given space, it influences how, how often, and among whom special interchange is carried on.

People everywhere organize themselves into functioning groups that are based on spatial proximity. Propinquity and relative frequency of communication affect much of the behaviour among kith and compatriots whether they are members of a band, a neighbourhood, a tribe, a village, a city or a nation.

The idea of region is often used as a fundamental territorial category, both for the practice of social relations and for the study of them. It broadly means a territory inhabited by socially distinct groups who share some common interests and cultural characteristics. These shared traits may facilitate joint action although, in themselves, they do not guarantee any. Beyond this bare definition, the idea of human region has been used in many differing ways. Saberwal's essay on the subject concludes with the observation that the term region has been applied in 'a protean variety of meanings' (1971: 95).

It is useful to sort out two kinds of meanings. One kind refers to a human region as defined by outside observers, based on their perceptions and taxonomy, using general 'etic' criteria to specify its nature and extent. The other kind is the region as perceived by participants in it, as understood in the insiders' 'emic' perceptions, as engaging their sense of identity and attracting their loyalties. A person normally identifies with several regions, of narrower or wider scope; one's loyalties are made manifest to a closer region in a particular situation, to a broader one in a different context.

Each of these two types of region takes in a different kind of reality, but not totally separate realities. People who share the same terrain and climate, who cope with like environmental conditions, who have similar cultural proclivities, have some common ground for developing a degree of common identity. That degree is enhanced if they see themselves as together confronting a mutual threat, whether of nature or of another set of people. True, spatial propinquity has often enough bred hostility and cultural proximity is no bar to enmity. Yet, on balance, both conditions generally favour the development of social and cultural bonds.

The peoples and the terrain of the Nilgiri Plateau have long attracted interest because of their unusual characteristics. Throughout three principal periods—aboriginal, colonial, national independence—the Nilgiri region has constituted a singular and singularly instructive enclave, a distinctive locale as perceived by observers as well as by its inhabitants. It is clearly an enclave in the sense of having special natural and human characteristics, markedly distinct from those of the surrounding lower lands. Though relatively small in area (12,549 sq. km. in the Nilgiri District) and population (630,169 in 1981), it has been given the broader term *region* by most who have written about it.

The first report by a European, Father Jacomo Finicio's relation of 1603, tells of the remarkable groups who shared the Nilgiri uplands (Rivers 1906: 719-30). There were, and still are, four peoples, Todas, Kotas, Kuṟumbas, and Badagas. Each speaks a different Dravidian language. Together they formed a social system that was similar to, and yet very different from, the caste societies of the surrounding plains. The Todas were and are pastoralists, the Kotas were artisans, musicians, and cultivators. The Kuṟumbas, on the jungle edges of the plateau, were gatherers, hunters, and swidden cultivators. The Kuṟumbas also provided magical services to the other three groups and were much dreaded by them as sorcerers. These tribesmen knew that together they had been the sole inhabitants of the plateau until the Badaga agriculturalists arrived as refugees, beginning in the sixteenth century. The Badagas became the most populous group by far and provider of foodstuffs to the others.

The colonial period began when British officials discovered the plateau and settled there in the 1820s. They quickly transformed the locale into an integral part of the British imperial régime. It became, for six months of every year, the seat of the government of the Presidency (province) of Madras. An important military base was built and extensive plantations established. Missionary stations were opened. Before long, the plateau area became a principal hill station, the main health and recreational resort for Europeans in South India. It was a central place for major components of the British rule, an administrative, military and societal headquarters.

Immigrant Indians, from the surrounding plains, settled in the Nilgiris in numbers. Some were brought up as plantation labourers, others came to work directly for the British or to provide services for them, still others immigrated in search of work or land.

This influx affected the older inhabitants in different ways. The Kuṟumbas in their remote hutments were little touched by it. Todas and Kotas adjusted to it but maintained their traditional culture with relatively little change for many decades. The Badagas flourished. Their population grew rapidly, their agricultural productivity increased, their local influence mounted. A good many Badagas took advantage of the educational opportunities offered by the missionaries and the government. While they steadily maintained their language and identity, more than the other groups they adapted effectively to the new régime.

When British officials demarcated the Nilgiris District as an administrative unit of Madras Presidency, they included adjoining tracts at lower elevations, mostly jungle and sparsely settled. The tribal inhabitants of these tracts, Iruḷas, Mullukurumbas, Naikens and others came tenuously under the jurisdiction of the officials of the Nilgiris District. The village of Gudalur was located on travel routes leading up to the plateau. The chapter by T. Adams tells us how Gudalur emerged as the central town for this part of the district when malaria was controlled there and new settlers flocked in.

Since independence in 1947, changes have come apace. The residents of the Nilgiri uplands have benefited from expanded governmental services, as in health provisions, education, transportation and agricultural assistance. The total population of the district has greatly increased, with all the immediate difficulties and long-term dangers such increase brings. There has been an upsurge in governmental projects, in administrative interventions, in modern party politics, and in the secular pilgrimages of tourism. The region, its inhabitants, its resources continue to be of special interest to others in India and to observers from many parts of the world.

What has given the Nilgiris as a human region its distinctive character throughout is the highly unusual natural environment. Its physical and biotic attributes are summarized in the chapter by von Lengerke and Blasco. The Nilgiri massif thrusts high above the plains, cloaked on its sides by dense jungle growing on steep slopes, surmounted by the hilly plateau of grassland, woodland, savanna. The climate of the plateau is very different from that of the adjacent plains; above all it is much cooler. It is like, in these authors' words, 'a cold tropical island rising above the warm tropical sea of South India'. The plateau, roughly a triangle some 32 by 56 km, supports a diversity of indigenous plants 'found nowhere else in the world', and a large number of endemic animal species as well. For the six species of nonhuman primates, as Poirier relates, it was an 'incredibly rich habitat'. Plants and animals introduced from distant parts of the world have flourished in the equable climate, ample rainfall, and productive soils of the Nilgiris. Each of the peoples of the

Nilgiris, during every period, has made its own particular use of their common, benign environment.

The Aboriginal Enclave

One effect of that environment on all the aboriginal inhabitants was to keep them isolated from regular relations with the people and cultures of the plains below. The isolation was not absolute, but the contacts before the nineteenth century seem, except for the coming of the Badagas, to have been brief and sporadic. The ascent was difficult and dangerous. Lowlanders had to traverse the Mysore Ditch on the northern edges of the massif or the bottomlands around the southern perimeter. The Nilgiri uplands were surrounded by an unbroken thicket belt made formidable by fearsome animals and quick-striking diseases. Once across these obstacles, travellers had to clamber up long precipitous slopes. On reaching the plateau, plains people, as Finicio's account attests, were eager to be gone. The unaccustomed cold struck through their thin clothes and chilled them to grim shivers.

There were some visitors nonetheless. The archaeological evidence summarized in the chapter by Noble shows that, for centuries, mortuary sites were built and stone structures erected, mostly on hilltops. But the evidence of these sites and of the artifacts found in them shows no relation with the known ways of the indigenous inhabitants. Neither their traditional tales, nor the accounts of European observers, nor anything in their culture indicates that they produced bronze utensils, made effigy pots, manufactured weapons, or practised the customs depicted on the sculptured stones. In all likelihood, then, the sites were the work of parties who came up to the plateau to perform mortuary and perhaps other rites. These purposes accomplished, they probably hurried down again.

Soldiers and raiders occasionally put in an appearance. There is an inscription dated AD 1117 which tells of a military presence on the plateau. There are a few remains of fortifications, but no evidence of any lengthy military occupation. There was very little atop the plateau then to attract soldiers or marauders.

Trading parties of hill people, according to tribal tales, ventured to plains markets to obtain cloth and other goods. These forays were apparently brief because the hillmen, for their part, felt uncomfortable in the heat and strangeness of the plains.

There are only bits of indirect evidence about the condition of the Nilgiri tribes before the Badagas became part of their exchange system. The linguistic evidence noted in Emeneau's chapter indicates that the Toda and Kota languages were separated from ancient Tamil about the time of, or earlier than, the first recorded Tamil texts of some 2000 years ago. The botanical evidence given in the chapter by von Lengerke and Blasco indicates that the Nilgiri grasslands are at least 3000 years old. If the Toda practice of maintaining these grasslands by periodic burnings is an ancient one, and was essential for the maintenance of grassland cover, this indicates some three millennia of such practice. These authors conclude, on the basis of botanical data, that permanent tillage was introduced only some three to four hundred years ago, a time period compatible with the presence of Badagas.

After the advent of Badagas, an extraordinary relation among the four groups was established. It was glimpsed by Father Finicio, depicted in some detail in a number of nineteenth century reports, meticulously recorded by Rivers in *The Todas* (1906), and subsequntly described and analysed by several anthropologists. The unusual aspects, particularly of Toda culture, have often been noted in these accounts. Less often, as Anthony Walker properly remarks in his chapter on the Todas, has there been adequate discussion of the ways in which the Nilgiri groups, severally and together, resemble the peoples and cultures of the nearby plains. In these respects, they are like others of South India; indeed in certain fundamental ways they share in the cultural mainstream of the subcontinent.

Their languages, though not mutually intelligible, are all Dravidian, of the same linguistic stock spoken throughout southern India. Their kinship terminologies are all versions of the generic Dravidian type. Their emphasis on pollution and purity, in daily life as well as on great occasions, parallels that upheld in Hinduism generally, both as to what pollutes and what purifies. As elsewhere in India, each group practises its own version of the pollution-purity complex, and all agree that the groups are hierarchically ranked according to the purity or pollution of the traditional practices of each. Thus, the Kotas, whose old practices of eating bovine meat (now abandoned) and of providing music for the other groups (now largely relinquished) consigned them to the lowest rung of the hierarchy, vigorously protested the disdain of the other groups, and proclaimed their own estimable virtues. By contrast, the Todas, as strict vegetarians engrossed with ritual practice, were considered by their neighbours to be a people of purity, worthy of respect.

The relations among the four upland groups, their economic and ritual specializations, their patterns of exchange, were much like those carried on among the *jatis* (castes or subcastes) of a locality throughout village India. These relations were in the characteristic *jajmani* mode by which a family of a specialist *jati* has regular interchange with specific families of a food-producing *jati*. Each specialist family provides goods and services to associated agriculturalist families, and in turn, receives a share of the harvest. There was a continual exchange of products and services among the four groups, though not of spouses. There was no intermarriage among them. The parallels to the caste system characteristic of Indian peoples, as noted in the chapter on Kotas later in this volume, are many.

Despite such similarities, in other respects they differed markedly from caste societies on the plains. There was no legitimation of caste relations in scripture, for they knew none. There were no Brahmins to serve as exemplars; they shared no priestly elite. Their religions were separate; each group worshipped its own gods. There were no explicit legal standards; they had virtually nothing of the apparatus of a state. There was not even marked economic inequality among the groups. The Kuṟumbas probably were not as well nourished as the others, but mainly because of their modes of subsistence.

Further, each of the four upland groups diverged from common South Indian practices in its own ways. The Badagas, removed from caste society by only a few centuries, were most like the villagers of their former homelands. As Paul Hockings notes in his chapter on the Badagas, their religion differed little from that of South Indian villagers generally. They formed a kind of *jati* cluster. That is, a set of groups who see themselves as socially distinct, one from another and as hierarchically ranked. But in the eyes of their neighbours, they are quite alike in culture, are ranked as one, and are often referred to by a single *jati* name (Mandelbaum 1970: 19-22). Because they are seen as one entity by others and because they do share similar attributes, the groups of a cluster sometimes unite in joint effort, especially to achieve higher ranking. The Badaga cluster presents certain uncommon features. Of the eight communities who speak the Badaga language, practise Badaga customs, and identify themselves as Badagas of the Nilgiris, only two—the highest and the lowest in rank among them—are firmly endogamous. Members of the other six may intermarry according to rules that allow for both hypergamy and hypogamy. Hypergamy, under which a woman may marry into a family of somewhat higher rank within the *jati* than that of her natal family, is common throughout India. Hypogamy, under which she may marry into a family of lower rank, is much less common. There are few, if any, *jati* clusters other than the Badagas within which women may be married either higher or lower according to specified circumstances.

The case of the Kotas also presents some unusual sidelights. The Kotas ate the meat of deceased or sacrificed buffaloes, a practice viewed as polluting by their neighbours, as the eating of bovine flesh is generally regarded among Hindus. Kotas customarily provided music for ceremonies; in the Nilgiris, as elsewhere in India, this service, especially the playing of music at funerals, is considered to be a lowly occupation. Members of *jatis* whose tradition includes the eating of degrading food and the providing of demeaning services have generally tried to abandon such practices when they had the wealth and power to do so. In the 1930s some Kotas began to drop these practices and almost all have now done so. But for a very long time, Kotas happily held to these practices, even though there was no interior need for the tainting food and not much external power that could force them to play funeral music. When I was with the Kotas in the 1930s, I learned that most of them, even then, just liked to eat meat and enjoyed playing music. They were independent enough (as Indian artisan *jatis* often are) to indulge these tastes despite the social consequences that they resented.

The Kuṟumba, Naiken, and Iruḷa tribelets, as described in the four chapters by Kapp and Hockings, Bird-David, Misra, Jebadhas and Noble, practised means of subsistence that were varied in mode but sparse in yield—foraging, hunting, trapping, and hoe cultivation of swidden fields. Their population was thin on the ground. Not only was their food supply meagre during much of the year, they were also exposed to mortal dangers from attacking animals, especially tigers, and virulent diseases, especially malaria.

Their cultures were correspondingly not elaborate. They differed one from the other in particulars but all upheld similar principles of ritual, pollution, purity, and social hierarchy. And, though their subsistence economy differed from that on the plains, their social and religious life was basically like that of the low impoverished *jatis* in caste villages on the other side of the jungle. For the most part they had little to do with villagers of the other side. Some Iruḷas did have closer contact when pilgrims came to the shrine served by them. Kuṟumbas occasionally traded with plains villagers, exchanging jungle produce for salt, cloth, and tools.

But three Kuṟumba tribelets, those closest to the plateau proper, were continually involved in the exchange system of the upland peoples. Kuṟumbas were seen both as wielders of magical powers and as guardians against such malevolent forces. A Kuṟumba family would have *jajmani* relations with a settlement of the other groups, as village watchmen for Badagas and as hired protectors against witchcraft for all upland groups. The relationship was not an easy one. The unexpected sight of a Kuṟumba was enough to stir forebodings in men and to strike dread in women and children. Upon the appearance of a Kuṟumba, women and children would scurry into their houses; if afield, scatter in fright. Kuṟumbas were allowed to come into upland settlements only for specific purposes, as participants in certain rituals, as suppliers of jungle products, and as protectors against the magical depredations of other Kuṟumbas. No Kuṟumba was allowed to enter a house, yet no household could do without their protective services, because the fear of Kuṟumba witchcraft was pervasive, powerful, and constantly imminent.

It is not unusual in South Indian villages that men of a low but feared *jati* be employed as watchmen on the principle that one sets a thief to catch a thief. Watchmen are often recruited from *jatis* which have local reputation for criminal tendencies. But, rarely are the fears of such purportedly dangerous people as imperatively strong and so closely riveted upon a single group as they are among Todas, Kotas, and Badagas.

One consequence was that men of the three upland groups banded together from time to time, as recounted in a later chapter to maim or kill suspected Kuṟumba sorcerers. Knocking out a suspect's incisors was one way of halting future misdeeds; he could not then properly enunciate magical spells. Thus detoothed, he was disarmed.

Within each upland society, there was little accusation of malevolent acts within the community, though one could employ a Kuṟumba to wreak evil on a fellow tribesman. What the Kurumbas think of these matters has not yet been thoroughly explored and explained.

The Todas were, and at this writing still are, the most distinctive in appearance, economy, and social organization. As pastoralists, each Toda settlement depended on associated families of cultivators for their principal foods, apart from the buffalo milk and ghee which they consume in prodigious quantities. Their economic production came entirely from their herds, but the buffalo were much more to them than an economic mainstay. As Anthony Walker tells us in his chapter, Toda religion was primarily concerned with elaborate rituals of caring for the buffalo, taking their milk, and preparing ghee from the milk.

The temples are the dairies of the sacred herds, the dairymen in them are the consecrated priests. Buffaloes figure critically in the major ceremonies of the life cycle; in funerals buffalo sacrifice is a pivotal act. At ceremonies, men circle in a simple, stately dance; two men in the circle improvise song-poems according to complex rules and the other dancers chant a refrain (Emeneau 1971b). The poetic structure is replete with formal phrases about buffaloes; the song contents are mainly about the beauty of buffaloes and beauteous relations between Todas and buffaloes. In the settlement of disputes and other jural procedures, decisions are not binding unless there is payment in buffaloes. Important social transactions among Todas, whether transfer of women or ritual transfer of rights, must be accompanied by transfer of buffaloes.

Women can have nothing to do with sacred buffaloes, or with the rituals and temples for their care. They can have little to do with ordinary buffaloes either, so the women take very small part in economic production or religious practice. But Toda women do not therefore feel demeaned or deprived. Quite to the contrary, they believe, as do the men, that the women provide the cohesion that binds the tribesmen together. One reason given for that belief is the practice of polyandry under which several men can share one woman as wife. In earlier generations, this arrangement was fostered by the custom of female infanticide. That was prohibited by the British and has almost disappeared. The incidence of polyandry has also been lessened because housing is now readily available. Separate residence for a newly united couple is now made easy because housing (of a standard and inefficient kind) is abundantly supplied by government agencies. Nonetheless, many Toda men of the current generations have shared a wife at some period in their lives, usually as one of a set of brothers or clan brothers. The experience of such sharing, Todas explain, teaches the men to avert rivalries, to live together harmoniously, to co-operate in all spheres of life.

Co-operation among Todas is more than a frangible ideal of fraternal concord. It is built into the structure of their society, in the pattern of co-operation between members of the two endogamous moieties. In a caste society, endogamy sets social barriers that discourage rather than encourage mutual support between two endogamous groups. But members of each Toda moiety are expected to give close support to members of the other in times of need, and they do so. At a funeral, when all of one moiety are mourners in some degree, the physical tasks involved in the rite, including the preparation of the corpse for cremation, are done by men of the other moiety. Bonds between men of the two divisions are further strengthened, in the Toda view, by the formal arrangement whereby a man and woman of different moiety may have a long-term sexual relationship. The man receives permission for the liaison from the father and husbands of the woman; he gives her certain gifts to symbolize their connection. He can never be the legitimate social father of her children; she can never be his legitimate wife. His relationship with her and with the men of her family is expected to be, and often is, one of friendship, mutual support, freely proffered aid. When one of this pair dies, the survivor has a special honoured role to perform in the funeral ceremony.

This is not to say that life among Todas is unruffled or unstrained. There are quarrels aplenty. The pressures of outside forces have generated tensions among the tribesmen and extremely grave problems have risen from the overuse of alcohol and opium. But quarrels are rarely protracted, grudges seldom held. And when a tribal council is called to adjudicate a dispute among members of one moiety, the council is conducted by men of the other moiety who take the lead in smoothing over animosities between the disputants.

Todas enjoy their way of life; they say so and act so. They have taken great pride in it, and are still sure that they are a superior kind of people. Their neighbours and rulers have long admired their qualities, but were also exasperated by them. Badagas supported them economically for centuries, though they had no temporal need to do so and received very small returns. British administrators expressed their impatience with Todas in reams of reports, complaining about their refusal to become revenue-producing cultivators, deploring their sexual customs, disapproving of their indifference to official orders. Yet Todas were generally favoured by British officials. Indian officials have repeated similar expressions of annoyance and have also continued a flow of special benefits to Todas through modern welfare agencies. As they did earlier, many Todas have managed to rid themselves quickly of most incoming monies and have succeeded in remaining true to their ancient character.

The Toda case presents some extreme variants of general Indian and South Indian practices. The deep Hindu reverence for cows is exalted among Todas so that buffalo cows (bulls are of little account) are the central concern of their economy, society, polity, art, and religion. Great as is orthodox Hindu reverence for cows, it is surpassed in the Toda absorption with their buffalo cows.

In kinship relations, Todas use the basic Dravidian terminology; yet they accommodate it to polyandry, a practice repugnant to most other Dravidian speaking groups. We have noted how endogamy, which usually serves to separate two otherwise close *jatis,* is directed by Todas to help enhance tribal solidarity.

Although the Todas are numerically one of the very smallest of the world's societies, their case has sometimes been cited in discussions of panhuman characteristics. Toda polyandry, for one example, is among the best known instances of this relatively rare form of marriage. It has been cited to demonstrate that male sexual possessiveness and jealousy, so common among men elsewhere, need not be taken as a universal characteristic of mankind. Todas themselves are well aware of their peculiarity in this respect. When men or women have explained to me Toda attitudes about sex, they have generally dilated on the reason and advantage that lies in their position.

The coherence of Toda culture is another noteworthy feature. To constitute a society at all, every set of people must act together in many ordered ways, sharing commonly accepted priorities. Few, if any societies are as tightly ordered around and pertinaciously focused upon a single element as is Toda culture on the buffalo.

Toda relations with their indigenous neighbours offer an instructive example concerning the transfer of cultural elements. The four groups lived in close proximity and had frequent contact, yet each took relatively little from the others in language, technology, or religion. Cultural diffusion was remarkably scant among them. Members of the four groups had considerable social interaction, not much cultural interchange. The Nilgiri groups also illustrate that the entities called tribes and those called castes represent ranges along a scale rather than absolute categories. They conformed to the profile of tribal groups in most respects (cf. Mandelbaum 1970: 576-85), yet they also carried on a system of exchange and of specialized production quite like that of a local caste system. Not least among the instructive aspects of these societies is the absence of warfare. They had no implements for combat, no memory or tradition of lethal hostilities directed by one whole group against another society. To the assertion that war is an inalienable feature of all human life, the Nilgiri case presents one refutation.

In sum, these indigenous groups formed a human enclave. It was a distinctive combination of peoples, cultures, and exchange system that was quite isolated by geography and also separate in culture from the speakers of Tamil, Kannada, and Malayalam who bordered the uplands on three sides. But no human enclave is an island entire unto itself. Despite the isolation, there were and are deep-rooted cultural linkages with other regions and with the wider civilization. The Badagas, removed from the Kannada-speaking area by only a few centuries, retained much of the village culture of that region. Todas have cultural affinities with certain practices that were known in Kerala, as in polyandry and life-cycle rites. Kota technology and religion find parallels among Tamil-speakers. Kuṟumba culture bears basic resemblance to that of other hunter-gatherers throughout South Indian jungle habitats.

All the Nilgiri groups share in the generic Dravidian language and kinship patterns of South India. They also share in basic tenets of Indian civilization, as in their concern with social hierarchy, graded according to criteria of pollution and purity. The indigenous peoples recognized the plateau as a distinctive enclave; Toda and Kota origin myths suggest this, and their trading forays to the lowlands impressed the fact upon them. But their ancient tribal horizons were too constricted to let them glimpse affinities beyond their hills.

The Colonial Communities

Within a few decades of the first British settlement, the Nilgiri region became the scene of activities that were bound into far-flung networks of government and commerce. The chapter on the English by Paul Hockings tells how the British established themselves once John Sullivan had prepared the way. He founded the principal town, Ootacamund; he recommended the shift of the capital of Madras Presidency to Ootacamund during the hot months on the plains. The large military base near the second town, Coonoor, was built. Great tracts were cleared and covered by tea and coffee plantations. The earliest British who ventured up to the plateau foresaw that it could become, as it rapidly did, a seasonal refuge for Europeans in India, a blissfully cool, homelike haven for their rest and relaxation.

By 1834 even the Governor-General, the highest official of British India, spent several months in the Nilgiris. In that year Thomas B. Macaulay put into Madras on his voyage to the then capital of British India, Calcutta, to take the high office as the legal member of Council. He was diverted to Ootacamund where the Governor-General and members of his council were conducting the affairs of state in its salubrious climate.

The British régime existed in the Nilgiris for about a century and a quarter. Through the first hundred years of that period, the indigenous exchange system was maintained in quite the pre-colonial form. The four groups were influenced by the new establishments, but only slowly were they deeply affected by them. Cultural diffusion from English sources to them was, for decades, as limited as diffusion had been among them. Only in the last two decades of the colonial period was the indigenous system impaired; many Badagas stopped participating, in part because the old relations reminded them of what they wanted to forget—that they had practised customs they now regarded as inferior and had acknowledged closer ties with lowly folk than they now deemed suitable.

The society of the British in India can be viewed from particularly good vantage in the Nilgiri setting. All sections of the ruling class were represented and functioned there. Elsewhere in South India they were scattered in small upcountry stations doing lonely jobs or surrounded in cities by a sea of non-English. Here there was a visible and viable community of their own kind. They were in a climate, social as well as physical, that was almost like home. Almost, because not even in the Nilgiri atmosphere could any of them afford to forget to behave in ways proper to members of the governing class, of the paramount varna. Many a sensitive observer has depicted that role. Leonard Woolf's classic version (1961) for colonial Ceylon applies quite as well to the British in colonial Ootacamund. Mollie Panter-Downes' memoir, *Ooty Preserved* (1967) is a beautifully written, insightful account of colonial Ootacamund and its heritage in the 1960s.

I saw something of British colonial society during my first fieldwork in the Nilgiris in 1937-8. My central concern then was with the Kotas, but the British residents also caught my anthropological interest. Most of them had come to India when the imperial (then a most honourable term) cadre and society were in full, successful career. By the late thirties, that society was coming under serious challenge, but like the indigenous Nilgiri exchange system, the parts were still in place and in use.

British society in the Nilgiris, as throughout India, comprised several sections, each clearly bounded and hierarchically ranked. The rankings were made and enforced among the British themselves; few Indians had much grasp of the niceties of that social order. The higher ranks enjoyed much more power, prestige, and perquisites than did the lower. Relations between members of different sections were limited to certain spheres of conduct. Interdining and intermarriage between those of the higher and those of the lower echelons of British was strongly disapproved and rarely occurred. The ranks and rankings are more fully discussed in Paul Hockings' chapter. I summarize here the order as I saw it in 1937-8.

Highest were the civil and military officers. Both were recruited from the same social classes, sometimes from the same families. But the conditions of army life tended to keep the military in their cantonments, removed to some degree from the other English. The civilian officials of the top administrative corps, the Indian Civil Service, generally led local British society; officials of the ancillary services, such as police, irrigation, forestry, upheld the standards and ranked just below the I.C.S.

A wife assimilated the rank of her husband. Children figured little in social matters. As among the upper classes in England, they were largely given over to nursemaids (perhaps somewhat less in India than in England) and removed at an early age to schools (a more continuous and complete separation from parents in India than in England). A few English professional people worked in the Nilgiris, several civilian doctors, lawyers, schoolmasters and schoolmistresses, clergymen of the Church of England and bankers. An American dentist practised his profession in the social season. These were ranked with the highest section.

Planters fell into a separate section, clearly lower. Though most came from respectable enough class backgrounds in England, the nature of their work and their involvement in a commercial enterprise placed them in position inferior in power and prestige to the officials and officers. They were, however, full members of the club, as the lower-ranking English were not.

Wherever several families of the British were stationed in India, they set up a club as a centre of their social activities. Such clubs were found in each of the three Nilgiri towns; the one in Ootacamund was one of the finest in India, complete even to an organized hunt with horses, hounds, horns, master, and all but foxes—jackals had to do. The club provided facilities for interdining and interdrinking, for outdoor sports and indoor recreations, for ritual celebrations, especially in the Christmas season. It also provided opportunities for those informal social relations that are needed in any society and were particularly important for administering the British Raj. Elsewhere Englishmen might meet, at the club they mingled. There, a junior officer could find friends, be sized up unobtrusively by his seniors and, in turn, assess them and their wives. Wives could find the company there that was lacking in houses full of servants but empty of kin.

British of lower sections were rarely seen in the club. Those of the other ranks in the army could no more be members there than they could be in the Officers' Mess of the Regiment. There were in Ootacamund a few British tradesmen, a haberdasher, a jeweller, a tailor, a bookseller and the manager of a general store. They did not qualify, by occupation and education, for the club, would not have felt comfortable in it, nor were they encouraged to do so.

Missionaries were a section apart. They marched to a different beat which one scarcely heard in the club, and did so in separate, denominational platoons. They were a disparate set in terms of class background, education, sectarian affiliation and even nationality. Some lived in the Nilgiris the year round, most came in numbers during the hot weather from the plains. Some were known and respected by those in the club, but for several reasons they were not ranked with other sections of the ruling British. For one reason, their purpose in India was not primarily to uphold that rule or provide services to the rulers, but to implant Christian religion and transmit British civilization. The latter was done in the service of the former but, as it turned out, they were far more successful in their schools than in their churches. Many Indian students learned English language, ideas, manners in the mission schools and went on to become members of the rising national elite. Not many of them became converts.

Missionaries laboured long years among Badagas, Kotas, and Todas. Viable Badaga Christian communities now exist, but in terms of numbers and influence they are small. Yet the number and local influence of the non-Christian Badagas who attended mission schools and whose parents and grandparents did, is large. Kotas loudly cursed and barked to drown out the preaching of an early missionary of the Basel Mission (Metz 1864: 134-5), and they have turned a deafened ear to such evangelizing ever since.

A redoubtable woman, Catharine Ling, worked unsparingly among the Todas for some forty years. She converted several young Todas, found spouses for them and set them up as farmers in a colony. The colony survives flatly; Miss Ling's spirit has long gone out of it. The more enterprising descendants have mostly left the colony and adopted other identities. One, Evam

Piljain Wiedemann, has remained a Toda Christian and has also become a strong influence in aiding the tribal Todas and inspiring them to follow their ancient tradition. Not an outcome that Miss Ling would have foreseen, but nonetheless a fruitful consequence of her missionary efforts.

Other missionaries in the Nilgiris directed their efforts to the Indian immigrants. One of them, as Paul Hockings relates, never once ate a meal with Indians during a six-month period of his diary which was closely examined, and probably rarely or never did so during his twenty years as a missionary in the Nilgiris. It seems likely also that he never once came to grasp the thinking and culture of those whose lives he was trying to turn. If missionaries in the Nilgiris, as throughout India, did not convince or convert a large proportion of Indians, they did enable many to acquire skills for good employment, social mobility, and an entry into English culture.

Another community that became established in the Nilgiris was that of the Anglo-Indians, people of combined British and Indian descent. As Hockings notes, the term had earlier been used for English stationed or domiciled in India. By the 1930s, it meant only Eurasians of 'mixed blood', who identified themselves with the English. The wealthier and better educated among them merged, if they could, with the British. The others, English in speech and Christian in religion, held an ambivalent social position, not accepted as fully British, not then considering themselves as fully Indian, having the worse of both worlds, the better of neither. Yet, in the Nilgiris at least, the core of this community had secure jobs in the railways and other agencies, found a secure place in church activities, lived relatively secure lives in the lee of the British Raj.

As each section of the British settled in the Nilgiris, its members brought along attendant and associated Indians. Civil officials could not operate without numerous clerks, constables, and other Indian functionaries. The Military brought up regiments of sepoys. Planters imported workers from the plains. As British power and pleasure centred in the Nilgiris during the hot weather, a number of Indian princes flocked there. They built palaces, kept racehorses, and in time entered into the whirl of the season's social engagements. Missionaries, for their part, were accompanied by Indian Christian teachers and other missionaries. Each attendant group of Indians kept itself, or were kept, as rigorously apart from other Indians as did the club from the non-club British.

All sections needed markets for shopping. Merchant and artisan communities, Muslim shopkeepers among them, arose in the three towns. Each of the three was reputed to have a particular strain of British residents. As Hockings notes, Ootacamund was the locale for government, military, and retired military, Coonoor was the planters' centre, Kotagiri the missionaries' haven.

For Indians in the Nilgiris, the British loomed even larger than they did in the cities and villages of the plains. They were more concentrated there, especially during 'the season'. Their activities impinged more directly on Indian occupations, their presence was more visible. Still, most of the inhabitants had little direct contact with them. Thus the plantation workers, brought in by British companies, overseen by British managers, did their work and lived their lives in a separate social world of their own. Immured in lines of huts on remote estates, they had very little contact with other Indians. The number of immigrant residents of the district soon became larger than that of the older inhabitants and the preponderance has steadily increased.

Badaga villagers benefitted from the British régime with regard to their agriculture, education and government services. They took the modernizing Hindus whom they met as models of change. Kotas were too enamoured of their indigenous life to pay much attention to the new opportunities until a reformer rose among them (Mandelbaum 1960). Todas went their own pastoral way, accepting the British as another source of benefits, to be treated with proper respect, but not with particular subservience.

To caste Hindus in the Nilgiris as in the plains, British rule carried a certain legitimacy beyond that imposed by force of arms. The British were seen as a kind of *jati* cluster of the Kshatriya varna, the social division of Hindu society whose members were traditionally warriors and rulers. While Kshatriya tradition had flourished in North rather than in South India, there had been *jati* clusters of warrior-rulers in the South for as long as scripture and memory could reach. Orthodox Hindus might abhor British food habits and other polluting customs; they could well understand British restrictions on interdining, intermarriage, and their other caste-congruent kinds of conduct.

For their part, the British understood enough about caste from their own class perspective to make use of the system and generally to avoid egregious infringements on caste sensibilities. While class distinctions within English society in India were sharp, English women and men responded as one in the face of any threat, real or imagined, from Indians. If any one of them received what was perceived as an insult from an Indian, all rallied to redress the slight. No matter how eccentric, indolent, hapless, or dissolute an Englishman might be, he was still an Englishman; he still warranted protection not only as a fellow countryman but perhaps even more as a representative of the rulers and of the rule. Conversely, no Britisher was allowed to breach certain standards in public, such as getting drunk with Indians, without incurring potent disapproval. In such matters, they were indeed *the* British.

Their life in the Nilgiris was easier, more comfortable, more interesting than was their everyday life on the plains. The hill stations provided a central place for their society, as well as for administration. The concept of a central place, especially with respect to the market towns, is important in the study of regions (cf. Smith 1976: 7-9). Here we see that a region can itself serve as a central place for people of other regions. The Nilgiri enclave was a central place for the colonial British society in South India; like other such places across the British (and other) Empire, it was all the more needed because men and women of that society commonly lived and worked among subject peoples who had to be kept, so they believed, at a clear social distance. Each English couple, moreover, had usually to rely for emotional support and personal reward on friends among their countrymen and compeers; their children and closest kin were half the world away. Friends and companions were more readily at hand in the Nilgiris than in their duty stations. And, in the buoyant, holiday ambiance of the Nilgiris (Richard Burton to the contrary notwithstanding) companionship was more forthcoming, friendship could be more intense. For the British, the Nilgiri enclave was a region of the mind and spirit as well as a district of land and government. It was a useful, perhaps essential, social node for the British in India that provided, as did other hill stations, for reinforcement of social standards, raising of morale, relaxation from the travails of loneliness, alienation, discomfort, and disease.

The indigenous peoples for their part were experiencing, during the colonial period, the first dozen decades of a continuing encounter with both Indian and western civilization, and with the interplay of the two. We know something of how they managed and were managed by these encounters. Most Badagas readily adopted the ways of their counterparts among Tamil and Kannada villagers. The Kotas fought a narrow, nostalgic rearguard action and then followed the Badaga course. The Kuṟumbas remained quite isolated in their remote hamlets until some were drawn into plantation labour. The Todas kept generally aloof from it all, thanks in part to subvention by others, though they came to the towns regularly and availed themselves there of goods, tools, foods, drugs, and alcohol.

There is little information about the other Nilgiri communities, the immigrants who formed a large majority of the district residents by the end of the period. They were plantation workers, town artisans and tradesmen, landless labourers, and owner-cultivators. They were of diverse *jati* affiliation and preponderantly from the three adjacent language areas. Most immigrants maintained ties with kinfolk in the old home, bonds that weakened as an immigrant community became a viable social entity in itself and as succeeding generations of the Nilgiri-born came of age.

With independence, the colonial period ended, all but a few of the British departed from the Nilgiris. The institutions they founded have been carried on, some attenuated (the Anglican churches), some little altered (the military base), some expanded and flourished (plantations and new industry).

After Independence

Since 1947, the central administration of the state, now called Tamil Nadu, is no longer shifted seasonally to Ootacamund. But the framework of state and district administration has remained much as it had been, with the critical addition of democratically elected officials to whom administrators are responsible. The elected officials press constantly to assure that district administrators respond to their needs and constituents' desires, especially to carry out the modern mandate that the government actively promote public welfare and diminish inequality.

The people of the Nilgiri region have experienced the same kinds of gains and losses as have those of other districts, but in greater degree than most. Population in the Nilgiris District has increased at a higher rate than the state and national average. Many new government projects have been installed, among them vast hydroelectric schemes, afforestation programmes, a large factory producing film, even a radio-astronomy observatory. Cash crops now dominate agriculture, potatoes and cabbage are grown for a national market, as well as tea for world sale. There are many more schools and health facilities than before.

Still, the prospect of the Nilgiri landscape is recognizably the same as it was when I first saw it in 1937. The terraced fields around Badaga villages march up the hillsides as before, but now they reach higher and over more hills. The public buildings of the colonial period remain in place and in use, with newer ancillary structures clustered about them. The spacious homes that British occupied now hold Indian families of the business and professional classes. The rural hutments of landless labourers are no less impoverished, but much more numerous. The three plateau towns are bigger, busier, with more housing for the poor and more households that are very poor.

The social landscape, too, has been altered, but not vitally amended. In administrative matters, the Collector, now from the Indian Administrative Service, remains the all-purpose chief administrator of the district. The continuing character of this office can be illustrated by an episode that occurred at a dinner given in 1979 at which the Collector was a guest. It was given in the ornate former summer palace of the Maharaja of Mysore, a rambling edifice that is now one of a hotel chain. As the first course was being served, the District Superintendent of Police appeared and discreetly whispered something to the Collector who rose at once, apologized hastily, was off in a moment. It turned out that two factions of Badagas had come to blows, blood had been shed, the local police were unable to quell the violence. The Superintendent had decided that stronger measures might be required. Before the police could use their guns, we were told, the order had to be given on the scene by the highest civil official of the district, a regulation carried over from the colonial régime. The factional conflict among Badagas too was in an old pattern, that of competition for precedence in the status hierarchy of ranking among Badagas.

Nonetheless, regulations and social patterns have also been modified to fit new conditions. The Collector's authority has become subject to intervention, official and unofficial, by elected legislators and the office holders appointed by them. Badagas have recognized that voting power is now a pivotal force. Factional disputes among them are played out in electoral contests and are about issues of power and material benefits as well as about symbols of status prestige.

If dire poverty has not been abolished, if colonial procedures are still enforced, if factional conflicts still consume much human energy, the outlook is not as sombre nor is society as stagnant as is sometimes depicted. Most people of the region, it is my firm impression, are better nourished than were their parents half a century ago, better clothed, better educated, enjoy better health and longer life. While people at all social levels are quick to point out grave defects in the social and political order, they also share a view that something can and should be done to remedy the defects, a notion that was much less prevalent in earlier times.

These impressions are not yet supported by detailed studies. Except for the statistical analyses of the census reports, very few studies have been conducted of the modern social condition of the Nilgiri region; for that matter, not many have been done of the modern development of any Indian region or district.

A Danish geographer, Steen Folke, described some characteristics of the Nilgiri region as of the mid-sixties in two papers (1966, 1967). He found that the usual hierarchy of central places does not obtain, each of the three towns is a central place for somewhat different networks. Bus and other road transport is well developed and intensely used. Cultivators have relatively easy access to markets; factory workers can readily commute to their work from outlying sections. Plantations too have benefitted from good transport; plantation workers were about 25 per cent of economically active persons in the district. The average per capita annual income for the district was higher than that of any other district of the state.

Within the district, the town and environs of Gudalur had the greatest growth rate of all. As T. Adams tells us in his chapter, the Gudalur tracts at the northern foot of the hills were morbid places because of malaria until that scourge was lifted in the 1950s. Immigrants from the nearby language areas then poured in and settled in mixed array. The social divisions and rank orders of their home localities were of small consequence in this frontier milieu. Even their speech reflects a diminished emphasis on traditional hierarchy. A new hierarchy may well emerge in Gudalur as new secular criteria for ranking become established.

The old relations among the four indigenous groups of the uplands are only tenuously maintained. New interests and influences have overshadowed traditional exchanges. Paul Hockings notes that most Badagas are now growing potatoes, cabbages, and tea for a national market. They are concerned with such matters as supplies of credit and of fertilizers, with the hiring of day labourers, with governmental programmes rather than with ancient links and prerogatives (cf. Hockings 1980a: 213-45). Those Badagas who are in the professions or in government service have even less interest in the old, and for them, totally obsolete exchange system.

For a dozen years in the 1960s and 70s, Badagas were the principal beneficiaries of the Indo-German project, an agricultural aid programme in the Nilgiris that was sponsored by the Federal Republic of Germany. While the

German agricultural experts repeated many of the mistakes made by the US and other aid programmes in India, this project did do much to improve Badaga agricultural practices. Todas and Kotas also benefitted from this project. As before, the Todas caught the interest of the officers of this project who treated them with a particularly generous hand. The benefits of that favour were offset by encroachments on pasture land and growing problems of addiction to drink and opium.

The Nilgiri uplands have been a favoured region quite simply because of good climate and rich landscape. The residents of the region are well aware of this advantage; those who immigrate stay. In the study of Indian regions, the geographic uniqueness of the Nilgiri region makes it less typical than, say, the adjoining district of Coimbatore within the larger region of Kongu. That region has been intensively examined by Brenda Beck and her associates (Beck 1972, 1979). And yet, as we have noted, the people of the Nilgiris, past and present, are clearly part of the culture of South India and of the civilization of India. In certain ways, understanding the Nilgiris as a region enhances and clarifies our understanding of these greater entities.

The Nilgiri Environment

HANS J. VON LENGERKE
and FRANÇOIS BLASCO

2

Ecology is the science of the interrelations between living organisms and their environment, of the structure and function of nature, it being understood that mankind is a part of nature.
E. P. Odum (1971: 3)

The Nilgiris as Habitat

For reasons not yet known, by mere accident, curiosity or, more likely, due to external pressure and beginning in prehistoric times, the Nilgiris or 'Blue Mountains' of South India have developed into a niche for several ethnic groups. On leaving the surrounding plains and on ascending the densely wooded slopes of the mountains they encountered an unfamiliar, partly hostile environment. This not only called for an initial process of adaptation, an adjustment of their way of life to nature as perceived, but it also inspired the peoples of the Nilgiris to manipulate actively the existing ecosystem according to their needs. Thus human settlement and land use have transformed the original, natural environment into an intricate man-environment interactive system, the present-day Nilgiris habitat that, of course, has to be seen as a transitory stage in a continuous dynamic process.

The approach of this book is ecological to the extent that it incorporates an opening chapter on environmental conditions before entering upon the main part which comprises essays on individual ethnic groups inhabiting the Nilgiris. Since man is an integral part of the complex Nilgiri ecosystem, not a single environmental factor is without some relevance to human aspects, and *vice versa*. Yet this chapter can only present the very basics, an outline of environmental conditions; for, apart from the limited space which excludes a comprehensive treatment, not all environmental elements have been studied sufficiently well; nor have their manifold interrelations within this ecosystem whether in quantitative or in qualitative terms. Therefore this chapter only tries to familiarize the reader with the spatial context of the Nilgiris and their topography, introduce the main environmental controls, describe some salient features of environmental conditions, such as climate, soil and vegetation, that we consider to be important with respect to human activities and that also serve well to emphasize both the individuality of the Nilgiris within South India and their internal environmental differentiation in space and time.

For further details the reader is invited to consult the references listed at the end of this book and in Hockings' (1978) exhaustive bibliography which covers all scientific fields.

Topography

The Nilgiris (Fig. 2.1) are located between 11°10′ and 11°30′ N latitude and between 76°25′ and 77°00′ E longitude at the junction of the Eastern Ghats and the Western Ghats, or Sahyadris, the two prominent mountain ranges that run almost parallel to the coastlines of Peninsular India. The approximate distance from Ootacamund—with a population of 78,277 in 1981, the largest town and administrative headquarters of the Nilgiris District (Tamil Nadu)—to the Arabian Sea (Malabar Coast) is 100 km, to the Bay of Bengal (Coromandel Coast) 350 km and to Cape Comorin (or Kanyakumari) 400 km.

Geologically the Nilgiris belong to the Archean continental landmass of the Indian peninsula, composed of pre-Cambrian, mainly metamorphic rocks (gneisses, charnockites, crystalline schists). Due to continental drift of the 'Indian shield'—which until late Jurassic times was a part of the ancient Gondwanaland—and coincident with the Himalayan orogenesis during the Cretaceous and Tertiary periods, geotectonic movements in the southern Deccan resulted in its fragmentation and in vertical dislocations along faults that are oriented in three main directions, viz. NNW-SSE, NE-SW and W-E, and that recur in the morphological boundaries as well as in the courses of many streams and rivers of the Nilgiris.

Thus, the triangular-shaped mountain block of the Nilgiris was formed by the phase-wise uplifting of a portion of the Deccan. This horst is almost entirely made up of more or less garnetiferous, acid hypersthenic charnockites (Holland 1900) with a general NE-SW to ENE-WSW strike of foliation (Eastern Ghats trend) and traversed by doleritic and quartzitic dykes (Fig. 2.2). It is slightly tilted towards the east—like the entire Deccan Plateau—and has a base size of roughly 2,400 km^2, of which 40 per cent rises above 1,800 m in the central Nilgiris Plateau (which falls off steeply on all sides). It culminates in Dodabetta, or 'big mountain', with 2,636 m elevation above mean sea level (m.s.l.). It is the second highest peak of Peninsular India, only surpassed by Anaimudi (2,695 m) south of Palghat Gap where the other three 'Southern Blocks' (Spate *et al.* 1967: 683) merge, the Anaimalais, the Palnis and the Cardamom Hills.

It is the close spatial association of two contrasting morphological zones or relief types—the 'mature' landforms of the elevated plateau and the 'juvenile'

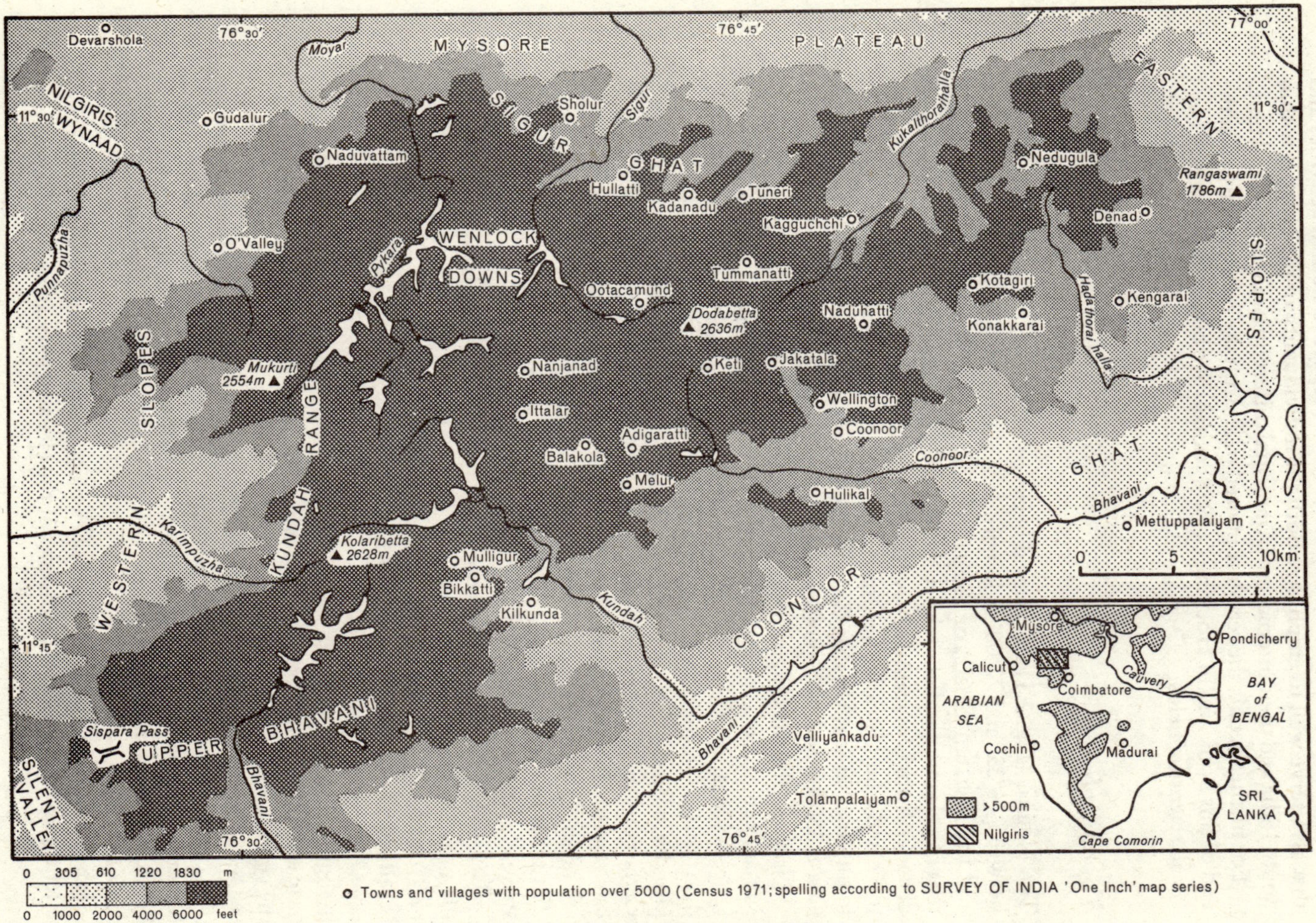

FIG. 2.1 The Nilgiris: Topographical Outline.

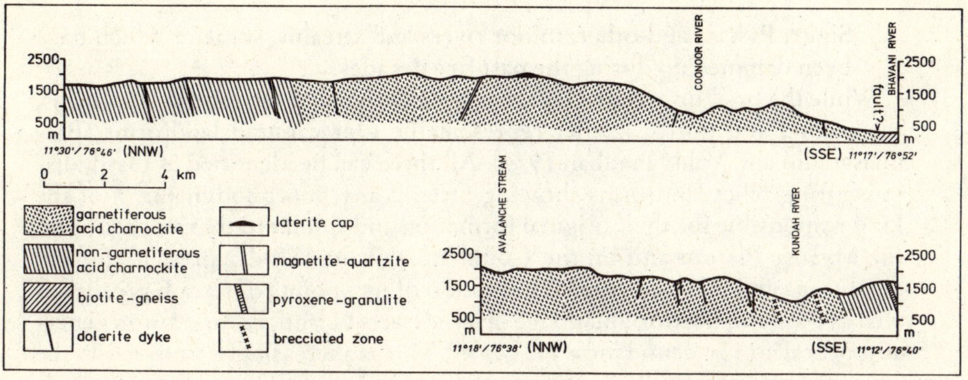

FIG. 2.2 Two Geological NNW-SSE Profiles of the Nilgiris.

escarpments surrounding it—from which the Nilgiris derive their compact physiognomy, which was certainly not without influence on their settlement history. Whether approached from the Malabar Plains (m.s.l. – 100 m), the Coimbatore Plains (300-400 m) or the Mysore Plateau inclusive of the Nilgiris-Wynaad (800-1,000 m), i.e. on different altitudinal levels, the Nilgiris invariably appear like an invincible fortress, protected by high-rising, wall-like flanks: the Sigur Ghat, the eastern slopes, the Coonoor Ghat and—most impressively—the western slopes (Pl. 1). Through a distance of only 15 km the latter rise by more than 2,400 m from the Nilambur Basin to the towering Kundah Range with its highest peak Kolaribetta (2,628 m); first gradually, then with increasing gradients and finally with perpendicular and at places even overhanging rocks, e.g. along the horseshoe-shaped recess just south of Mukurti Peak (2,554 m).

Despite its great petrographic homogeneity the Nilgiris Plateau is far from being a monotonous horizontal plane. It exhibits a remarkable diversity of morphological features that clearly points to its polycyclic development—controlled not so much by 'petro-variance' but mainly by a combination of tectonic activity and local climatic changes due to uplifting and large-scale paleoclimatic fluctuations. However, little is known about the relative contribution and the succession of events during the geological past. In any case, at least three 'high levels' (Demangeot 1973, 1975) can be recognized:

the crest-level of the convex monadnocks, or Inselbergs, above 2,400 m, particularly along the western edge of the Nilgiris Plateau (Kundah Range) and in its centre, including Dodabetta and other prominent peaks around Ootacamund;

the undulating remnants of a denudational surface between about 2,200 m and 2,400 m; this pediplain is best preserved on the western plateau (Wenlock Downs, Upper Bhavani); it has been more or less dissected into isolated ridges by

several valley systems between 1,800 m and 2,200 m elevation, drained by the Bhavani, Kundah, Coonoor, Hadathoraihalla, Kukalthoraihalla,

Sigur, Pykara and other, minor rivers and streams, some of which have been dammed up during the past five decades.

While the first, uppermost level has been designated Dodabetta landforms, the second and third together represent the Ootacamund landforms (Pardhasaradhi and Vaidyanadhan 1974). All three can be identified as fossil, disintegrating relict landforms showing little, if any, morphodynamism of the kind responsible for their original formation and similar to that still active on the Mysore Plateau and on the Coimbatore Plains. Under present climatic conditions and due to upstream migration of nickpoints of stream profiles as well as the deforestation and tillage of large tracts of land, linear erosion generally prevails over denudation processes. This is increasingly so towards the edges of the plateau where numerous streams and rivers leave the upper Nilgiris in torrential rapids and picturesque cascades, with heights of up to 120 m, which emphasize the youthfulness of the Nilgiri mountain block as a morphological unit.

Weather and Climate

CLIMATIC CONTROLS AND SEASONS

The spatial and temporal differentiation of climatic conditions in the Nilgiris is basically controlled by five interacting factors, namely, their latitudinal position close to the equator, where the annual variation of day-length and solar radiation is small (*thermic diurnal climate*) compared with middle and high latitudes (*thermic seasonal climate*); their longitudinal position in that part of the tropics where the monsoon phenomenon, i.e. the seasonal alternation of winds and associated weather patterns, is most characteristically developed; their position relative to the Indian Ocean, the main source of precipitable water vapour, their proximity to the Arabian sea and their considerable distance off the Bay of Bengal; their specific relief features (these have been outlined in the preceding paragraphs); and human interference with the ecology of the mountain massif, particularly the destruction of the natural vegetation cover over large tracts of the elevated plateau, which affects local climatic and hydrological conditions.

Contrary to a widespread misconception, viz. the 'monotony' or 'uniformity' of equatorial or tropical climates in general, climatic conditions in the Nilgiris are characterized by a pronounced seasonality—not in the sense of extratropical thermic 'season' but mainly with respect to hygric conditions, i.e. the absence or presence of water in terms of humidity in the air (water vapour, cloud and fog droplets, raindrops), surface water, soil moisture and groundwater.

From the meteorological point of view—considering upper-air dynamics, surface winds and weather patterns including rainfall—four seasons or periods can be distinguished in South India: the Northeast (NE) Monsoon period from December through March, the First Intermonsoon period in April and May, the Southwest (SW) Monsoon period from June through September and the Second Intermonsoon period in October and November.

This seasonal division must not be understood as a fixed schedule of atmospheric events. It only represents the average annual succession of typical associations of aerological and surface weather conditions which, however, may be subject to large interannual variations 'hidden' behind the mean values shown in Figs. 2.3-2.12 such aspects have been discussed extensively by von Lengerke (1977).

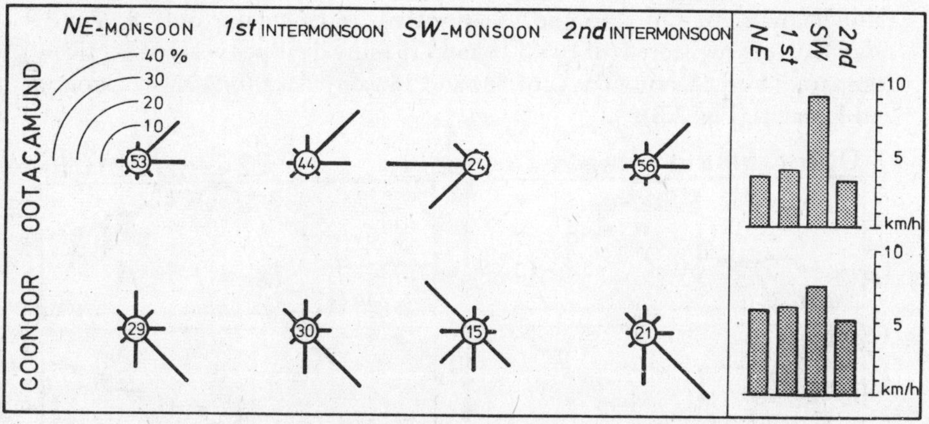

FIG. 2.3 Seasonal Changes of Wind Direction and Velocity.

THE NORTHEAST MONSOON PERIOD (December-March)

During this four-month period South India entirely falls within the northern hemispheric trade-wind belt with light but steady north-easterlies in the frictional layer of the lower troposphere. The aerological and dynamic characteristics of the trade-wind régime are large-scale upper-air subsidence, low-tropospheric divergence and the formation of the 'trade inversion' at 1.5-2.5 km above m.s.l. These result in fair weather conditions over the greater part of Peninsular India, including the Nilgiris (Fig. 2.4); low cloud amounts (2.5-3.5 oktas of sky cover; generally *cumulus humilis* clouds), long duration of sunshine (8-10 hours/day or 75 per cent of maximum possible sunshine duration), low humidity (50-70 per cent) and high evaporative power of the air (see pp. 40-3).

The moisture picked up by the NE Monsoon current from the Bay of Bengal is either precipitated in connection with synoptic disturbances and depressions—over the southeastern Tamil Nadu Plains and along the Coromandel Coast—or due to uplifting of air above the exposed flanks of high mountains where increased cloudiness, supported by diurnal up-slope winds, may trigger scattered orographic rainfall—as on the Eastern Slopes, along the Coonoor Ghat and on the eastern Nilgiris Plateau. Here NE Monsoon rainfall well exceeds 250 mm, locally even 500 mm, and thus accounts for more than one-fourth of the annual average (1 inch of rainfall =-25.4 mm) (Figs. 2.7, 2.10A). The rest of the Nilgiris and the surrounding plains, however, experience prolonged rainless periods, particularly the leeward slopes and the western Nilgiris Plateau. At Naduvattam, for instance, the average rainless period between mid-November and mid-April lasts for 69 (consecutive) days (maximum: 123 days, minimum: 28 days; observation period: 1941-70) as against 37 days (94 days, 19 days) at Havukal Tea Estate between Kotagiri and the Hadathoraihalla. In the northwestern parts of the Nilgiris less than 100 mm during the NE Monsoon (or 25 mm/month) and local minima as low as 50 mm in the Nilgiri-Wynaad are on record; i.e. less than 2 per cent of annual rainfall totals. At Ootacamund, Naduvattam and Gudalur only 9, 6 and 4 rainy days are registered (of 105, 128 and 115 rainy days per year, respectively) compared with 15 rainy days (of 86) and 16 rainy days (of 100) at Coonoor and Kotagiri (Fig. 2.8).

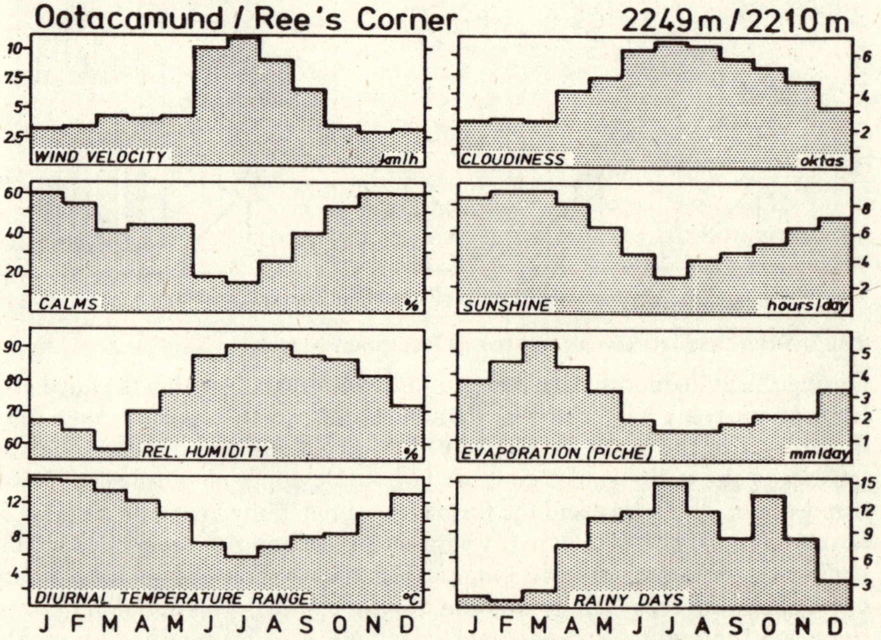

FIG. 2.4 Average Annual Course of Six Atmospheric Elements.

The spatial differences of air temperature largely reflect altitudinal differences and they correspond to a general vertical decrease of temperature, or negative lapse-rate of 0.5°-0.7°C/100 m depending on the season and the exposure to the prevailing, rain-producing winds (*foehn* effect). During the NE Monsoon period average temperatures range between 22° and 28°C on the plains and plateau surrounding the Nilgiris and 12°-17°C on the Nilgiris Plateau, with the lowest averages either in December ('winter' solstice) or in January (Fig. 2.5).

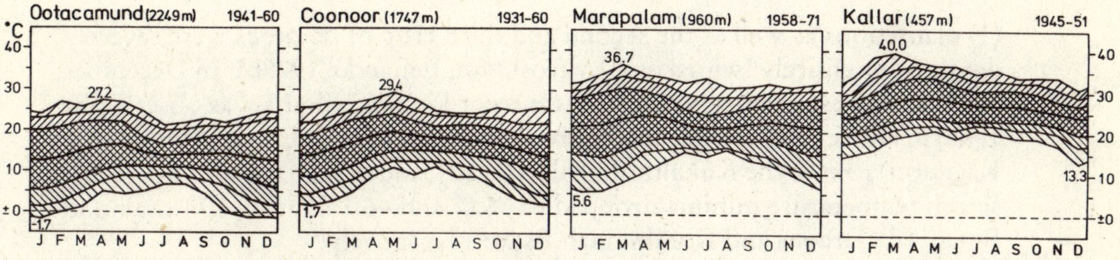

FIG. 2.5 **Annual Course of Screen Temperatures at Selected Stations.**

While these temperatures deviate but little from annual averages and fall short of the highest monthly averages (in April or May) by only 4°-6°C, daily temperature amplitudes (even in the Stevenson screen, 1-2 m above the ground) show a marked annual variation. During the NE Monsoon or fair weather period the sharp contrast between intensive short-wave solar radiation during daytime and almost unrestricted nocturnal long-wave reradiation of energy from the ground during calm and clear nights results in average diurnal temperature amplitudes of more than 12°C in the western Nilgiris (e.g. Marapalam 14.8°, Ootacamund 14.0°). Somewhat lower values are registered in the eastern and southeastern parts (e.g. Coonoor 10.6°, Kallar 8.3°) where increased cloudiness, humidity and air turbulence exercise a dampening effect on radiational processes.

Amplitudes between actual temperature maxima and minima, however, are considerably larger: averages of monthly extremes reach values above 20°, while absolute extremes on record even surpass 25° within 24 hours, and the occurrence of frost on the upper Nilgiris is not uncommon. Over large tracts of the central and western Nilgiris Plateau, above 1,800-1,900 m, night minima at or below the freezing-point of water and hoar-frost formation during early morning hours, before sunrise, are regular, annually recurrent climatic phenomena: wherever radiational cooling is not sufficiently compensated by controlled drainage of cold air it is impeded by topographical depressions, in 'frost hollows' and 'frost basins' (Pl. 10), and by man-made barriers.

Although it is most frequent in December and January, on the western Nilgiris Plateau the phenomenon of night frost may occur from the third week of October throughout a potential 'frost season' of almost six months until the second week of April. Maximum frost season frequencies of 46, 52, 58, 63 and 67 nights with hoar-frost on the ground or negative screen temperatures are on record for Dunsandle Estate (south of Sholur village; in 1964/65), Ootacamund Botanical Gardens (1964/65), Lovedale Estate (between Ootacamund and Ketti, 1963/64), Nanjanad (1971/72) and Korakundah Estate (8 km SW of Kilkundah; 1971/72), respectively (Fig. 2.6). High frequency and/or high intensity of frost heavily affect agricultural and silvicultural activities—as was the case during the exceptional, devastating 1971/72 frost season when thousands of hectares of tea and young bluegum (70)* and wattle

*Numbers referring to species are identified in the Appendix to this chapter.

(1) plantations as well as the second and third crop of potatoes were severely damaged or entirely 'wiped out' by frost (von Lengerke 1978b). In December 1971 the lowest occurrence of frost on record in the Nilgiris was observed at 1,420 m above m.s.l. on the lower division of Marvahulla Estate (northeast of Kaggucci) next to the Kukalthoraihalla (Pl. 12); and in January 1972 nocturnal screen temperature minima dropped to –6.0° and –6.7°C at the Ootacamund Botanical Gardens and Korakundah Estate respectively.

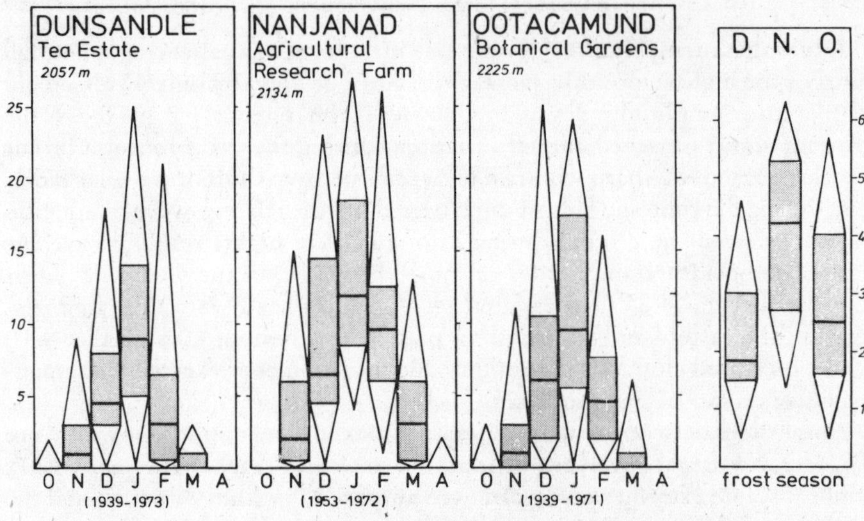

FIG. 2.6 Frost at Three Selected Stations.

On account of their equable microclimate frost does not occur within the sholas, the remnants of the montane evergreen forest (climax) vegetation, despite the fact that they are more often than not confined to ravines and valley bottoms, i.e. to topographical situations where frost is most severe as far as grassland, exposed soil or fields covered with low-growing crops are concerned. Thus the phenomenon of frost in the Nilgiris provides a good example of man's impact on his environment and secondary, in this case adverse, effects. The relationship between its spatial distribution and the line of causes is obvious: low rainfall amounts and prolonged 'dry spells' during the NE Monsoon period, low moisture content of plant tissue, high inflammability and improved applicability of 'slash and burn' techniques, destruction of forest vegetation and extension of savannas; these in their turn catch fire even more easily—accidentally or deliberately set by migrating Toda herdsmen searching for fresh grass (cf. Noble 1977)—and therefore reduce the chances of forest regeneration apart from inhibitory factors such as frost and exposure to strong winds during the SW Monsoon period that restrict arboreal dynamism, with the exception of the hardy Nilgiri rhododendron (154), the 'pioneer tree' of the montane forest vegetation (Pl. 5). Frost and fire are also responsible for the striking absence of a 'tension belt' between sholas and savannas (Pl. 3).

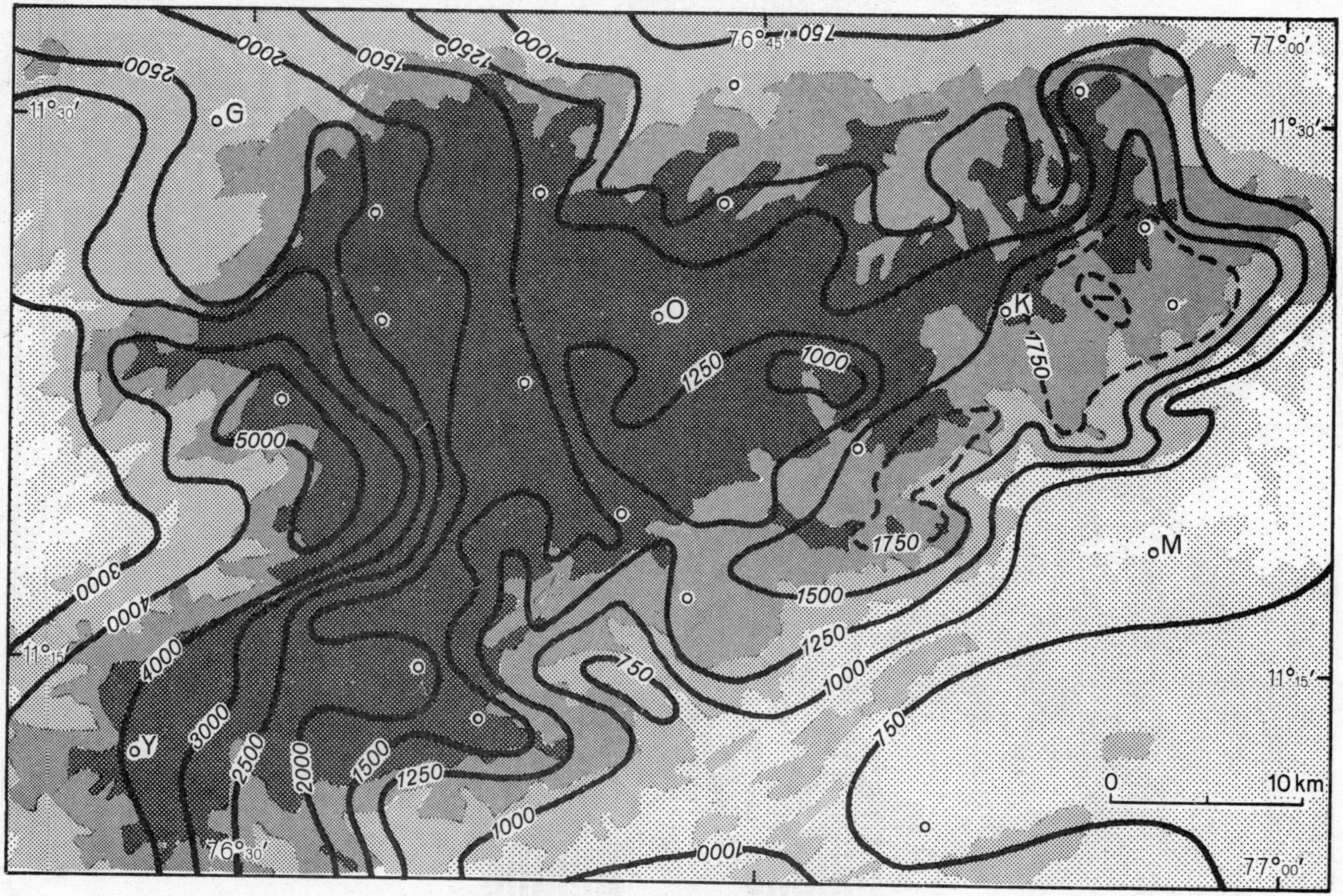

FIG. 2.7 Average Annual Rainfall.

FIG. 2.8 Average Annual Number of Rainy Days.

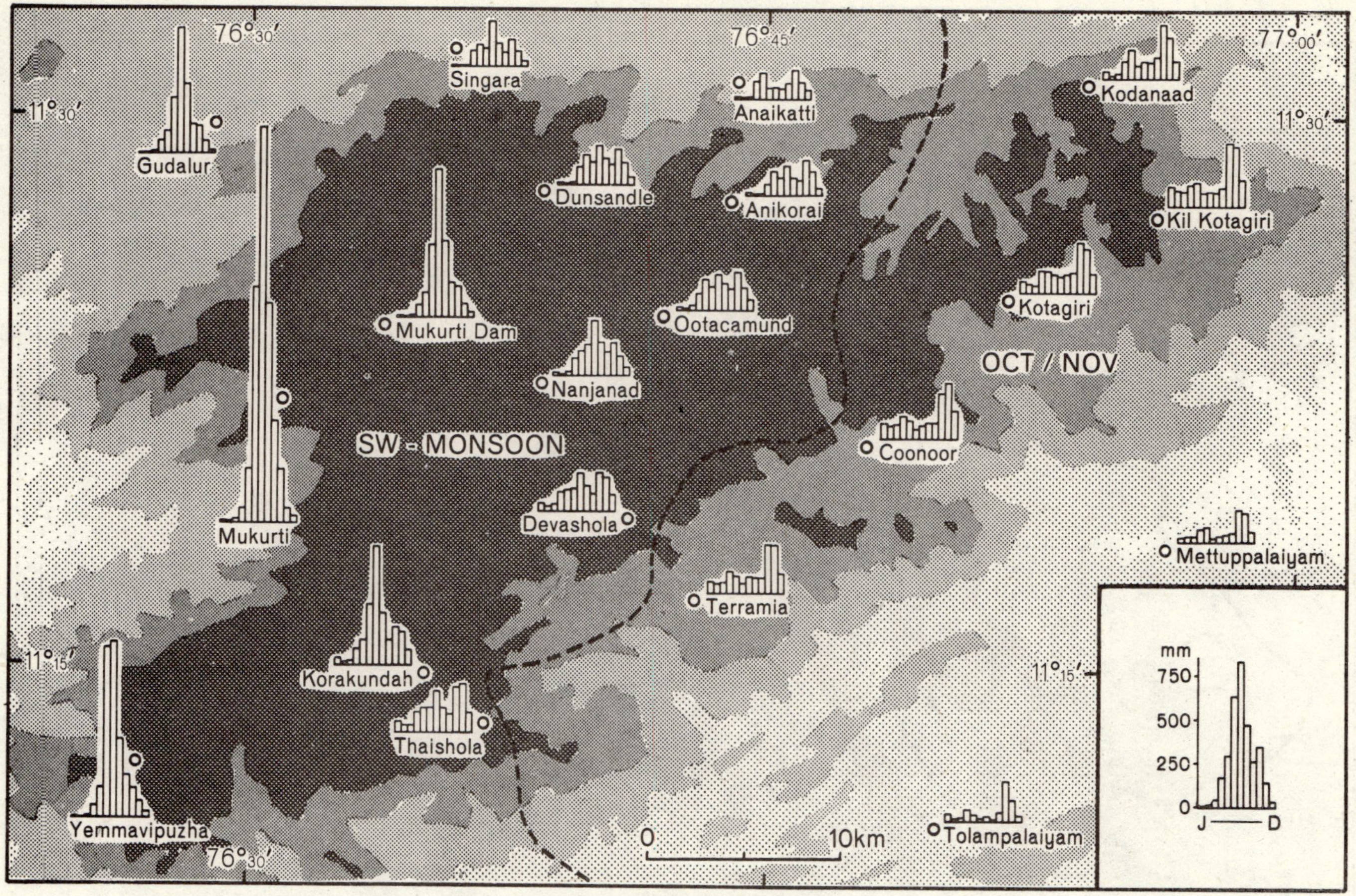

Fig. 2.9 Average Monthly Rainfall.

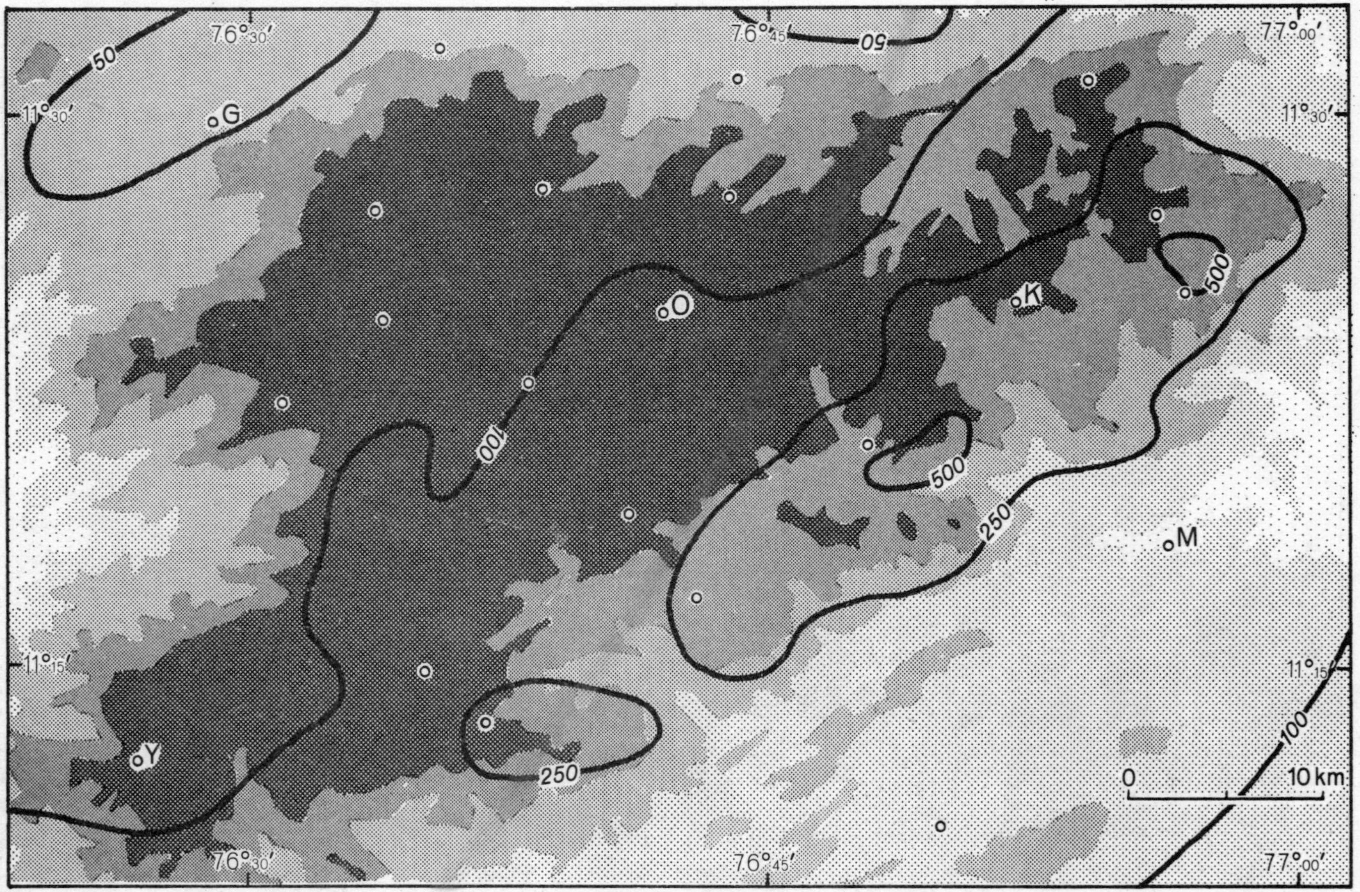

FIG. 2.10A Average Monsoon Rainfall (December-March).

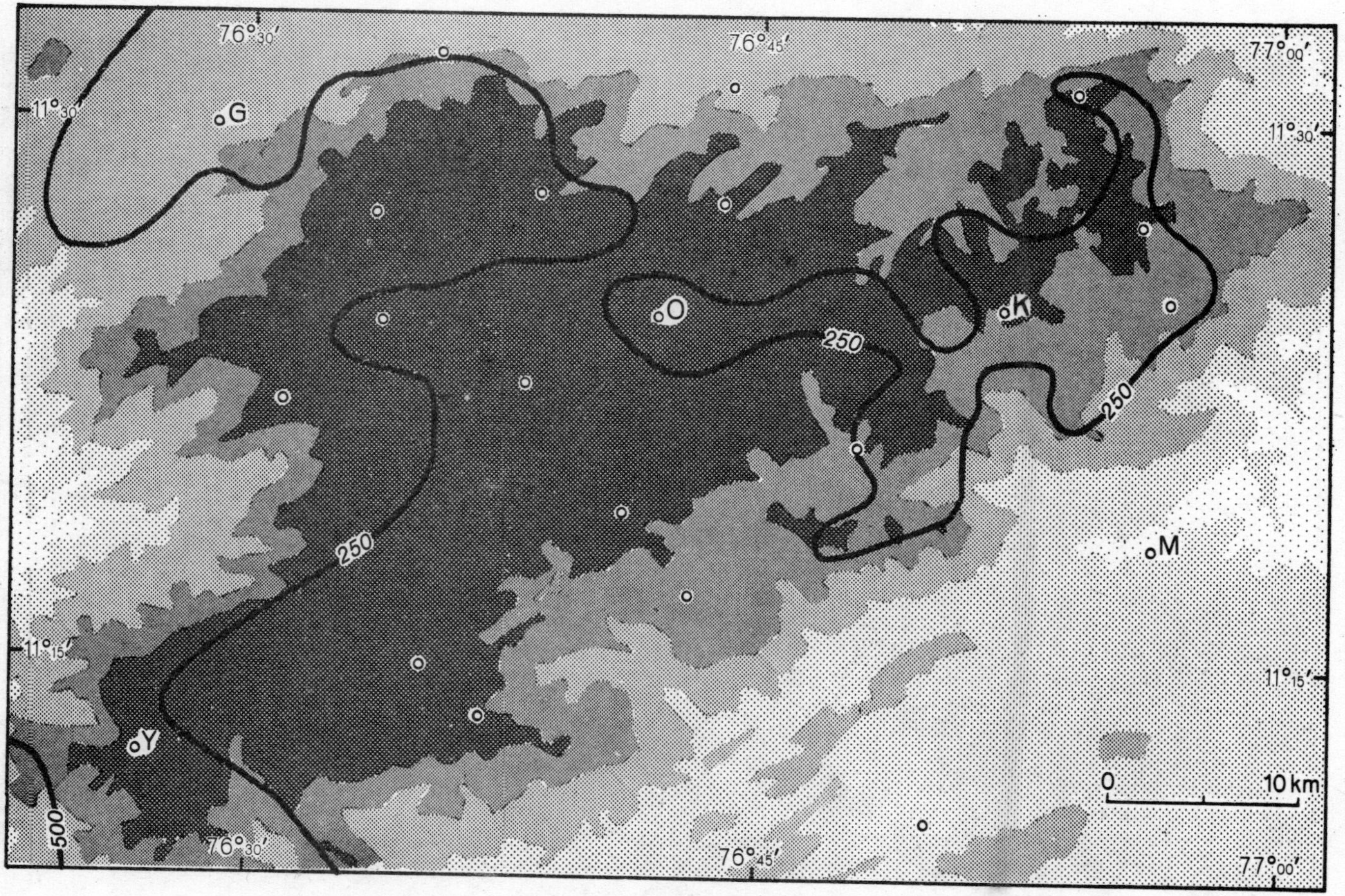

FIG. 2.10B Average Monsoon Rainfall (April-May).

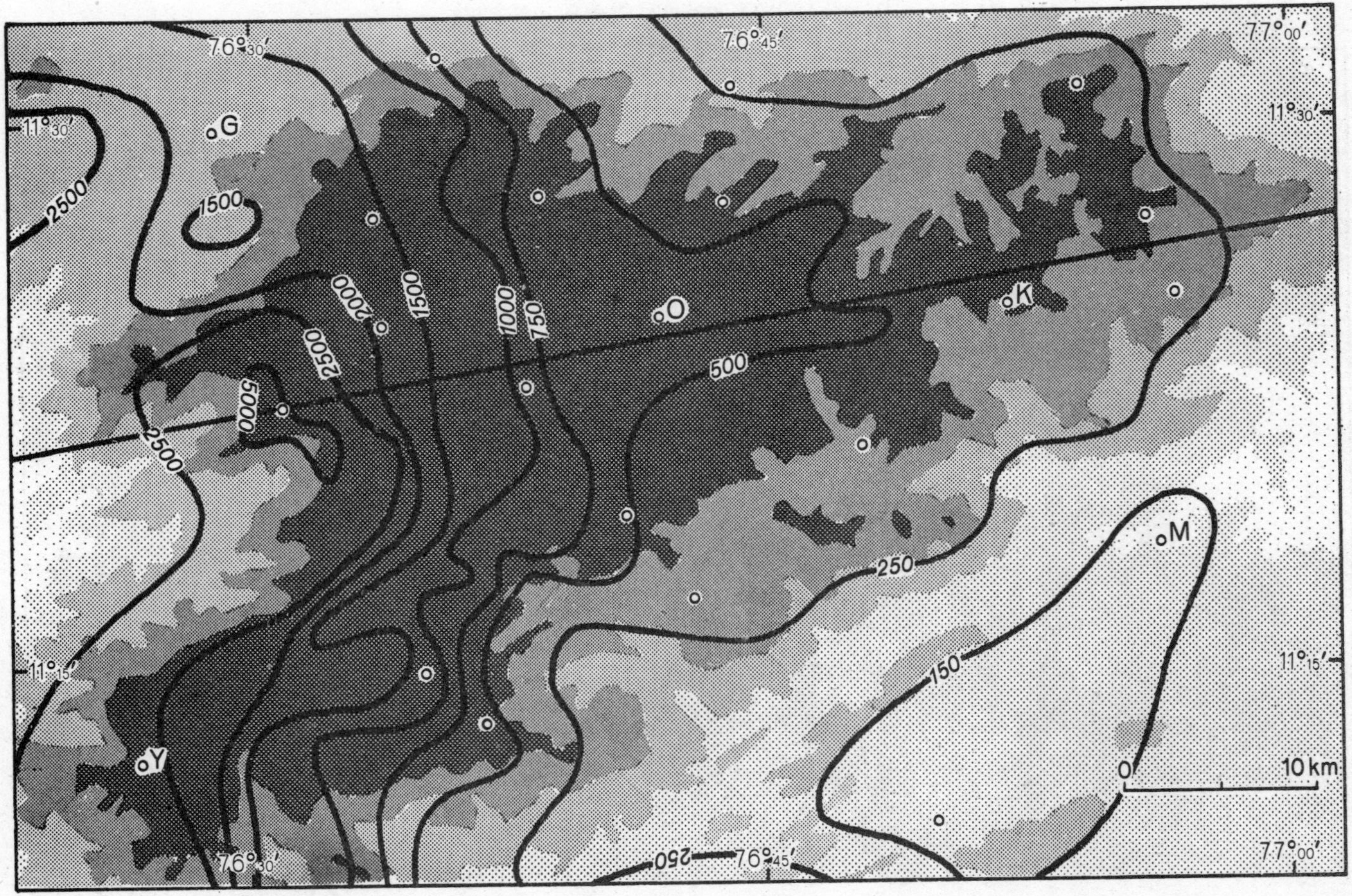

FIG. 2.10C Average Monsoon Rainfall (June-September).

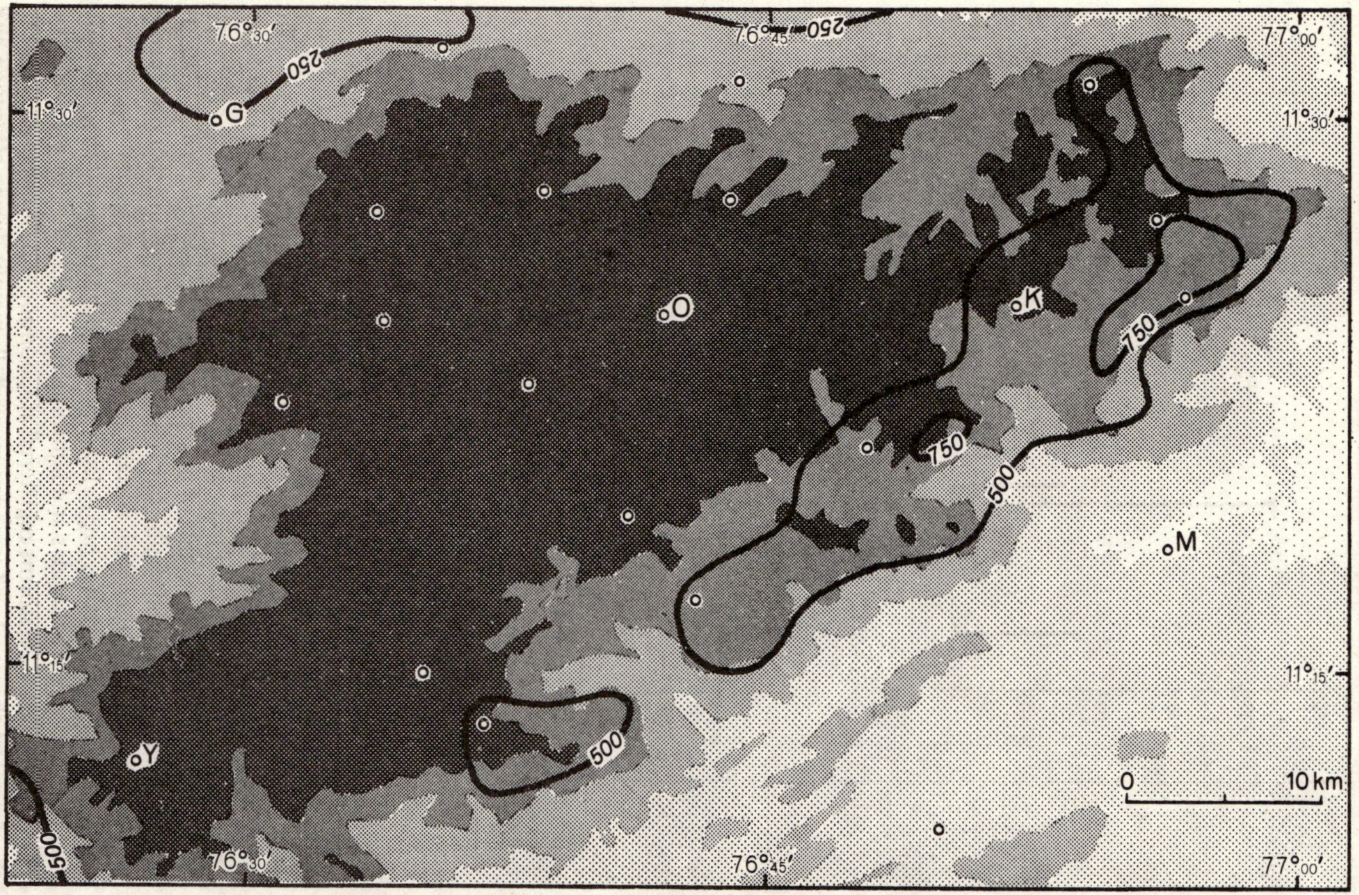

FIG. 2.10D Average Monsoon Rainfall (October-November).

FIG. 2.11 Average SW Monsoon Rainfall.

THE FIRST INTERMONSOON PERIOD (April-May)

From the synoptic-meteorological point of view this period is transitional in character because in April the tropospheric dynamics and weather conditions typical of the NE Monsoon gradually disintegrate, while the SW Monsoon régime is only established in one out of three years before the end of May.

Except for the upper Nilgiris where easterlies and, locally, southeasterlies (e.g. at Coonoor due to topographical deflection) continue to prevail (Fig. 2.3), winds at lower elevations become more variable with increasing westerly components due to the development of intensive thermal low pressure systems over the Tamil Nadu Plains and the Mysore Plateau during afternoon hours and the consequent influx of maritime air from the Arabian Sea. Upper-air subsidence ceases and the higher moisture content of the shallow diurnal westerly current together with surface overheating result in atmospheric instability, local convectivity, high-rising clouds (of the *cumulus congestus* and *capillatus* types) with thunder and lightning (on 10-15 days, or more than 50 per cent of all thunderstorms registered at Coonoor and Ootacamund during the course of the year) and to showery rainfall, mainly during afternoon and early night hours. Rainfall distribution is fairly uniform and varies between

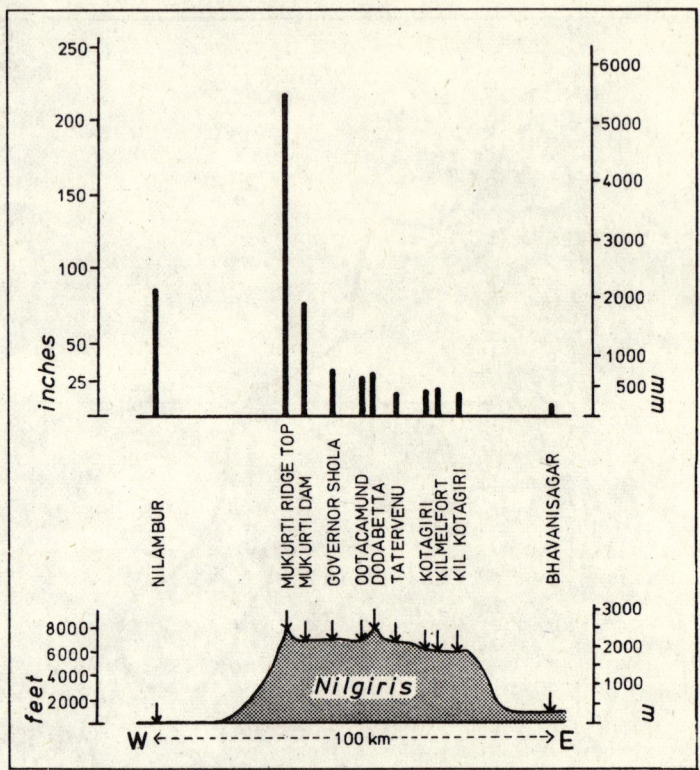

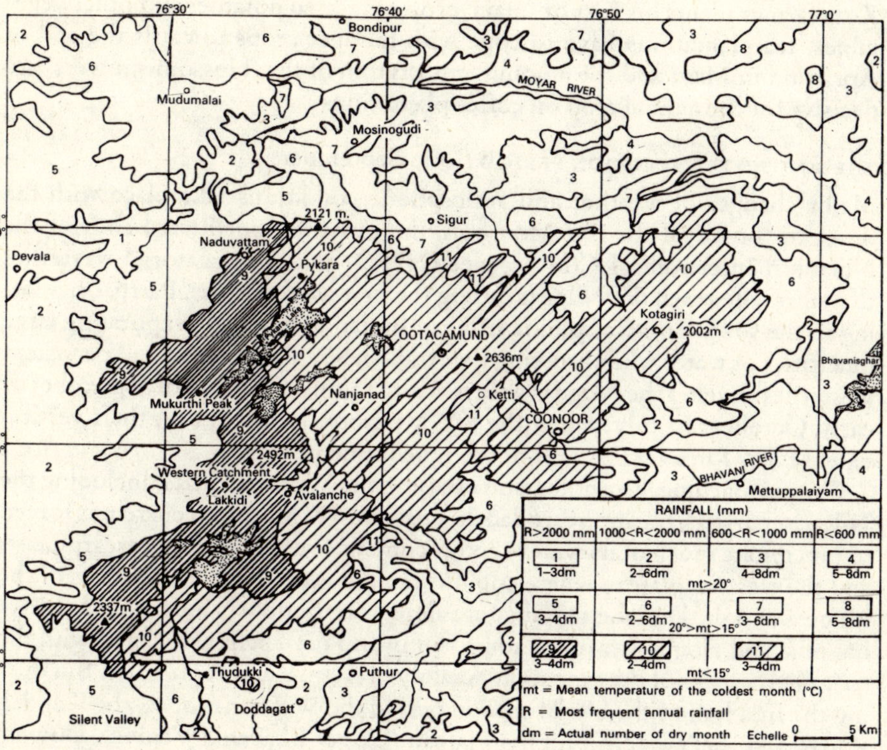

FIG. 2.12 Bioclimatic Map of the Nilgiris.

150 mm and 300 mm, on an average, or 10-20 per cent of annual totals. Averages of more than 500 mm are confined to the southwestern spurs of the Nilgiris (Fig. 2.10B).

The pre-SW Monsoon period is also referred to as the 'hot season'. On the Malabar and Coimbatore Plains average daytime temperature maxima reach 35°C and even on the Mysore Plateau, at 900–1,000 m altitude, they still exceed 30°C, e.g. at Marapalam in the Nilgiri-Wynaad. Absolute extremes close to 40°C in the shade are on record, while nocturnal cooling is much reduced. Naturally such thermal conditions keep human activity to a minimum; and thousands of people who can afford it try to escape the 'dog days' on the plains by spending their annual vacation or at least a couple of days on the cool heights of the Nilgiris and other mountains in South India, where temperatures are very agreeable during the day—seldom above 25°C—and the nights are not quite as chilly as during the previous months (Fig. 2.5). This is 'The Season' for a number of hill stations and tourist resorts, including Ootacamund, Coonoor and Kotagiri in the Nilgiris.

After the long rainless period the first April showers are a long-awaited event: the brown colours of the savannas turn into lush green, the Badaga or

Kota farmer plants his first or 'main' crop of rainfed potatoes and other vegetables, tea plantations have to cope with bumper crops towards the end of April and in May, and the amount and rhythm of the 'blossom showers' are decisive for the annual crop on coffee plantations.

THE SOUTHWEST MONSOON PERIOD (June-September)

Major changes of weather and atmospheric conditions take place with the onset or 'burst' of the SW Monsoon, the pulsating northward shift of the Northern Intertropical Convergence (NITC) and the equatorial westerlies that normally reach the Nilgiris during the first week of June. Partly originating in the southern hemispheric tropical Indian Ocean and evaporating large quantities of water while crossing the Arabian Sea, these strong and persistent westerlies envelop the Nilgiris with maritime air until the end of September or early October. In July, when the SW Monsoon is at its peak, their vertical extent is 6-7 km over this part of the subcontinent.

On approaching, ascending and crossing the Western Ghats, including the Nilgiris 'barrier', the moisture-laden, unstable SW Monsoon current is forced to precipitate much of its water vapour content over the exposed escarpments and the western plateau where copious rainfall is received. These parts of the Nilgiris and the Dodabetta Range are almost continuously hidden under driving mist and clouds—save for some sunny and dry spells associated with the so-called 'breaks of the monsoon', mainly in August and September. But during the first half of the SW Monsoon period the Western Slopes, the Kundah Range and the Western Nilgiris Plateau (Upper Bhavani, Wenlock Downs) are characterized by overcast, windy and extremely wet conditions.

Even at Ootacamund, in the centre of the Nilgiris Plateau, cloudiness exceeds 6 oktas of sky cover during this four-month period and sunshine drops to a minimum of 2.8 hours per day in July (Rees's Corner data). Naturally, visibility is also considerably reduced: in only 6 per cent of all (morning and afternoon) observations from Ootacamund Observatory the Kundah Range (at a distance of about 20 km) can be seen in July, while it is visible in 65 per cent of all cases from January through March. Wind speeds are more than twice as high as during the rest of the year; and despite the lack of anemometer data from the westernmost parts of the plateau, the strength of the westerlies that blow over these tracts and that are channelled through various gaps of the Kundah Range is indicated by the wind-shaped growth and deformations of many isolated trees and exposed groups of trees, mainly Nilgiri rhododendron (154), dotting the extensive savannas. The low, stunted growth of the few remaining sholas is also conspicuous. The extreme humidity, prevailing during the SW Monsoon period (80-90 per cent relative humidity, on an average) is revealed by an abundance of mosses and bearded lichens even covering the trunks of isolated trees outside the perhumid microclimate of the sholas and unprotected against the desiccating NE Monsoon winds.

To the leeward of the Dodabetta Range, the main climatic divide of the Nilgiris Plateau, in the Ketti Valley and other sheltered locations of the eastern Nilgiris as well as on the adjacent Coimbatore Plains and along the lower course of the Moyar River, weather conditions are rather different, much brighter and not as dull and uncomfortable as on the 'hostile', westward exposed parts of the mountain block. This spatial contrast (also reflected by the migratory pattern of the Todas) is caused by the well-known *foehn* effect and it is best revealed by the steep negative W-E rainfall gradients that appear on both the SW Monsoon isohyetal map and the profile from Nilambur (Malabar Plains) across the upper Nilgiris to Bhavanisagar Colony, at the confluence of the Moyar and the Bhavani below the Nilgiri Eastern Slopes (Figs. 2.10C, 2.11).

On a distance of only 40-45 km SW Monsoon rainfall averages decrease from more than 5,000 mm (90-100 rainy days) in the northern Kundah Range around Mukurti Peak to values below 550 mm in the Mettuppalaiyam-Tolampalaiyam area south of the Coonoor Ghat, with less than 20 per cent of annual rainfall received on only 10-15 rainy days; whereas rainfall amounts registered to the west of the line connecting Thaishola, Devashola, Ootacamund, Dunsandle and Singara (Power House) not only exceed those during any other season (Fig. 2.9), but they even account for more than 50 per cent of annual averages. On Mukurti Ridge Top, at 2,545 m, 86 per cent are on record for the normal (30-year) observation period 1941-70, i.e. 5,479 mm of an annual average of 6,330 mm. This even surpasses the long-term value for Mahabaleshwar in Maharashtra (70 km SSW of Poona), which so far has been considered to be the 'rainiest' spot along the Western Ghats and, for that matter, of Peninsular India.

The excessive rainfall around Mukurti and in the three 'Western Catchments' of the Kundah Range (where the Hydro-Metric Survey, Madras, maintains several raingauges in very remote locations) is due to the specific configuration of the upper Western Slopes. Here the moist air has to rise to considerable heights and to converge at the same time. Therefore July, SW Monsoon and annual rainfall averages exceeding 2,000 mm, 4,000 mm and 5,000 mm respectively are on record for this area—with individual extremes of more than 4,900 mm in July, 9,000 mm from June to September and 10,000 mm (10 m!) in a single year (1961).

However, such extremes are confined to the very edge of the Kundah Range, while to the east of its crestline a sharp decrease of rainfall is noticed within a short distance: between Mukurti Ridge Top (5,479 mm) and Mukurti Dam (1,932 mm), for instance, it amounts to 500 mm/km and it is still above 100 mm/km between the latter station and Ootacamund, 16 km further east where 10-12 mm per rainy day are recorded against 40-45 mm per rainy day around Mukurti Peak. It appears that on the Nilgiris Plateau rainfall amounts and intensities during the SW Monsoon period are a function of the distance from the Kundah Range. This tendency is only interrupted by the Dodabetta Range and its rain-shadow towards Kotagiri.

THE SECOND INTERMONSOON PERIOD (October-November)

Dynamically this transitional period appears to be a 'mirror' of events in April and May. Yet the shift of the NITC from North and Central India towards the equator and the consequent retreat of the equatorial westerlies are not quite as rapid as their northward progression in late May/early June. Apart from that, their vertical depth is much less than during the 'burst' of the SW Monsoon. Therefore, tropical (upper) easterlies reappear on the Nilgiri Plateau at the beginning of October, while at lower levels the transport of moist air from the Arabian Sea continues, but towards the end of October and in November only diurnally, in connection with afternoon heat lows on the plains and very strong sea-breeze activity along the Malabar Coast.

Humidity of the air is higher than during the First Intermonsoon period, not only because temperatures are lower, but mainly due to the additional supply of water vapour from the local moisture storage in the soil and its vegetative cover. Therefore, atmospheric instability and afternoon convectivity result in higher rainfall amounts throughout the area. But just as in April/May, its distribution is rather uniform and does not show such spatial contrasts as during the NE and SW Monsoon periods when large-scale advective processes interacting with the relief produce the pronounced luff-lee rainfall patterns.

On an average of 15 to 20 days 250-500 mm of rain are registered over most of the Nilgiris and their surroundings (Fig. 2.10D). Only the Coonoor Ghat and the easternmost parts of the Nilgiri Plateau receive more than 500 mm, locally between 750 mm and 800 mm. These tracts and the Ketti Valley, the Kukalthoraihalla catchment area and the Coimbatore Plains record 34-45 per cent of their annual rainfall within two months, which must be attributed to the high frequency of cyclonic disturbances and storms during the Second Intermonsoon period which form over the south-eastern Bay of Bengal. A 'cyclone' (the South Asian equivalent of a hurricane in the Caribbean) usually intensifies on its two- or three-day travel towards the Coromandel Coast before recurving in a northerly or northeasterly direction or else crossing the southern Indian peninsula. In the latter case rainfall over the Tamil Nadu Plains and the tracts of the Nilgiris mentioned earlier is particularly high and wide-spread floods occur in the plains, while landslips are frequent in the mountains.

WATER BALANCE AND CLIMATIC CLASSIFICATION

Besides solar energy, water is one of the most decisive environmental factors for the existence and functioning of terrestrial ecosystems. The availability of water largely determines the ecological potential of an area, its carrying capacity with respect to floral, faunal and human populations. It is therefore important to consider the aspect of drought which may be expressed in descriptive terms as the low or subnormal quantity of water within the hydrological cycle at a given place and time or, if perceived as a state of deficiency, as the imbalance between available moisture and specific water requirements.

From the plant physiologist's point of view drought is characterized by a gap between actual evapotranspiration (evaporation from the soil + transpiration by plants) and optimum or potential evapotranspiration of a natural or cultivated plant community under given atmospheric conditions (solar radiation, temperature, saturation deficit and movement of air, etc.). For the hydrologist and water manager drought is the reduction of surface (or underground) water supply below the amount required for drinking water, for irrigation (to bridge the gap between actual and potential evapotranspiration) or for the generation of hydro-electric power, to name only a few examples relevant to the Nilgiris.

On account of the homogeneity and impermeability of the rock base the drainage density is very uniform throughout the Nilgiris, but the spatial and temporal differences of stream-flow, or run-off, are considerable. Since no aquifers exist—such as sandstone or calcareous formations—run-off mainly represents 'direct' or 'prompt subsurface' run-off (precipitation excess) and 'base run-off' of water that infiltrates the soil and is released after a more or less extensive time-lag, depending on the slope and water retention capacity of the soil; deep percolation and groundwater run-off must be considered negligible in the Nilgiris. Therefore, annual run-off amounts and run-off regimens are highly correlated with annual rainfall and rainfall patterns.

Within a distance of 30 km average annual run-off (expressed as areal run-off depth) varies between more than 3,000 mm in the Kundah Range and less than 400 mm in the upper Kukalthoraihalla catchment. These amounts represent over 75 per cent and under 30 per cent of rainfall received in the respective catchment areas. However, hydrological data from Mysore Plateau areas, below the Sigur Ghat, indicate that average run-off/rainfall ratios may even drop to 15 per cent, and individual figures suggest that in some years these areas do not contribute to the stream-flow of the Moyar at all. In other words, local precipitation as well as additional water consumed by the gallery forests and bamboo thickets along the banks of several perennial streams descending from the upper Nilgiris are 'lost' by evapotranspiration. This is potentially high: at Arakadupatti, for instance, on the Mysore Plateau 12 km north of Tuneri (just outside Fig. 2.1) open water pan evaporation ('Class A'-type instrument) amounted to 2,059 mm/year during the observation period 1958-66 versus 555 mm of annual rainfall (imbalance: –1,538 mm); and at the foot of the Sigur Ghat, next to the river of the same name, at 990 m elevation, the respective figures were 1,841 mm/year (1957-65) versus 838 mm (–1,003 mm). Similar conditions prevail on the Coimbatore Plains below the Coonoor Ghat.

The situation is quite different on the Nilgiris Plateau, particularly its central and western parts, due to higher rainfall amounts and lower pan evaporation and potential evapotranspiration, which may reach 80-90 per cent of the former where soil moisture supply is unlimited and dense vegetation with high transpiration capacity covers the ground. While the imbalance between pan evaporation and rainfall is –312 mm/year (1957-66) in the Kukalthoraihalla Valley (at 1,500 m), positive annual balances of 100-150 mm are found in the Ootacamund–Melur–Coonoor triangle; and for the Upper Bhavani area (at 2,450 m) and Silent Valley (at 900 m) values exceeding 1,200 mm and 2,600 mm respectively are on record.

Nevertheless, on the central and western Nilgiris Plateau and even in the high-rainfall areas of the Kundah Range, Upper Bhavani and Western Slopes—with annual averages of more than 2,500 mm and locally above 5,000 mm (Fig. 2.7)—physiological as well as hydrological droughts are annually recurrent phenomena. This is due to the prolonged rainless period or frequent dry spells during the NE Monsoon period coincident with high potential evapotranspiration that reaches its maximum in March, i.e. towards the end of the dry season when both run-off and soil moisture content are at their minimum, particularly where the evergreen montane forest vegetation with its 'sponge effect' has been cleared and replaced by grassland, shrubs or open farmland. Apart from the small acreage under irrigation along perennial streams producing a third crop of special, remunerative agricultural and horticultural produce (besides potatoes: cabbages, beans, peas, cauliflower and other 'European' vegetables as well as such specialities as strawberries), most fields lie fallow during this period with soils unprotected against intensive solar radiation and dry winds.

Contrary to this shortage of water we find the seasonal concentration of precipitation either during the SW Monsoon period (unimodal rainfall pattern and run-off regimen) or—along the Coonoor Ghat and Eastern Slopes—during the First and Second Intermonsoon periods (bimodal rainfall pattern and run-off regimen), i.e. during months when the evaporative power of the air, or rather its saturation deficit, is low (Fig. 2.9).

This alternation or periodicity of 'water surplus' and 'water deficit' has during the past hundred years inspired engineers to build a number of storage reservoirs in order to reduce the run-off fluctuations and to secure the water supply during the lean months. Several drinking water reservoirs were constructed in the vicinity of Ootacamund, Coonoor-Wellington and Kotagiri to meet the demand of an ever-increasing urban population as well as of the thousands of visitors and tourists who start to crowd these towns and push up water requirements at the beginning of 'The Season' in March.

Further west, in the almost uninhabited Wenlock Downs and in the Upper Bhavani area where precipitation excess is high during the SW Monsoon period and the danger of silting by eroded top-soil from arable land is small, large reservoirs were built during the last fifty years. They facilitate a regular supply of water to the generators of the Kundah and Pykara-Moyar Hydro-Electric Power Schemes, the two major sources of electricity in Tamil Nadu. All lakes shown in Fig. 2.1 are man-made, recent artificial additions to the Nilgiris environment; and the temporal as well as spatial redistribution of water by discharge control and diversion through tunnels and penstocks across natural watersheds has substantially altered the hydrological pattern of the western Nilgiris. This interference with nature has, without any doubt, been a great stimulus to the economic development of the district and its inhabitants, but the long-term geoecological implications might bring about secondary adverse effects that are not yet fully apprehended.

While man is capable of manipulating the hydrological cycle according to his needs, at least to a certain extent, natural vegetation and soil formation are fully subject to the vicissitudes of nature, including droughts and resultant 'drought stress'. In order to show the spatial differences of physical conditions and potentials an attempt at a 'bioclimatic' regionalization of the Nilgiris and their immediate surroundings is presented in Fig. 2.12. It is based on three interrelated parameters: the amount of annual rainfall to be expected in three out of four years (lower quartiles), the average temperature of the coldest month, and the actual number of dry months per year according to the pluviothermic index by Bagnouls and Gaussen (1957) defining a month as dry if 2 x t (t = average monthly temperature in °C) is less than p (monthly rainfall in mm). These environmental factors are important controls of biological and chemical processes within local soil-plant ecosystems and their combinations are therefore closely related to the composition and distribution of natural vegetation formations. But at the same time they serve as indicators for the cultivation potentials with respect to native as well as 'exotic', introduced agricultural and silvicultural plants.

According to this classification 11 'bioclimatic types' or combinations can be distinguished that emphasize the remarkable diversity of environmental conditions within the small study area—ranging from *humid* (1-2 dry months) to *semi-arid* (6-10 dry months) and from *warm* (average annual temperature above 18°C, no frost; Classes 1-8) to *cold tropical* conditions (Classes 9-11).

Soils

Similar to the polygenetic character of landforms, the soils found in the Nilgiris also show signs of a temporal succession and spatial superimposition of different formative processes on account of climatic changes. But, while the surrounding plains and plateaux were mainly subject to palaeoclimatic fluctuations and consequent minor changes of weathering and soil formation, the original low-level soil cover of the Nilgiri area was uplifted into an entirely different climatic realm.

Soils on the Nilgiri Plateau today are therefore basically relicts of a formerly much thicker soil cover that developed under pedo-dynamic conditions prevailing from late Jurassic to early Tertiary times (Subramanian and Murthy 1976) and that were similar to those still 'active' at lower, *warm tropical* levels.

In the Nilgiris this soil cover has, however, been exposed to a *cold tropical* and partly moister climate throughout the Quarternary, and it has been more or less altered from above by soil formation processes that correspond to present climatic conditions. Microscopic analysis of thin sections of soil samples, for instance, reveals that even soil particles from undisturbed black top-soils are identical in colour and in chemical composition with particles from the unaltered, fossil subsoil (Beckmann 1972).

No detailed soil map of the Nilgiris exists, which might be explained by the difficulty of surveying and presenting cartographically the extreme variation of soil conditions within small areas in the upper Nilgiris. This diversity, however, does not reflect a regular topographical succession of recent soil formation products in terms of natural 'catena', but it is mainly caused by anthropogenic disturbances of the soil cover—following the introduction of permanent land tillage by the Badagas some three to four hundred years ago and its extension over the major parts of the central and eastern Nilgiri Plateau during the past centuries.

Prior to these fundamental changes—which were further stimulated and carried out by the Europeans during the nineteenth century—it would have been quite easy to draw a soil map of the Nilgiri Plateau; for—apart from variations of the depth of soil covers overlying the rock base (up to 40 m) or of individual soil 'horizons', and excepting some exposed high-rainfall areas with excessive natural denudation and scattered rock outcrops (Pl. 4)—the entire plateau could be classified under 'humic ferralitic mountain soil' (Gaussen *et al.* 1962) on account of the great uniformity of top-soil as well as soil profile characteristics.

This 'normal', undisturbed soil type is found under forests and savannas alike (with only minor quantitative differences); it is basically composed of different layers or horizons—a black to dark grey top-soil, 10-70 cm in depth, with high humus content and rich in organic matter (carbon generally 3-10 per cent, locally even exceeding 20 per cent of soil material), due to the low decomposition rate under cool climatic conditions; in badly drained places peaty top-soils have developed and many of these bogs were dug for fuel during the early nineteenth century (Francis 1908: 461); at some places a whitened or light greyish eluvial horizon (tendency towards podzolization), usually a thin layer, hardly any organic carbon present and maximum leaching; brown to yellow-brownish layer of strongly altered fossil subsoil, with gravelly bands of disintegrating quartzitic veins or base-rock remnants, mineral accumulations and concretions; yellow-orange to reddish unaltered fossil 'red earth' or 'lateritic clayey loam'; scarcely any roots of present vegetation are found except for some tap roots of arboreal species; on some hilltops this fossil lateritic or lateritoid soil forms hard, morphologically resistant caps (Fig. 2.2) due to exposure to the air and seasonally dry atmospheric conditions, probably dating back to late Tertiary times, and many of the ancient cairns found on the Nilgiris were built with this material; fossil weathering horizon, reddish in colour, with spheroidally weathered boulders of crystalline rock of varying size, sometimes exceeding 1 m in diameter; underlying, unweathered charnockitic rock base with clefts here and there due to tectonic strains and decompression.

Therefore, the four soil types distinguished according to the colour of the present soil surface, viz. black, brown, yellow and red (e.g. Jeyadev 1957: 3), represent nothing but the more or less 'eroded phases' (Govindarajan and Datta Biswas 1964) of the above profile structure; and it is interesting that some Badaga toponyms refer to the red colour of fields characteristic of the advanced stage of man-induced erosion and soil deterioration (Hockings 1974).

However, apart from these anthropogenic alterations of soil conditions, even the undisturbed 'humic ferralitic mountain soil' shows some spatial variations of physical and chemical properties that should not be left unmentioned because they are significant in questions of agricultural land-use and soil conservation. According to Mahalingam and Durairaj (1968) there is a trend of increasing clay, iron, alumina, carbon and nitrogen content with increasing elevation (or decreasing temperature) and rainfall; on the other hand, silica content and cation exchange capacity are negatively correlated with these environmental factors, and due to the very low alkali earth base acid reaction of soils prevails throughout the upper Nilgiris—the pH-values generally below 5.5. As far as the textural and physical properties are concerned, three main altitudinal levels or groups can be distinguished—below 900 m: loamy sand, 900-1,500 m: sandy loam, above 1,500 m: sandy clay-loams.

The fertility of the black top-soil is high, but easily deteriorated and it invariably lacks in lime; therefore the application of fertilizers is essential wherever intensive crop cultivation is practised, and the response to fertilization is generally good.

One of the most aggravating problems in the upper Nilgiris is soil erosion. It is most severe where agricultural land-use with annual crops and intermittent periods of fallow has been indiscriminately extended to steeply sloping terrain even without slope terracing. It was only during the 1950s that soil conservation became a high-priority agronomical and political issue that enforced the programme of the Soil Conservation authorities, including the recommendations (Govindarajan and Datta Biswas 1964:124) given on p. 46.

These soil conservation measures have by now become a widely applied routine; and it seems that most Badaga and immigrant Kanarese farmers realize the benefit derived from both the prevention of soil erosion and the improvement of the micro-scale and local water balance.

Flora

Since the first account of the flora of the Nilgiris by Wight (1845-51) many botanists have been attracted by the very peculiar floristic properties of this area. During the second half of the last century Colonel Beddome in particular took a prominent part in its botanical exploration and description. Later on these works were revised and completed by two famous botanists at the beginning of the twentieth century, Gamble (7 vols., 1915-25) and Fyson (2 vols., 1915-20).

Land characteristics	Soil conservation measures needed
valley bottoms and gently sloping valley sides	provision of drainage, raising rotational crops, application of lime and fertilizers
slopes up to 10 per cent	contour farming, graded type contour bunding, with outlets and disposal drains, rotational cultivation, strip cropping, liming, manuring, etc.
slopes from 10 per cent to 15 per cent	contour and graded trenching with disposal drains and drop pits and all agronomical soil conservation measures including liming and manuring
slopes from 15 per cent to 33 per cent	bench terracing at 4 to 6 ft. vertical intervals with disposal drains, drop pits, other agronomical soil conservation measures including liming and manuring

Today the flora of the Nilgiris is well known in spite of its extreme diversity and complexity. Even a century ago Beddome could write that 'with the exception of the dense evergreen moist forests on the western slopes, the whole area has been well explored by Botanists, and it is probable that there are no plants now botanically unknown on the plateau and the deciduous forests of the slopes' (1876:17). It is generally accepted that two distinct main groups of plants compose the flora of the Nilgiris, according to their thermic requirements. Up to about 1,800 m above m.s.l. the dominant plants are 'megatherm'. Above 1,800 m, on the plateau, we find a distinct group showing European or north temperature affinities (valerians, violets, anemones, pimpernels, barberry, etc.).

The outstanding floristic peculiarity of the Nilgiris is due to the exceptional amount of endemic plants side by side with an unceasing increase of introduced, 'exotic' trees, shrubs, herbs, etc.

ENDEMIC PLANTS

The following are the plant species that are only found on the Nilgiris Plateau and nowhere else in the world:

Botanical	*English**	*Badaga**
ACANTHACEAE		
Andrographis lawsoni Gamb.		
A. lobelioides W.		
A. stellulata Cl.		
Leptacanthus amabilis (Cl.) Brem.		
Mackenziea violacea (Bedd.) Brem.		
Nilgirianthus papillosus (T. And.) Brem.		
N. wightianus (Nees) Brem.		
Phlebophyllum lanatum (Nees) Brem.		
Pleocanthus sessilis (Nees) Brem.		
AROIDEAE		
Arisaema tuberculatum C. Fisch.	cobra flower	na:gara hu:
A. tylophorum C. Fisch.	cobra flower	na:gara hu:
ASCLEPIADACEAE		
Brachylepis nervosa W. & A.		
Caralluma nilagiriana Kim & Rao.		
BERBERIDACEAE		
Berberis nilghiriensis Ahrendt.	common Nilgiri barberry	jakkalu
CAPRIFOLIACEAE		
Viburnum hebanthum W. & A.	guelder rose, wayfaring tree, hairy yellow-flowered laurustinus, snowball tree	ka:ḍambu
CELASTRACEAE		
Microtropis ovalifolia W.	holly-flowered spindle tree	ottane
COMPOSITAE		
Anaphalis neelgerriana DC.	Nilgiri pearly everlasting; cudweed; cudwort; buffalo carolah	pu:ṇḍu
A. notoniana DC.	everlasting, cudweed, cudwort	ottu giḍu
Helichrysum wightii Cl.		
Senecio kundaicus Fisch.	Nilgiri ragwort	kuri giḍu
S. lawsoni Gamb.	Nilgiri ragwort	" "
S. lesingianus Cl.	Nilgiri ragwort	" "
S. polycephalus Cl.	Nilgiri ragwort	" "
Yungia nilgirriensis Bab.		
CONVOLVULACEAE		
Argyreia nellygherrya Choisy.	silverweed	mi:nige

(continued)

Botanical	*English**	*Badaga**
CYPERACEAE		
Ascopholis gamblei C. Fisch.		
Carex pseudo-aperta Boeck.	sedge	go:re
Fimbristylis latinucifera	sedge	go:re
F. latiglumifera	sedge	go:re
ERIOCAULACEAE		
Eriocaulon pectinatum Ruhl.	hat-pin flower	
E. robustum Steud.	white-tailed hat-pin flower	
EUPHORBIACEAE		
Dalechampia velutina W.		
GENTIANACEAE		
Swertia trichotoma Wall.		
GERANIACEAE		
Biophytum polyphyllum Munro.		
Impatiens beddomei Hk. f.	balsam	kiyu: giḍu
I. debilis Turcz.	balsam	" "
I. laticornis C. Fisch.	balsam	" "
I. lawsoni Hk. f.	balsam	" "
I. neo-barnesii C. Fisch.	balsam	" "
I. nilgirica C. Fisch.	balsam	" "
I. orchiodes Bedd.	balsam	" "
I. rufescens Benth.	pink marsh balsam	to:ṭa giḍu
I. tenella Heyne.	balsam	kiyu: giḍu
GRAMINEAE		
Arundinaria wightiana Nees. var. *hispida* Gamb.		
Eriochrysis rangacharii C. Fisch.		
Garnotia geniculata Santos.		
Helictotrichon asperum (Munro) Bor var. *polyneuron* C. Fisch.		
Isachne deccanensis Bor		
Poa gamblei Bor		
HYPERICACEAE		
Hypericum japonicum Thunb. var. *major* Fys.	marsh St. John's wort, tutsan	honne
LABIATAE		
Leucas rosmarinifolia Benth.	dead-nettle	tumbe
Orthosiphon rubicundus Benth. var. *hohenackeri* Hk. f.		
Pogostemon nilagiricus Gamb.	patchouli, patchouly	koḍaṅgu
P. paludosus Benth.	" "	koḍaṅgu
Teucrium wightii Hk.	wood germander	

Botanical	*English**	*Badaga**
LAURACEAE		
Cinnamomum perrottetii Meissn.		
LORANTHACEAE		
Dendrophthoe neelgherrensis Van. Tieghem var. *clarkei* Hk. f.		
Loranthus recurvus Wall.	loranthus, Indian mistletoe, honeysuckle mistletoe	ottane
Viscum orbiculatum W.	mistletoe	
MELASTOMACEAE		
Memecylon flavescens Gamb.		
MIMOSACEAE		
Acacia hohenackeri Craib.	wattle	banni
MYRTACEAE		
Syzygium montanum Gamb.	Nilgiri mountain black plum, Nilgiri jaumoon	pu: na:ge
ORCHIDACEAE		
Cirrhopetalum acutiflorum A. Rich.	tree orchid	arkattu hu:
Coelogyne odoratissima Lindl. var. *angustifolia* Lindl.	sweet-scented plantain orchid	" "
Habenaria fimbriata W.	ground orchid	" "
Liparis biloba W.	(orchid)	" "
PAPILIONACEAE		
Alysicarpus beddomei Schindl.		
Crotalaria barbata Grah.	crotalaria, rabbit's ears, rattle-pod, rattle-wort	tuggili
C. candicans W. & A.	" " " "	"
C. formosa Grah.	" " " "	"
Dalbergia gardneriana Benth.	rosewood	
PIPERACEAE		
Piper pikarhense C. DC.	pepper	so:lekuḍi
P. ootacamundse C. DC.	pepper	
ROSACEAE		
Pygeum sisparense Gamb.		
Rubus rugosus Sm. var. *thwaitesii* Focke.	bramble	muḷḷi

(continued)

Botanical	*English**	*Badaga**
RUBIACEAE		
Oldenlandia hirsutissima O. Kze.		
O. sisparensis Gamb.		
Ophiorrhiza pykarensis Gamb.		
Pavetta hohenackeri Brem.	pellet shrub	
RUTACEAE		
Melicope indica W.		
SYMPLOCACEAE		
Symplocos microphylla W.	sweet-leaf	
UMBELLIFERAE		
Bupleurum plantaginifolium W.	giant hare's ear	
Heracleum hookerianum W. & A.	cow parsnip, hogweed	nare

*These Badaga and English equivalents have been supplied by Paul Hockings, who advises that they are tentative in many cases

The Nilgiris appear as the most important centre of speciation in South India, next only to Travancore (now southern Kerala). When comparing them with the Palnis, for instance, we are surprised to find only nine plants endemic on the Palni Hills.

INTRODUCED OR EXOTIC PLANTS

Nobody knows their exact number: it probably exceeds 400. Many of them are of great economic interest, and have contributed decisively to the development of the district and to its comparative prosperity. (Note, for example, the New World domesticates in the list given by Jebadhas and Noble.) The following are the commonest introduced plants.

Wattles: common tan wattle, silver wattle, black wattle (1-5). They have played an important economic role since the cessation of import of the bark from South Africa (about 1940). The Australian blackwood is used in construction (Pls. 4, 6). Introduced as early as 1832, wattle trees now make up two-thirds (16,000 ha) of the forestry plantations on the plateau.

Eucalyptus: among these bluegum (70) is the best known (Pl. 10). It was introduced from Australia in 1842 and then planted on a large scale; today it occupies approximately 8,000 ha. Many other species of *Eucalyptus* (probably more than 40) have been introduced for timber, firewood, oil, or as ornamental trees, etc.

Conifers: the introduction of these is due mainly to the Forest Department (Jeyadev 1954), not a single conifer being spontaneous in this region. The largest area planted with conifers bears several species of pines: *Pinus roxbur-*

ghii (135) from the Himalayas, Monterey pine (134) from California, Mexican yellow pine (133) from Central America; during the 1970s, the last-named tree was planted on more than 600 ha of frost-prone terrain where bluegum and wattle had proved unsuccessful in previous afforestation attempts. Other conifers like the Himalayan cedar (34), the Californian cedar (165), and cryptomeria (44) are also frequently planted.

Alder: (9) has also been introduced by the Forest Department owing to its soft wood and good yield.

Fruit trees: most of the common European fruit tree species have been tried in the Nilgiris, particularly plum, peach, apple and pear (144-5, 151-2).

Fodder grasses: among these the most popular is certainly the Kikuyu grass (127), easy to recognize everywhere, being a creeping and closely matting species.

Weeds: these are very numerous. None is as abundant as the lantana (99), introduced from South America. It became a pest throughout much of India in the first half of the last century. Dispersed by birds, it is a wayside weed forming dense thickets mainly on poor, eroded soils. Another widespread introduced weed is the American ironweed (73). Very conspicuous on walls and along road-sides, another North American weed, Canadian fleabane (69), became familiar all over the inhabited places of the Nilgiris. Two escaped European garden plants must also be mentioned, both perfectly naturalized: the yellow broom (161) and the gorse (192).

European vegetables: horticultural plants from mid-latitude regions, such as cabbage, knolkohl, radish, carrot, beetroot, cauliflower, pea, bean, Brussels sprouts, form an important asset to the tropical mountain economy, as does the South American potato (169) which John Sullivan introduced in 1822. Potatoes are now cultivated on about 9,000 ha of generally sloping arable land, at places yielding three crops in a year; they are major agricultural produce of the district, in value second only to tea. Potatoes from the Nilgiris are exported to all parts of Peninsular India and occasionally even to the distant Andaman Islands.

Coffee (mainly *Coffea arabica,* 41): after successful trials in the vicinity of Coonoor in 1838, the coffee plant soon spread to the eastern and northern slopes and to the Ouchterlony Valley. The boom lasted until 1880, when over 10,000 ha were under coffee in the district; a rapid decline, however, was caused by the Ceylon coffee-leaf disease *Hemileia vastatrix* and by severe competition from Brazil. Within the next 15 years the acreage under coffee dropped by over 50 per cent (1896: 4,375 ha). It was only after 1930 that the industry recovered; at present the acreage is close to 8,500 ha, a third of this in the Nilgiri-Wynaad alone.

Tea (28): the first seeds were brought to Ootacamund in 1832-3 and propagation of tea was carried out in Ketti from 1835 onwards; yet another 20 years passed before commercial-scale tea planting commenced in the Nilgiris.

By 1880, the tea plant had conquered 1,800 ha and began to replace coffee on many estates. During the past fifty years, its acreage has quadrupled and now amounts to over 22,000 ha, i.e. 40 per cent of the district's agricultural land-use area.

Cinchona (37): also known as 'Peruvian bark' tree, this quinine-producing genus was introduced by Sir Clements Robert Markham from the high Andes *via* Kew Gardens. In late 1860, he brought seed material to Ootacamund where the first cinchona plantation of the British Empire was established above the Botanical Gardens, soon to be followed by plantations in the Naduvattam area (Brockway 1979). Here a factory for the processing of the bark started to operate in 1870, extracting the valuable anti-malarial substance. Private estates also took up cinchona planting during the coffee crisis, but the peak of 4,000 ha cultivated with cinchona had already been reached in 1882, when competition from Ceylon and Java became too strong and tea more remunerative. In 1974, a mere 400 ha remained, all under Government control.

Many more useful as well as useless or even notoriously 'aggressive' plants have been introduced into the Nilgiris which cannot however be listed here; Krishnamurthi (1953) and Matthew (1969) have compiled most of them. In any case, man has truly created a cosmopolitan floristic assemblage in the Nilgiris during the past century and a half.

Vegetation Types

In spite of the enormous extent of deforestation, illicit cutting, selected felling, extraction of valuable timbers, etc. it is still possible to find good examples of the main forest formations that together with the savannas of the Nilgiri Plateau make up the four vegetation types to be distinguished, viz.

(1) the moist evergreen forests of the slopes and at low elevations,
(2) the dry deciduous forests and thorny thickets,
(3) the sholas or low forests of the plateau,
(4) the grasslands of the plateau (Bellan 1985).

THE MOIST EVERGREEN FORESTS

Before the introduction of tea (28) in the 1830s and its large-scale cultivation since the 1850s, these forests covered vast areas of the western and southern slopes of the Nilgiris, up to 1800 m elevation. Today they are confined to small protected areas. One of the best preserved is that of Silent Valley (in Palghat Forest Division). This region was practically uninhabited until the end of the last century. Although its climate has been classified as semihumid, with a dry season of two to four months, the air humidity inside the forest is rarely less than 80 per cent. The forest is best preserved at about 1000 m elevation, where the dominant stratum ranges from 30 to 40 m high. Most of the trees are large evergreen species, with dense foliage and fairly straight boles, often buttressed. They show a clear tendency towards gregariousness.

The following trees are very conspicuous: rudraksha (62), the most handsome species among Indian *Elaeocarpus;* unfortunately its very soft wood is practically useless; wild durian (45) is probably the most common tree of these jungles, but it is not a timber of great commercial interest; the Indian copal tree (195); the poon spar tree (27); the black dammar tree (29) which yields a viscous resin, the black dammar of commerce; and the Nahor oil tree (110) which has a very hard wood, extracted for railway sleepers, posts etc. In the Silent Valley, rosewood (50) and white cedar (58) are rare nowadays. Both these species are of great value on account of their ornamental heartwood. Most probably they were already exploited before the end of the last century.

In the secondary stratum of the forest, we generally find a great density of crowns belonging to species reaching their maximum size at a level of 10 to 20 m: longan tree (116), wild cinnamon (38), several clove trees (180-2) and wild nutmeg (115) are present.

The shrubby stratum is very irregular; it can be identified, however, by the profusion of rattans (26) and the numerous clumps of the graceful cane betel palm (132).

On the ground small herbaceous plants and woody undershrubs entirely cover the soil. Here the fern flora, balsams, and *Neurocalyx* (117) are extremely diversified; but the most abundant are certainly the strobilanths (171-4), shrubs of the underwood which flower after a growth of six years or more and die thereafter; they reproduce by seeds. Another typical shrub of this herbaceous stratum is the umbel spiked pepper (82) of 1 to 2 m height, easy to recognize with its large leaves (100-50 cm), deeply cordate. Everywhere climbers (136), fine creepers, epiphytic orchids, ferns, and mosses can be seen naturally. In boggy places typical trees are the Indian gutta-percha (126), sometimes tapped for gutta-percha production. In the same localities, important stands of a shrubby bamboo (120) are usually very conspicuous. At medium elevation, towards 1500 m, the dominant stratum is lower (20 to 30 m), buttresses are smaller, trunks are shorter and not as straight at those found below 800 m elevation. Species common at lower elevations generally disappear, almost without exception above 1800 m (Pascal 1984).

THE DRY DECIDUOUS FORESTS

These rather dry forests cover large areas of the Mysore Plateau to the east of the Mudumalai Wild-life Sanctuary. They are designated in India as the 'Madras dry deciduous teak-bearing forests' or 'South Indian dry deciduous forests' (Champion and Seth 1968). Teak (185) is their most valuable tree species. It thrives well only below 1100 m. All intermediate forms exist between dense and clear tropophilous forests, savannas and shrubby savannas.

The long dry season on these tracts, including the lower Sigur Ghat, has greatly facilitated the regressive evolution of these forests into savannas, wooded grasslands, that are regularly destroyed by fires.

The upper canopy of the forests is not very tall (15-25 m). Among the most common species besides teak (185) are the button tree (16), the common myrabolam (186), the Indian kino tree (148), emblic myrabolam (67), and several bamboos. Sandalwood (159) has been planted in the foot-zone of the Sigur Ghat. All these species are demanders of light, fire-resistant and deciduous from December to March.

Where drought periods are shorter and less severe, for instance between Mudumalai and Naduvattam, benteak (98), conspicuous with its white boles, becomes abundant among giant bamboos (23). The herbaceous cover, sometimes tall and continuous, is mostly composed of various grasses (48, 189).

The region still possesses vestiges of more or less dense thickets of little economic interest. Many species are armed with thorns: the prickly brasiletto climber (149), Bengal currant (31), prickly nail dye (24), bell vervein (80), several jujubes (200), etc. The grass cover is very poor, often overgrazed and discontinuous.

THE FORESTS OF THE MONTANE ZONE

Their ecological conditions differ entirely from the previous forest types. In the montane zone of the Nilgiris, above 1800 m, the cool average temperatures of the coldest months (10-15° C) together with the frost phenomenon during the NE Monsoon period imply a strict selection of botanical species. Another important bioclimatic factor is the extremely variable humidity of the air in the non-forest areas: it varies within very short distances between saturation point and 30 per cent or less.

Despite the spatial climatic differentiation over the Nilgiris Plateau described earlier in this chapter, the vegetation climax is that of an evergreen montane forest type which, however, is now confined to depressions. These relic forests are present in almost all basins of the Nilgiris as well as on the Palnis and Anaimalais. They occupy very restricted and often isolated areas along the valley bottoms. Therefore they are essentially edaphic formations, characterized by rather rich soils, good drainage and relatively high soil moisture content throughout the year.

The residual montane forests show a high degree of structural and physiognomic uniformity in spite of their different exposure to the elements, particularly wind and rainfall. Their growth height varies from basin to basin and generally decreases from the western Nilgiris towards the eastern, more sheltered parts. Nevertheless, they all come under the type of evergreen montane forest and they are unanimously referred to as 'sholas' by the peoples of the Nilgiris.

A 'shola' (Badaga *so:le*—Fig. 2.13) is an evergreen forest of the elevated Nilgiri Plateau, located along stream banks or in hollows and surrounded by large tracts of savanna. It is essentially composed of medium-sized (15-20 m) and small trees (7-15 m). Its timbers are of much less value than those of lowland forests. Generally speaking, the timber produced by native trees is used very little. A 'shola' has two or three woody strata; undisturbed, it is rich in epiphytes (mosses, orchids, ferns). On the ground the herbaceous cover varies considerably according to soil moisture, thickness of wood cover, etc. (Interesting examples can be observed near Avalanche Reservoir, north of Mulligur.) The dominant stratum is rather irregular, mainly consisting of medium-sized and some scattered emerging elements (20-28 m); the average density of these emerging trees is about 70 per ha. Their diameter at breast height rarely exceeds 50 cm.

However, ever so often one finds specimens of big cinnamon (38), jambu (180), and the saw-leaved olive linden (59) having a regular trunk and a great diameter, exceeding 100 cm. They have voluminous, more or less hemispherical crowns, bearing small but very dense leaves (Pl. 8). The average height of the other species composing this stratum varies between 12 and 20 m, viz. the wild olive (122), kidney plum (150), Nilgiri elm (35), and shepherd's panax (162-3).

With the exception of three species all trees are evergreen. The Nilgiri magnolia (111), the Nilgiri elm (35), and the spiraea tree (109) may be partially or totally deciduous during the dry season, in January and February. They are 'pioneers' at the edges of the 'sholas' or in forest clearings. (Good examples are found around Nanjanad.)

The secondary stratum of the 'sholas' comprises small gnarled, tortuous trees with an average height of 7 to 12 m. The stratum is entirely composed of microphyllous species with less dense foliage, grouped at the tip of the branches. The following species are typical: the elliptic-leaved holly (86), *Rapanea wightiana* (153), wild tea (187), sweet-leaf (176), and holly-flowered spindle tree (113).

The shrubby stratum, if undisturbed, is very dense, with densely branched stems of five species of strobilanthes (171), which outnumber other species such as bastard ipecacuanha (146) and offal shrub (101). Typical species found close to the rivers and streams are: oblanceolate-leaved wild tea (74), yellow milkwort (142), guelder rose (197), etc. There is no large tree fern (47) in the montane zone of the Nilgiris: the only tree fern found in abundance is the grove fern (10; Pl. 7).

No tree or shrub of the 'sholas' tolerates poorly drained, water-soaked soil, with the probable exception of the hairy yellow-flowered laurustinus (198), endemic in the Nilgiris, and the Indian willow (158), a species completely deciduous for two or three months.

The herbaceous stratum is more or less continuous, rarely dense. It is floristically rich, chiefly in Urticaceae such as the harmless nettle, (63-4), bastard nettle (131), Nilgiri nettle (100), orchids, mints, balsams, and ferns.

Epiphytes are very numerous on the dominating strata, essentially along the watercourses. They are found in all humid forests of this zone. The most

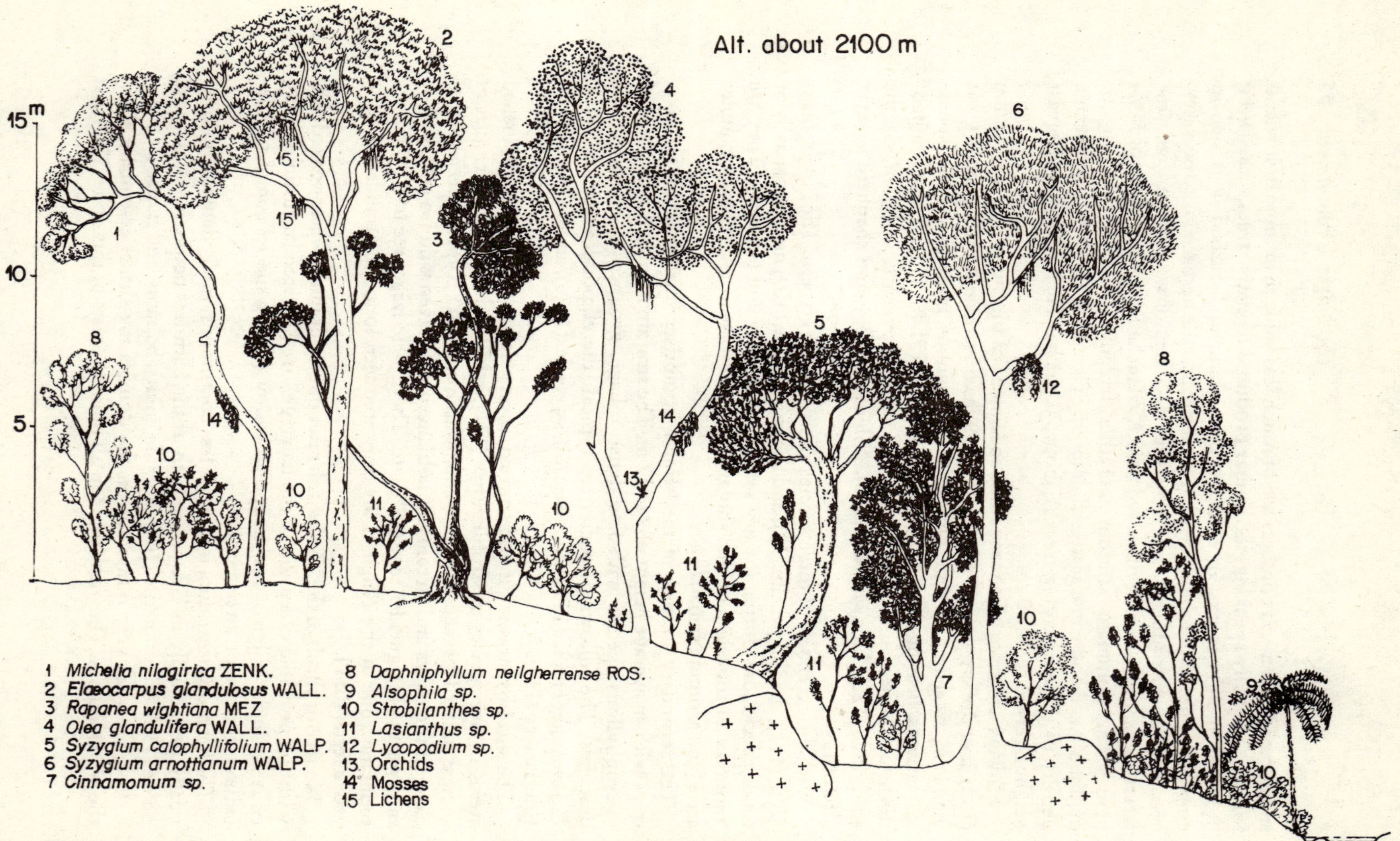

FIG. 2.13 Profile of a Shola.

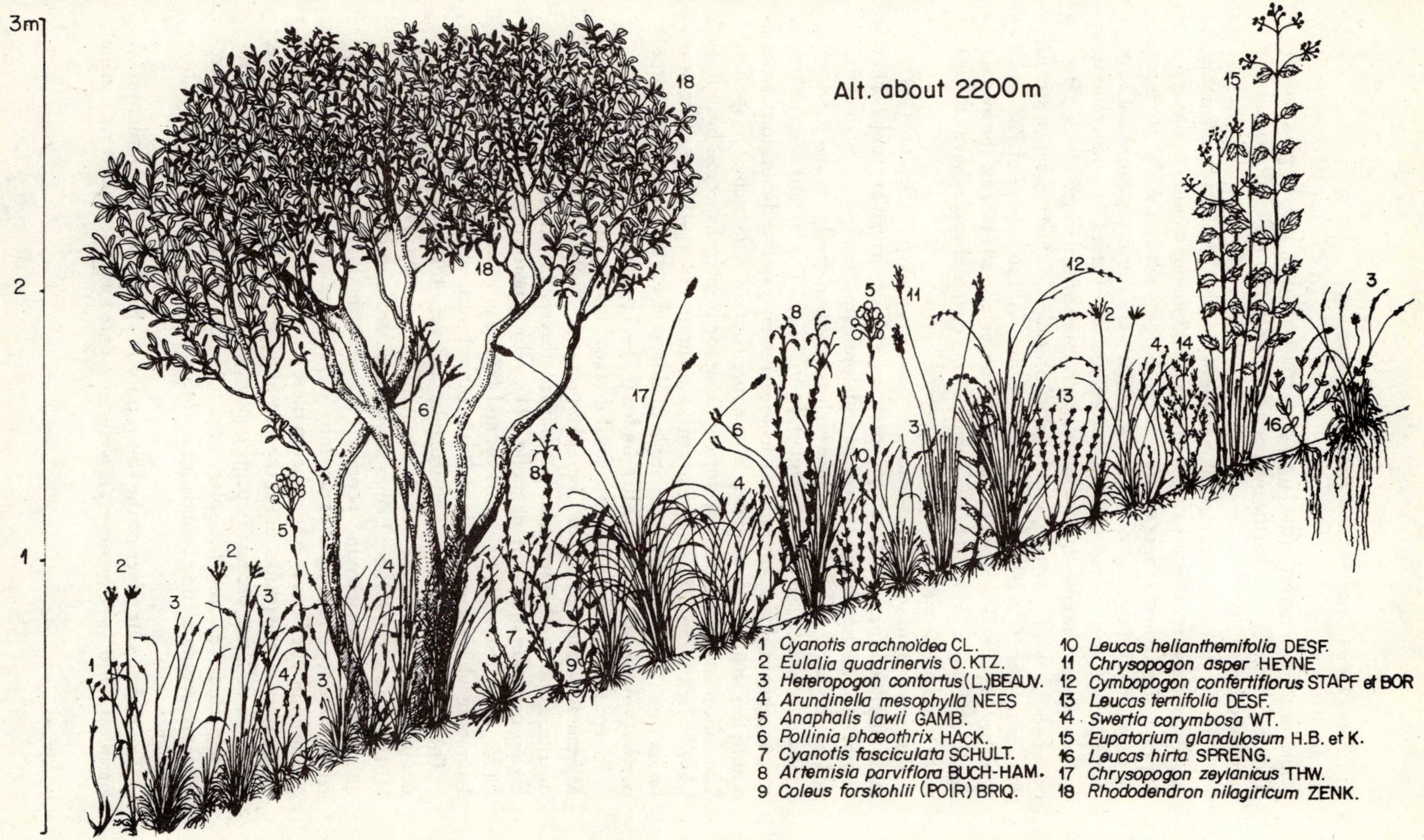
3m
2
1
Alt. about 2200m
1 Cyanotis arachnoïdea CL.
2 Eulalia quadrinervis O. KTZ.
3 Heteropogon contortus (L.) BEAUV.
4 Arundinella mesophylla NEES
5 Anaphalis lawii GAMB.
6 Pollinia phaeothrix HACK.
7 Cyanotis fasciculata SCHULT.
8 Artemisia parviflora BUCH-HAM.
9 Coleus forskohlii (POIR) BRIQ.
10 Leucas helianthemifolia DESF.
11 Chrysopogon asper HEYNE
12 Cymbopogon confertiflorus STAPF et BOR
13 Leucas ternifolia DESF.
14 Swertia corymbosa WT.
15 Eupatorium glandulosum H.B. et K.
16 Leucas hirta SPRENG.
17 Chrysopogon zeylanicus THW.
18 Rhododendron nilagiricum ZENK.

FIG. 2.14 Profile of a Shrub Savanna

common are two orchids (the larger pink tree orchid, 8, and the pink sac-orchid, 157), one Piperaceae (the pepper elder, 128) and one Polypodiaceae (the polypody fern, 139).

Naturally there are altitudinal variations of the 'shola' physiognomy, mainly on account of different exposure to the dominant winds. On the upper slopes of the highest ranges of the western Nilgiris where the SW Monsoon blows most violently, we only find low 'sholas' very dense, not exceeding 10 m in height and constituted of small-leaved species of trees. (Good examples are found near Lakkidi on the southwestern spurs of Kolaribetta Peak.) Here dwarf-sholas with only two woody strata prevail, the upper of which attains 8 m, rarely 9 or 10 m. They are extremely dense, with an average of 3000 small trees per hectare. The diameter of trunks never exceeds 60 cm; the average diameter is about 15 cm, at breast height. 34 per cent of the trees have a diameter of less than 10 cm; only 6 per cent exceed 30 cm.

The most voluminous trees belong to beautiful-leaved black plum (181), a microphyllous species with short, tortuous and very irregular trunks. They number 450 per hectare on an average and dominate other species by their size and by the width of their crowns which are comparatively regular in shape, broad and densely leaved at the end of the branches. The crowns of other tree species, always closely intermixed, have extremely variable dimensions, from 6 m to less than 1 m in diameter. This upper stratum is the richest in species. The following is the result of our counting for principal species: beautiful-leaved black plum (181)—26 per cent; cinnamon (38-9)—15 per cent; umbrella tree (180)—5.5 per cent; apple bladder-nut (191)—5 per cent; gutta-percha (96)—4 per cent; and olive-linden (60)—3 per cent.

Except for the dominance of beautiful-leaved black plum (181—and of the Myrtaceae in general) and of *Cinnamomum* spp. (38-9), the other species are distributed in a rather irregular way in these forests. Here and there appears a high proportion of Nilgiri magnolia (111), a shrub noteworthy for its regular trunk, almost straight for 3 to 4 m, and Brazilian nutmeg (43), an endemic Lauraceae which is practically absent in other forests of the Nilgiris.

The secondary stratum is an irregular and very tortuous pole-formation dominated by the spindle tree (72); bastard ipecacuanha (146) and offal shrub (101) are also found here. Lianas are less numerous, but bastard privet (78) and common Ootacamund pepper (137) reach the highest crowns. At the edges there are also Indian ipecacuanha (81) and virgin's bower (40).

The epiphytic flora is even richer in these high-altitude 'sholas'. Mosses are found only on the trunks and at the base of principal branches. Besides these, the larger pink tree orchid (8), pepper elder (128), dagger orchid (119), pseudo-bulb orchid (68), and the spleenwort fern (21) are common everywhere. There is an outstanding abundance of club moss (106) right up to the highest branches.

The herbaceous cover of the soil is poor in species and discontinuous. A harmless nettle (64), however, is well represented with scabrid-leaved snake root (124), and *Polystichum aculeatum* var. *rufobarbatum* (a fern, 143) occurring along the shady watercourses.

The bamboo brakes are secondary riparian formations in which the shrubs have been destroyed by fire and replaced by a profusion of bamboos resulting in dense thickets of small hill bamboo (19) with maximum heights ranging between 3 and 5 m. At their outer limits, a small bushy Acanthaceae, *Nilgirianthus wightianus* (118) frequently forms a loose narrow belt. Very often we find *Moonia heterophylla* (114), a Compositae occupying moist and open places, jointly with the Acanthaceae.

Generally, the umbrella tree (180), Nilgiri privet (104), hill gooseberry (155), *Tarenna asiatica* (184) and the holly-flowered spindle tree (113) are found in the bamboo brakes. In the Sispara Pass region we encounter, in addition, a great number of *Photonia serratifolia* (130) and of the obtuse-leaved sweet-leaf (177). In the central Kunda Range (Western Catchments area) shepherd's panax (163) and madder (121), are found in abundance; at Mukurti the following occur: Indian cranberry (194), wild tea (187), umbrella tree (180), elliptic-leaved holly (86), *Rapanea wightiana* (153), Nilgiri magnolia (111) and beach-leaved melon featherfoil (79). In the Upper Bhavani area, the flora is further diversified by guelder rose (197) and another obtuse-leaved sweet-leaf (178).

THE SAVANNAS OF THE MONTANE ZONE

The vast extension of the grasslands and savannas on the Nilgiris Plateau raises the question of the role of biotic factors. Until now the mystery of the antiquity of human settlement remains unsolved (see Noble's chapter, below). When at the beginning of the nineteenth century the Europeans discovered the high plateau of the Nilgiris, the forests occupied only small areas. Thus Ouchterlony noted in 1847: 'The Neilgherries, or rather the plateau formed by their summits, are by no means densely wooded, the forest occurring in distinct and singularly isolated patches, in hollows. This absence of forest leads me to conclude that the vast tracts of primeval forest land must have been cleared to make room for cultivation at no very distant period' (*in* Baikie 1857, Appendix IX, p. 1).

On the slopes, the extensive felling of trees is of relatively recent origin. It was carried out mostly after 1850, i.e. after the introduction of the most important commercial crops, coffee and tea, and their cultivation in large plantations. From the end of the nineteenth century till now the demographic growth of this region has been accelerated by in-migration, and agricultural land-use has expanded considerably, including efforts to reafforest vast areas, mainly with Australian trees (*Acacia* spp. and *Eucalyptus* spp.). Little by little even the savannas are disappearing because of large silvicultural plantations that, during recent years, have even been extended to the edge of the western plateau (Pl. 4).

Two grassland formations, or savanna types, can be distinguished on the

Nilgiris Plateau: the 'common savanna' found around inhabited places and covering large contiguous tracts of land, and the peculiar, less extensive 'high rainfall savanna'. Regarding their ecological factors we can say that temperature conditions do not vary much from one place to another. However, diurnal variations are extremely high. Within a few minutes temperatures at sunrise from –1°C to +9°C have been registered on similar savannas on the Palnis from December 1967 to March 1968 (Legris and Blasco 1969). The diurnal amplitude, between minima around 6 a.m. and maxima around 2 p.m., is very high during the NE Monsoon period. Close to the ground it often exceeds 35°C. Such fluctuations are certainly unfavourable for the growth of most indigenous tree species.

Annual rainfall generally exceeds 1500 mm on the upper Nilgiris and it permits the development of sloppy soils bearing hygrophytes during the rainy season, from June to October.

The average leached ferralitic soils found under these savannas have three distinct horizons easy to recognize almost everywhere (Troy 1979):
surface horizon: dark, rich in organic matter, sometimes very deep, with a fine structure and containing great amounts of roots without preferential orientation;
gravelly horizon: with a lower amount of organic matter, but rocky elements representing at least 30 per cent, thickness between 20 and 40 cm;
clayey horizon: light orange in colour, sometimes reddish, almost constantly humid, practically without living roots, sometimes several metres thick.

Another important ecological factor responsible for the extent of these savannas, for a great floristic selectivity and adaptation and for the slow dynamism of woody elements, is the periodic fires which spread rapidly over practically all montane grasslands. Every year from December to April, fires are possible because savannas dry up after the Second Intermonsoon rains cease (Pl. 9). However, at the end of the dry season, with the beginning of the First Intermonsoon rains, almost all herbaceous species rapidly sprout again and blossom.

The common savanna (Fig. 2.14) is found almost everywhere on the Nilgiris, near villages, towns, or even around wattle and eucalyptus plantations. It is recognizable by the dominant tufts of the most common grass, *Chrysopogon zeylanicus* (36), conspicuous with its narrow leaves which are left ungrazed by cattle. On the Nilgiris, as well as on the Palnis and even in Sri Lanka (Ceylon), the development of this grass is associated with human interference. For this reason, these savannas cover many old plantations and arable land. Vast stretches are also encountered in the Mukurti region where *Arundinella fuscata* (20) abounds too. Among other common grasses, we find auburn tresses (141), *Themeda triandra* (190), purple grass (14) and *Ischaemum indicum* (95).

The most common shrubs and small trees are a strobilanth (172), common St. John's wort (85), Nilgiri rhododendron (154), Indian cranberry (194), and hill gooseberry (155). The floristic diversity of these savannas is usually very great: *Pleocanthus sessilis* (138), white dead-nettle shrub (102), cow-parsnip (83), ragwort (164), Nilgiri everlasting (11), etc. impart a floristic character unknown anywhere else. Many of the savanna species of the Nilgiris tolerate hydromorphic soils. Examples are balsams (88); common wild anemone (15) and teasel (57) are frequent in the savannas of Avalanche. Distinct floristic differences can be noticed according to locality.

The 'high rainfall savanna' is only found in the southwestern and westernmost parts of the Nilgiris, including the Western Catchments, the area west of Lakkidi and around Sispara Pass. Among the most essential ecological factors are extremely strong winds and excessive rainfall during the SW Monsoon period that are hostile to plants, animals and human beings alike. The ecology of this savanna type is not conducive to the growth of trees. Low temperatures during the dry season and then violent winds and rains during the SW Monsoon period result in poor, coarse soils that also render attempts at afforestation of these tracts of the Nilgiris a difficult task (Pl. 4).

This vegetation type consists of a more or less dense herbaceous cover, interrupted only by rocky outcrops and dotted with a few gnarled and stunted woody species: Nilgiri rhododendron (154), Nilgiri privet (104), and beautiful-leaved black plum (181). A number of dwarf undershrubs are encountered which are absent in the Palnis: two strobilanths (173-174), wood germander (188), white dead-nettle shrub (103), Nilgiri everlasting (11), pearly everlasting (12), and creat (13), among others.

The grasses well adapted to maintain themselves in these regions are *Themeda triandra* (190) and *Isachne kunthiana* (94). During the rainy season, a number of balsams may be collected: red liberty cap (93), and garden balsam (87, 89-92).

EVOLUTION OF VEGETATION

Since 1938, when Ranganathan published his noted 'Studies in the ecology of the shola grassland vegetation of the Nilgiri Plateau', two very different views have been expressed regarding the ecological status of the montane grasslands in South India. In the opinion of some foresters and scientists, grasslands are the natural vegetation type of the high plateaux (climax). Others consider them to be entirely artificial, due to the destruction of the 'sholas' and maintained by frequent fires.

If we accept Ranganathan's view that the effect of frost is a determining factor limiting the growth of woody species, then grasslands are to be considered as the best adapted to present ecological conditions and as *climaxic* (see Ranganathan 1938 and Meher-Homji 1965). If, on the contrary, we accept the commonly expresed view that fire is the principal agent responsible for the destruction of the forest and for the present stability of the savannas, then grasslands are only secondary formations due mainly to anthropogenic factors (see Bor 1938, Shankarnarayan 1958, Legris 1963, Gupta 1960, etc.).

The limit between 'sholas' and grasslands is generally well marked. Their floristic compositions also differ entirely from one another. Naturally both vegetation types are also different in their topographical location: the 'sholas' are only found in the bottom of the valleys and coombs, whereas the grasslands are found almost everywhere. Recent research has shown that almost all indigenous woody species of the Nilgiris Plateau are slow-growing plants. Moreover, the majority of trees are very sensitive to fire and were destroyed at the stage of seedlings or small bushes. Therefore biological reasons (slowness of growth) as well as direct or indirect human influences (fire and frost) explain the remarkable extent and actual perenniality of montane grasslands.

Regarding the early history of the evolution of the vegetation cover no clear explanations have yet been given. Several diagrams of fossil pollens have been studied, particularly for the Pykara, Parson's Valley, Kakathope and Rees's Corner areas (Vishnu-Mittre 1974). Until now the oldest reconstructable vegetation history dates back to about 38,000 years, in the vicinity of Kakathope (Wenlock Downs). It appears that even without human interference local vegetation has followed important changes in the form of progressive evolution, with a decline of the grasslands (35,000 years) and a rise of woody plants, reaching a dominance of 'shola' species (pollen) around 15,000 years ago. As far as we know most of the present grasslands of the upper Nilgiris are fairly old, at least 3000 years (Blasco and Thanikaimoni 1974).

These conclusions are provisional. The main reason for the uncertainty is the insufficiency of our knowledge regarding the representativeness of selected pollen spectra.

The 'Island' Character of the Nilgiris

On account of their altitudinal extent (2636 m) and their compact physiognomy interacting with the large-scale atmospheric control, the Indian Monsoon, the Nilgiris possess a unique environment within South India—in terms of horizontal as well as vertical diversity. The seasonal changes of upper-air dynamics, synoptic patterns and surface winds cause exceptional spatial contrasts between exposed and sheltered parts of the Nilgiris and consequent temporal alterations of wet and dry conditions. The pronounced luff-lee effect during the SW Monsoon period, from June to September, is the most striking example: while the Western Slopes and the Kunda Range receive excessive rainfall, severe drought conditions prevail some 40 to 60 km further east, in the dry valleys of the eastern Nilgiris and on the adjacent tracts of the Mysore Plateau and the Coimbatore Plains in particular.

However, the environmental peculiarity and individuality of the Nilgiris are even more accentuated by the thermal differentiation of the area due to the decrease of temperature with altitude. The present chapter, it is hoped, provides sufficient climatological, pedological and botanical evidence of the remarkable altitudinal variation of ecological conditions and of the particular physical environment of the elevated Nilgiris Plateau. The environmental conditions and ecological potentials of the upper Nilgiris are so different from those on the lower slòpes of the mountain block and its surroundings that the former could well be described as a cold tropical 'island' rising above the ecological 'barrier' of the warm tropical 'sea' of South India; and the high degree of floral as well as faunal endemicity found on the Nilgiris Plateau further supports this comparison with the specific character of isolated island ecosystems.

Appendix

Botanical Identifications of 200 Nilgiri Plants

	COMMON ENGLISH NAME	BOTANICAL	TAMIL
1	acacia, wattle, &c.	*Acacia* Willd., spp.	cīmai vēlam battai; vēla maram, &c.
2	white wattle; silver wattle; mimosa; bipinnate wattle	*A. dealbata* Link.	cīmai vēlam battai; vēla maram; veḷvēl
3	common tan wattle; green wattle; black wattle	*A. decurrens* Willd.	cīmai vēlam battai; vēla maram
4	black wattle	*A. mearnsii* Willd.	cīmai vēlam battai; vēla maram; cavukku
5	Australian blackwood; phyllode wattle; Tasmanian light wood	*A. melanoxylon* R. Br.	cīmai vēlam battai; vēla maram
6	rusty mimosa; copious narrow-leaved soap pod; fine leaved acacia	*A. pennata* (L.) Willd.	cīngai; cīngai muḷḷu; veḷḷai; iṇḍam būdai; iṇḍu; iṇḍamuḷḷu; kāṭṭiṇḍu; muḷḷuccīngai; pēyccīyakkāy
7	bear's breech family; bractwort family; justicia family	*Acanthaceae*	(numerous names)
8	larger pink tree orchid	*Aerides ringens* Fisch.	okkiṭṭu; pakaṭṭu vaṇṇamalarc caṭivagai
9	Nepalese alder	*Alnus nepalensis* Don.	pūrccamaram pōṉṟa maravagai
10	large tree fern; grove fern; common tree fern	*Alsophila latebrosa* Hook.	peruñcūral; maravaṭivac cūral
11	Nilgiri everlasting	*Anaphalis neelgherriana* DC.	
12	pearly everlasting	*A. wightiana* DC.	
13	creat	*Andrographis lawsonii* Gamb.	

	COMMON ENGLISH NAME	BOTANICAL	TAMIL
14	spear grass; purple grass	*Andropogon lividus* Thw.	paṉṟipul
15	common wild anemone	*Anemone rivularis* Ham.	
16	button tree	*Anogeissus latifolia* Wall.	veḷḷaināgai; namai; vekkāli; veḷḷai namai
17	alternate-leaved sugar palm; single-leaved sugar palm; wild coconut of Travancore	*Arenga wightii* Griff.	alāmbaṉai
18	wild jack; jungle jack; large jack; anjely	*Artocarpus hirsuta* Lamk.	akkiṇi; añcali; āyaṇibiḷā; āyini; kāṇḍambalā; kāṭṭuppaḷā; pēyppalā
19	common Nilgiri shrubby bamboo; small hill bamboo; hill reed	*Arundinaria wightiana* Nees.	mūṅgil
20	(erect grass)	*Arundinella fuscata* Nees.	pullu
21	spleenwort fern	*Asplenium falcatum* Lamk.	
22	margosa; neem; nim; ash-leaved bead tree	*Azadirachta indica* A. Juss.	aruḷuṟudi; congumaru; kimpam; kaḍuppagai; kiñci; mālugam; niriyacam; picidam; ukkiragandam; vēmbu; vēppu; varuṭṭam; vēppa maram
23	thorny bamboo; hollow large bamboo; female bamboo	*Bambusa arundinacea* Willd.	āmal; amai; periya mūṅgil; ambu; āmbal; aril; capam; cāṉagi; cey; kaḷai; kiḷai kāmbu; kīcagam; kuḻāy mūṅgil; muḷai; macukkaram; munduḻ; miruttucam; muḷḷu mūṅgil; muḍangal; mūṅgiṟkōl; nāḍi mūṅgil; neḍil; neṭṭil; pācu; palāndam; paṇai; pāṇḍil; peruvarai; taṇḍu; perumūṅgil; tiṇiyaṉ; taṭṭai; tuḷai; tūmbu; vaṉṉikaruppam; vedir; vēḻam; vāḻai; vēy; varaimūṅgil; vēṇu; vēral; veyal; viṇḍil; viṇṭu; iṟaivarai.
24	prickly nail dye	*Barleria* L., spp.	marudoṉṟi; muḷḷi; kuḍāṉ cem muḷḷi; ceṉguḍāṉ; koḍippacalai; kovindam; vaṟalmuḷḷi; veṭṭarguṟṟi, &c.

	COMMON ENGLISH NAME	BOTANICAL	TAMIL
25	dyer's berberry; Indian berberry; tree turmeric; Nilgiri barberry; yellow wood	*Berberis tinctoria* Lesch.	muḻḻukkaḷā; ūcikkaḷā
26	rattan çane; cane feather palm; &c.	*Calamus* L., spp.	pīrambu; cūral; cuvēdam; nīrvanci; vāṉiram; vēdacam; vēttiram; cuvēdakāṇḍam; piciṉ; mellicuppirambu; vañcikkoḍi.
27	poon spar tree	*Calophyllum* Bedd., spp.	pongu; pungu; malampuṉṉa; nāgam; cirupiṉṉai; namēru; puṉṉagam; puṉṉai; piṉṉai; kāṭṭupiṉai; viri
28	tea	*Camellia sinensis* L.	tēyilai; tēyilaic ceṭi
29	black dammar tree	*Canarium strictum* Roxb.	attam; karungundurukkam; karungungiliyam; karuppu dāmar; karuppukkungiliyam; kundurukkam; kungiliyam; kukkil; kungulu
30	honey cauray; spinous honey-thorn; common honey-thorn	*Çanthium parviflorum* Lamk.	cengārai; muḻḻu kārai; kaḍalattal; kārai; kuḍiram; nallakkārai; tèraṉai; tēravai
31	carissa; Bengal currant	*Carissa* L., spp.	kaḷā; kaḷākkāy; perungaḷā; kiḷākkāy; cirukkaḷā; ciṉṉakkaḷā
32	bastard sago; fern palm; kitul palm	*Caryota urens* L.	kūntaṟ paṉai kamuga; pūgam; konda paṉai; tippilippaṉai
33	downy rachis glandular senna; eared senna; tanner's senna; avaram bark	*Cassia auriculata* L.	āvaram; āvārai; āvirai; cadurgūli; cemmalai; cummai
34	Himalayan cedar; deodar; Indian cedar; sacred Indian fir; Indian sanobar; lancet tree; fountain tree	*Cedrus deodara* (Roxb.) D. Don.	tēvatāram; tēvatāri; tēvatāru; vaṇḍugolli

	COMMON ENGLISH NAME	BOTANICAL	TAMIL
35	Nilgiri elm; lotus tree; grecian honey berry; Roxburgh's Indian nettle	*Celtis tetranda* Roxb.	kuviyā; āḍa; kōṇa; murungaṉ; tavuḍi
36	(grass)	*Chrysopogon zeylanicus* Thw.	pullu
37	cinchona; chinchona; Peruvian bark; Jesuit's bark; Countess' powder; Cardinal's bark	*Cinchona* L., spp.	ilavangap paṭṭai; lavangam; kuḷimātti
38	big cinnamon; wild cinnamon; jungle lavangam	*Cinnamomum wightii* Meissn.	cembilā; ilavangam; kāṭṭukkaruvā; pulippalā; tāḷicappattiri; veṟṟam; ilavangap paṭṭai
39	Ceylon cinnamon; common cinnamon; true cinnamon; sweet bark	*C. zeylanicum* Breyn.	caṉṉalavangapaṭṭai; coracattōraci; ilavangam; ilavangappaṭṭai; karuvā; pulambiḷavu
40	virgin's bower; clematis; traveller's joy; old man's beard	*Clematis theobromina* Dunn.	taḻuviyēṟum koṭivagai
41	coffee; Arabian coffee	*Coffea arabica* L.	kāpi
42	wild taro	*Colocasia antiquorum* Schott.	caṉa tumpa
43	Brazilian nutmeg	*Cryptocarya lawsoni* Gamb.	pālai
44	cryptomeria; Japanese cedar; sugi pine	*Cryptomeria japonica* D. Don.	jappāṉiya tēvatāru maravagai
45	wild durian	*Cullenia excelsa* Wt.	āṉaippalā; malaikkōñcil; pālāvu; veḍpalā
46	Californian cypress; Monterey cypress	*Cupressus macrocarpa* Hartw.	cūram
47	large tree-fern; cup tree-fern	*Cyathea* Smith, spp.	peruñcūral; maravaṭivac cūral

	COMMON ENGLISH NAME	BOTANICAL	TAMIL
48	citronella grass	*Cymbopogon nardus* (L.) Rendle var. *confertiflorus* (Steud.) Stapf. ex Bor	pullu
49	broom	*Cytisus* L., spp.	āṭakam; āṭaki; tuvarai
50	blackwood; Malabar blackwood; dark blackwood; retuse leaved rosewood; Bombay rosewood; Bombay blackwood	*Dalbergia latifolia* Roxb.	īṭṭi; karundōrviral; karuntuvarai; nūkkam; nūkku; tōdagatti; tavaḍi
51	common yam	*Dioscorea bulbifera* L.	paṉṉukkiḻangu; kāṭṭu vaḷḷi; pūṇṭuk kāyvaḷḷi; pūṇṭuccērakavaḷḷi
52	common yam	*D. hispida* Dennst.	pey perendai
53	betel yam	*D. oppositifolia* L.	kiḻangukkoḍi; verōlaivaḷḷi tavaikaccu
54	prickly yam	*D. pentaphylla* L.	cedukkandi; vaḷḷaikodi;
55	prickly yam	*D. pentaphylla* L. var. *wightii* Prain & Burk.	cedukkandi; vaḷḷaikodi;
56	thorny yam	*D. tomentosa* Heyne	nalvēli kiḻangu; caval kiḻangu
57	teasel	*Dipsacus leschenaultii* Coult.	muḷḷi; iḻaiyūri
58	white cedar	*Dysoxylum malabaricum* Bedd.	veḷḷaiyakil
59	saw-leaved olive-linden	*Elaeocarpus glanduiosus* Wall.	kāṭṭukkārai
60	olive-linden	*E. recurvatus* Corner.	
61	saw-leaved olive-linden; paralysis seed tree; wild olive; rahoo seed tree	*E. serratus* L.	olangarai; sēlāmaram; uruttirāḍcam; uruttirākśam
62	downy-nerved olive-linden; utrasum bead tree; rudraksha	*E. tuberulatus* Roxb.	uruttirāḍcam; uruttirākśam; pagumpal
63	harmless nettle	*Elatostemma lineolatum* Wt.	muḷ illāta muḷceṭi pōṉraceṭi vakai

	COMMON ENGLISH NAME	BOTANICAL	TAMIL
64	harmless nettle	*E. lineolatum* Wt. var. *falcigera* Thw.	
65	(a herb)	*E. sessile* Forst. var. *pubescens* Hook. f.	
66	(small herb)	*E. surculosum* Wt.	
67	emblic myrabolam; featherfoil; gooseberry tree; awla tree	*Emblica officinalis* Gaertn.	āmalakam; indul; andakōram; attakōram; cirōṭṭam; kāṭṭunelli; nelli; perunelli; tāttiri; toppunelli
68	small tree orchid; pseudo-bulb orchid	*Eria nana* A. Rich	okkiṭṭu
69	Canadian fleabane	*Erigeron canadensis* L.	
70	bluegum; fever-gum; eucalypt	*Eucalyptus globulus* Labill.	karuppūram; astrēliyāvil vaḻarum
71	rose-apple; black plum; Eugene myrtle; fruiting myrtle; West Indian myrtle; jaumoon	*Eugenia* L., spp.	evujeṉiyenpavaḻ
72	spindle-tree	*Euonymus crenulatus* Wall.	vīṇi; malaikkurattai
73	ironweed	*Eupatorium glandulosum* H.B.K.	
74	oblanceolate-leaved wild tea	*Eurya japonica* Thunb., var. *nitida* Korth.	
75	fragrant quaternion tree	*Evodia* Forst., spp.	
76	ovate-leaved fig; Bengal fig; common banyan; Indian fig; grove tree; descending tree; wad; arbor-de-rais	*Ficus benghalensis* L.	āl; āla; ālam; ēguvācam; kā; kāṉ; kagavacugam; kavacugam; kōḷi; nekkurōdam; pāli; paḻu; pērāl; pūdam; pūdavam; pūdaviruḍcam; toṉ; vaḍam; vaḍavirukkam

	COMMON ENGLISH NAME	BOTANICAL	TAMIL
77	pipal; peepul; Indian aspen; jointed poplar-leaved fig; common long-tailed peepul; Indian poplar; sacred fig; Bo tree; bodhy; wisdom tree; Vishnu's tree	*F. religiosa* L.	acuvattam; accuvattam; aracu; arayāl; attiru; attugamāṉi; caladaḷam; caraṇam; cuvalai; ilaṇai; kaṇavam; kuñcarācaṉam; magādurnmam; marā; nārāyaṇam; pādarōgaṇam; paṇai; pittalam; taṇavam; tiru; vāṇagandi
78	bastard privet; stipulelined privet	*Gardneria ovata* Wall.	
79	beach-leaved melon featherfoil	*Glochidion fagifolium* Miq.	vayaṉṉaṉṉal
80	bell vervein; small Cashmere tree	*Gmelina asiatica* L.	arucāvirā; kaḍambal; kadukkumiḻ; kumiḻ; kumiḻmaram; kumiḻañ ceṭi; kumirkoḍi; nilakkumaḻā; ciṟukumaṭāṉ
81	Indian ipecacuanha; ipecacuanha swallow-wort	*Gymnema montanum* Hook. f.	ādikam; ayakam; amuduputpam; cakacaram; cirukuriñcā; kōkilam
82	umbel spiked pepper; fan-nerved leaved pepper	*Heckeria subpeltata* Kunth.	
83	cow-parsnip; hogweed	*Heracleum hookerianum* W. & A.	pacuttīvaṉac ceṭivagai
84	mountain tamana oil tree	*Hydnocarpus alpina* Wt.	āṟṟu cancalai; malaivaṭṭai; vaṭṭai
35	common St. John's wort; tutsan; Aaron's beard	*Hypericum mysorense* Heyne	mañcaḷ malaruḷḷa ceṭivagai
86	elliptic-leaved holly	*Ilex wightiana* Wall.	veḷḷudai
87	garden balsam; touch-me-not.	*Impatiens acaulis* Arn.	kācittumpai
88	" "	*I. chinensis* L.	"
89	" "	*I. clavicornu* Turcz.	"
90	" "	*I. crenata* Bedd.	"

	COMMON ENGLISH NAME	BOTANICAL	TAMIL
91	" "	*I. pusilla* Heyne	"
92	" "	*I. scapiflora* Heyne	"
93	red liberty cap	*I. tomentosa* Heyne	"
94	(a grass)	*Isachne kunthiana* W. & A.	pullu
95	(a grass)	*Ischaemum indicum* (Houtt.) Merr.	pullu
96	gutta-percha	*Isonandra candolleana* Wt.	kaṭṭiṟukka cāmpalniṟa marappāl piciṉ vagai
97	mint family	*Labiatae*	
98	benteak; crape myrtle; white teak; venteak	*Lagerstroemia lanceolata* Wall.	kaccaikkaṭṭai; veṇḍēkku; veṇveyilā
99	lantana; wild sage	*Lantana camara* L., var. *aculeata* Mold.	makkaḍam pū; uṉṉi cedi
100	Nilgiri nettle	*Laportea terminalis* Wt.	oṭṭarpalā
101	offal shrub	*Lasianthus capitulatus* Wt.	
102	white dead-nettle shrub	*Leucas rosmarinifolia* Benth.	caṉṉattumpai
103	" " " "	*L. suffruticosa* Benth.	"
104	Nilgiri privet	*Ligustrum perrottetii* A. DC.	puṉkaṉ; kōli
105	Australian holly gum; tallow laurel; grey mango laurel; betel-nut laurel; Ceylon laurel	*Litsaea* Lamk., spp.	
106	club moss; fir club moss; stag's horn moss	*Lycopodium hamitonii*	
107	holly-leaved berberry; Nepal barberry	*Mahonia* Nutt., spp.	kaṭa; muḷḷukkaḍambu; muḷḷukkaḷā; muḷḷumurunga

	COMMON ENGLISH NAME	BOTANICAL	TAMIL
108	cupid's favourite; cuckoo's joy; spring tree; mango; common mango	*Mangifera indica* L.	ādiceḷarayam; āmiram; āmbiram; cēdāram; cēgāram; cūdam; cuḷḷi; iradaṃ; keccakkār; kilimūkkumā; kōgilōrcavam; kokku; mā; mādi; mādudūdam; māgandam; māḷai; māmagam; māndi; maṉmadaṅgaṉai; mattiyagandam; mirudālagam; omai; palacirattam; palōrbatti; pigubandu; palōrbatti pigubandu; tēmā; tevam; tiḍalam
109	spiraea tree; honey-sweet tree	*Meliosma arnottiana* Walp.	kadiri; tagari; tuḷi; kuccavir
110	nahor oil tree; Ceylon iron-wood; serpent	*Mesua ferrea* L.	cirunāgappū; iruḷ maram; palai; nāgēcuram; nāṅgu; nākkēcuram; nāgacampikai
111	champak; white champak; Nilgiri magnolia; Nilgiri tulip-tree; Nilgiri champak	*Michelia nilagirica* Zenk.	cambagam; kāṭṭuccaṇpagam; Nilakiri caṇpakam; ceṇpagam
112	spindle tree	*Microtropis* Wall., spp.	katirkkōl ceyvataṟkup payaṉpaṭum mara vagai
113	holly-flowered spindle tree	*M. ramiflora* Wt.	" " "
114	(a herb)	*Moonia heterophylla* Arn.	
115	wild nutmeg; mace; jauty	*Myristica* L., spp.	cātikkāy; ātipalam; cāti; attikam; cālukam; citam; camuttirantam; ciṉēvam; cirukāriṭam; cīvikā; kōcam; civakaram; civikaram; tuvitāttumakam, &c.
116	longan tree; burdock soap-nut	*Nephelium* L., spp.	iliṭci; kāṭṭuppūvam; pirappiṉ; pūvam; puvatti; varattarpūvam; muḷḷai &c.

	COMMON ENGLISH NAME	BOTANICAL	TAMIL
117	(rubiaceae)	*Neurocalyx* Hook., spp.	
118		*Nilgirianthus wightianus* (Nees.) Brem.	
119	dagger orchid	*Oberonia brunoniana* Wt.	okkiṭṭu
120	shrubby bamboo; berry-bearing bamboo; reedy bamboo	*Ochlandra* Thw., spp.	īrttarkaḷḷi; irttal; kagamūngil; oḍai; karīrttal
121	madder	*Oldenlandia beddomei* O. Kze.	mañciṭṭi; mañcaḷ malarkaḷaiyuṭaiya pūṇṭu vakaikkoṭi
122	glandular-leaved olive; wild olive	*Olea glandulifera* Wall.	iṭalai; payir; kaṉaipporumpalu; kuraikkal
123	(item omitted)		
124	scabrid-leaved snake root	*Ophiorrhiza hirsutula* Wt.	cataicci; kīrippuṇḍu; kīrippuraṇṭāṉ
125	yellow wood-sorrel; Indian sorrel	*Oxalis corniculata* L.	
126	Indian gutta percha; Malabar caoutchouc; oval mohwah; pauchouty	*Palaquium ellipticum* Engler.	iluppai; madūkam; kāḍḍau ilupai
127	Kikuyu grass	*Pennisetum clandestinum* Hochst.	pullu
128	pepper elder	*Peperomia reflexa* A. Dietr.	
129	small date-palm; dwarf date; common hill dwarf date; long-peduncled dwarf date; straight stemmed glaucous dwarf date	*Phoenix humilis* Royle	ciruyiñcu; īccamā
130		*Photonia serratifolia* (Desp.) Kalm.	
131	bastard nettle; soft nettle	*Pilea* Lindl., spp.	

	COMMON ENGLISH NAME	BOTANICAL	TAMIL
132	feather palm; Western Ghats cane betel palm	*Pinanga dicksonii* Bl.	
133	Mexican yellow pine	*Pinus patula* Sch. & Cham.	
134	Monterey pine	*P. radiata* D. Don.	
135	Mexican yellow pine	*P. roxburghii* Sarg.	
136	pepper vine	*Piper* L., spp.	(many specific names)
137	common pepper of Ootacamund; glabrous loose-spiked pepper	*P. schmidtii* Hook. f.	
138		*Pleocanthus sessilis* (Ness.) Brem.	
139	polypody fern	*Pleopeltis lanceolata* Kaulf.	iḻattalari; kuppiyaḻari; kāḷḷimandārai; nāvillāvaḻari; perungaḷḷi
140	pagoda tree; jasmine spurge; temple tree; frangipani; Spanish American jasmine; goolicheen	*Plumeria acutifolia* Poir.	
141	auburn tresses (grass)	*Pollinia phaeothrix* Hack.	
142	yellow milkwort; red-eye; Nilgiri milkwort	*Polygala arillata* Ham.	naṅgai
143	(a fern)	*Polystichum aculeatum* Sw. var. *rufobarbatum* Wall.	
144	plum; prune	*Prunus domestica* L.	ālpokoṭā; pruṉiyē maṛattiṉ paḻam
145	peach	*P. Persica* Bth. & Hook. f.	mīṉ piṭittal
146	bastard ipecacuanha	*Psychotria elongata* Hook. f.	cīmūlappū
147	bracken	*Pteridium aquilinum* (L.) Kuhn. ex Deck.	kāṭṭupputar vagai; cūral vagai

	COMMON ENGLISH NAME	BOTANICAL	TAMIL
148	Indian kino tree; Indian dragon's blood; Malabar kino; kino rosewood.	*Pterocarpus marsupium* Roxb.	vēṅgai; acaṉam; caruvacātakam; kaṇi; carutākam; kuriñci; timil; pītacāralam; piracāram; pitakārakam; tamicu; taṉṉiṉi; timicam; timicu; udiravēnkai; vīcakā; vaṇḍunārmalar
149	prickly brasiletto climber; white brasiletto climber	*Pterolobium indicum* A. Rich.	cekappuccīyaku; cīyaku; īṇḍu; karīṇḍu; puliyīṇḍu
150	kidney plum	*Pygeum gardneri* Hook. f.	kāṭilai; kallumā; konkai; paḻankkacci
151	pear; wild pear; amrood; Keiffer pear	*Pyrus communis* L.	pēriyiṉakkāy vagai
152	apple; wild apple; crab apple	*P. malus* L.	āppiḷ paḻam
153		*Rapanea wightiana* Wall.	
154	Nilgiri rhododendron; large evergreen rhododendron; common rhododendron	*Rhododendron nilagiricum* Zenk.	aliṅki; eppōtum paccaiyāṉa
155	hill gooseberry; wild guava; woolly myrtle; tri-nerved myrtle	*Rhodomyrtus tomentosa* Wt.	kāṭukkoyyā; malai ocirukoyya; taviṭukkoyyā
156	bramble; blackberry; dewberry; raspberry; loganberry; vine-berry; Gowry's fruit	*Rubiaceae* L.	muḷḷi
157	yellow or pink sac-orchid; medium pendulous-stemmed orchid	*Saccolabium filiforme* Lindl.	okkiṭṭu
158	Indian willow; country willow; four-seeded willow	*Salix tetrasperma* Roxb.	vañji; nīruvañji; ārruppālai; cuvētam

	COMMON ENGLISH NAME	BOTANICAL	TAMIL
159	sandalwood; almug; algum; Malayam tree	*Santalum album* L.	srikantam; acam; anukkam; āram; cantaṉam; cantaṇi; cantaṉamppoṭi; cālēkam; cēlēkam; cēlōdam; ciciram; īkam; iṅgam; kandam; kōravāram; kuḷavuri; malaiveṭpu; malaiyāram; malaiyacam; pāṭīram; paṭīram; pāṭiram pītacāralam; pītacāram; ulōcidam
160	soapnut tree; monkey's blood	*Sapindus trifoliatus* L.	muṇippungu; neykkoṭṭāṉ; nīttavāñci; poṉṉāngoṭṭai; pūvandi; pūḍcikkoṭṭai; puṉalai
161	yellow broom	*Sarothamnus scoparius* (L.) W.D.J. Koch	
162	aralia; shepherd's panax	*Schefflera racemosa* Harms.	oppaṉaikkuriya koṭiyiṉa vagai
163	aralia; shepherd's panax	*S. rostrata* Harms.	oppaṉaikkuriya koṭiyiṉa vagai
164	ragwort; ragweed	*Senecio polycephalus* Cl.	
165	Californian cedar; Californian evergreen redwood; common Wellingtonia	*Sequoia sempervirens* Endl.	
166	China root; sarsaparilla; prickly ivy; rough bindweed; chobcheeny; wild sarsaparilla	*Smilax aspera* L.	
167	" " " "	*S. prolifera* Roxb.	
168	" " " "	*S. zeylanica* L.	kāṭṭukkoḍi
169	potato	*Solanum tuberosum* L.	
170	Johnson grass	*Sorghum halepense* (L) Pers.	pullu

	COMMON ENGLISH NAME	BOTANICAL	TAMIL
171	strobilanth; conehead; goldfussia	*Strobilanthes* Blume, spp.	kuriñci; kuruṅgai
172	" " "	*S. kunthianus* T. And.	kuriñci; kuruṅgai
173	" " "	*S. lawsoni* Gamb.	kuriñci; kuruṅgai
174	" " "	*S. wightianus* Nees.	kuriñci; kuruṅgai
175	Indian poison nut; nux-vomica; vomit nut; strychnine tree; false angostura bark; Krishna's arrow	*Strychnos nux-vomica* L.	citti; cuvācagam; eṭṭi; kāgōdi; kāgodi; kāḷam; kāñcigā; kāñciram; kāñcirai; karaḷam; kecamāmuṭṭi; kiruṭṭiṇābāṉam; kōbaguṇḍam; koḍaram; koṇḍaguḷam; kubacubā; maduragam; māvagam; muṭṭi; naccu; visamuṭṭi
176	sweet-leaf; lodh tree; Grenada tea; Californian quinine	*Symplocos* L., spp.	kambliveṭṭi; paraḷai
177	obtuse-leaved sweet-leaf	*S. microphylla* Wt.	kambliveṭṭi; paraḷai
178	obtuse-leaved sweet-leaf	*S. obtusa* Wall.	kambliveṭṭi; paraḷai
179	acute-leaved sweet-leaf	*S. spicata* Roxb.	kambliveṭṭi; paraḷai
180	umbrella tree; Arnott's mountain black plum; small jaumoon; jambu	*Syzygium arnottianum* Walp.	nāval
181	beautiful-leaved black plum	*S. calophyllifolium* Walp.	nāval
182	Nilgiri mountain black plum	*S. montanum* Gamb.	nāval

	COMMON ENGLISH NAME	BOTANICAL	TAMIL
183	tamarind; Indian date; chinch	*Tamarindus indica* L.	āmbilam; āmilam; āmiligai; cancivagaraṇi; carittarai; cevvārai; cincani; cindagam; cindam; cinduram; cīri; egiṉ; egiṉam; indam; kiñcam; mugiṇi; odīmaṇi; puḷi; tindiḍam; puḷiyam maram puḷi; tindiruṇi
184		*Tarenna asiatica* (L.) Saut. & Merch.	teraṇi
185	teak; Indian oak; ship tree	*Tectona grandis* L.f.	cāti; cāgam; caracam; kāḷinti; kumidigam; tēkku
186	common myrobolam; Indian myrabolam; chebulic myrabolam; black myrabolam; country gall nut; ink nut tree; Negro's olive	*Terminalia chebula* Retz.	akkantam; āmākōlā; arabi; aritagu; aritati; attaṉ; cēyā; kaṭukkāy; citēki; cinki; cirōṭṭam; cirrilai; civā; civanti; kaḍu; kakoṭakacinki; necci; pattiyam; piratamai; tatuvairi; tuvarccikai; urōgiṇi; vayataram
187	wild tea	*Ternstroemia gymnanthera* Bedd.	
188	wood germander	*Teucrium wightii* Hook.f.	
189	kangaroo grass	*Themeda cymbaria* Hack.	nosai palai pullu
190	(a grass)	*T. triandra* Forsk.	erikai tattu pullu
191	apple bladder-nut	*Turpinia cochinchinensis* (Lour.) Merr.	kaṉali; pampaveṭṭi
192	gorse; furze; whin	*Ulex europaeus* L.	
193	bearded lichens	*Usnea* spp.	marappāci; karpāci

	COMMON ENGLISH NAME	BOTANICAL	TAMIL
194	Indian cranberry; craneberry; farkleberry; Indian wintergreen; hill carondah; carbolic acid plant; whortleberry	*Vaccinium leschenaultii* Wt.	malaikkaḷā; kaḷā
195	Indian copal; piney varnish tree; Indian gum anime; white dammer; piney tallow	*Vateria indica* L.	attam; catākuḻai; kukkulu; kundurukkam; kuṅgiliyam; kuṅgulu; kukkil; tīpam; veḷḷaikuṅgiliyam; veḷḷaikkunturukkam
196	common winged-nut buckthorn climber; red creeper; pappily; vembaudam bark creeper	*Ventilago madraspatana* Gaertn.	ciṟu koṭi; cuṟuḷ; cuṟuḷbattaikkoḍi; pappiḷi; vēmbāḍai; vēmbāḍaṉ
197	guelder rose; snowball tree; mountain laurustinus	*Viburnum erubescens* Wall.	kōnarkāraṉ
198	hairy yellow-flowered laurustinus	*V. hebanthum* W. & A.	kōnarkāraṉ
199		*Xeromphis spinosa* (Thunb.) Keay	
200	jujube; cherry-plum; lotus; badary	*Zizyphus* Juss., spp.	koṭṭai; cūrai; attiram; civagam; atitāram; irati; ilantai; kōli &c.

The Non-Human Primates of the Nilgiris

FRANK E. POIRIER

3

From the day we entered the forest we began to collect specimens of the black langur..., which actually swarmed in the tree-tops wherever we went. We often saw more than a hundred and fifty in a day, and had we desired, might easily have killed fifty every week.... From first to last I shot about forty-five langurs, out of which I got twenty skeletons and eight skins. The tree-tops were so lofty I was obliged to shoot them all with my rifle, and in order to get a skeleton leaving no broken bones, I had to shoot one monkey through the head and take its body and legs, and shoot another of the same size through the body for the sake of its skull.... The black langur is a very handsome monkey.

William T. Hornaday (1885: 142-4)

Introduction

The Nilgiris is, or once was, a paradise for nonhuman primates. Six different nonhuman primate species lived within this incredibly rich habitat. Some members of this primate assemblage, such as the common grey or Hanuman langur (*Presbytis entellus*), the bonnet macaque (*Macaca radiata*) and the Nilgiri langur (*Presbytis johnii*) are relatively well known. Others such as the lion-tailed macaque (*Macaca silenus*), the Madras tree shrew (*Tupaia* or *Anathana ellioti*) and the slender loris (*Loris tardigradus*) are virtually unknown. (See notes 1-3).

The Nilgiri Hills: Ecological Overview

The following information is drawn from Poirier (1967) and Spate *et al.* (1967). The Anaimalai, Cardamom, Nilgiri, Palni and Shevaroy Hills are of similar origin. They were raised mainly during the post-Jurassic and late

---

I would like to dedicate this chapter to the people of the Nilgiris. Their graciousness and understanding made the time we spent in India most enjoyable. I hope that the habitat conditions I knew in 1965-6 will deteriorate no further and that future generations of Indians will have the opportunity to enjoy this beautiful region. If this volume and this article spur some to thinking about preserving the precious flora and fauna of the Nilgiris, then I will have partly repaid my debts.

Tertiary. The Anaimalai, Nilgiri and Palni Hills are considered subdivisions of montane subtropical forests. They are 'southern subtropical wet hill' on the lower slopes and 'southern wet temperate' above 1500 m. All have an annual rainfall usually exceeding 1500 mm, reaching 6000 mm locally, and a monthly mean temperature of 45–55°F with a maximum of 60–75°F. On the western slopes wind is often an important inhibitory factor. The result is a rich shrub savanna or parkland with occasional peat bogs. Vegetation is characterized by rather low, 12–18 m, forests with considerable undergrowth and many epiphytes, mosses and ferns.

The Nilgiri Hills can be roughly divided into four tracts (see map), each with its respective flora and fauna. Tract 1 is the deciduous forest growing along the slopes. The forests are basically deciduous in the dry months of January-March, but in this tract some trees are never entirely bare. Many tropical trees characteristic of Tamil Nadu grow in tract 1. Tract 2 is characterized by the moist evergreen forests which are most luxuriant on the western slopes at about 900-1200 m. Here trees of the top canopy occasionally grow to 60 or 90 m. Tract 3 holds the sholas or woods of the high plateau area. The grassy downs and scattered trees covering much of the plateau area comprise tract 4.

Except for occasional excursions into the Mudumalai Sanctuary in tract 1, my research was centred on tracts 2 and 3. These will be analysed in more detail. Tract 2 forests are purely evergreen and exceedingly moist from the

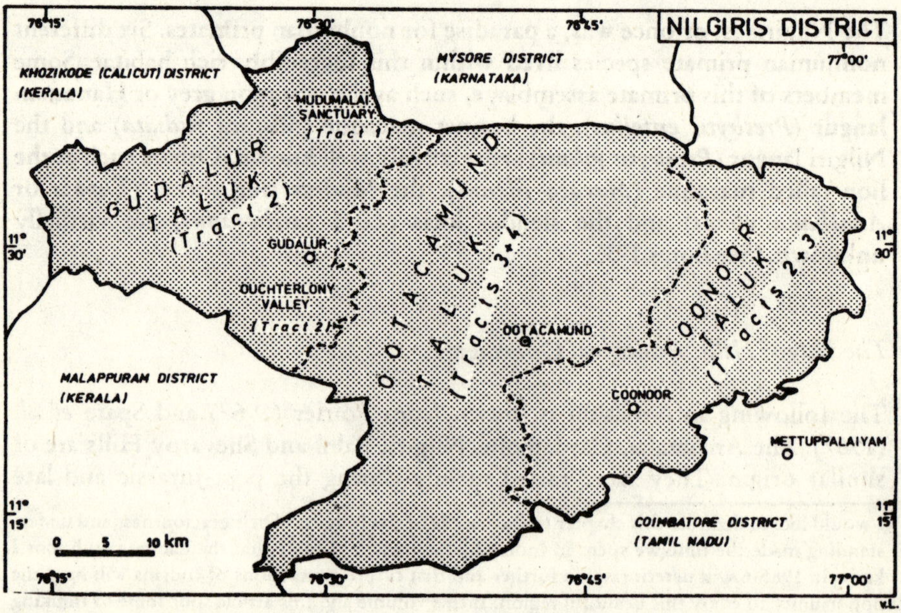

FIG. 3.1 Nilgiris District.

first rains in March until the end of December, when they abound with leeches. Trees are covered with epiphytic orchids, ferns, mosses and creepers. In the undergrowth *Strobilanthes* (171-4) and palms (*Caryota urens* and *Arenga wightii*; 32 and 17 in the above Appendix) are conspicuous.

Above 1200 m the forests of tract 2 begin to decrease in size. Towards the plateau they are gradually displaced by sholas. On the Malabar (western) side, forests reach down to the plains, as they do in parts of South Kanara, Coorg and Travancore Districts. Elsewhere, evergreen forests give way at approximately 300 m from the base to deciduous forests or tracts composed mainly of reed bamboo.

Tract 3 sholas are the major Nilgiri langur niche. A shola is a sloping strip of forest surrounded by grasslands with a narrow watercourse running through it. Sholas are found in the divided valleys dotting the surface of undulating hills characteristic of the interior of the plateau. They are most common in wind-sheltered valleys (see pp. 54-5).

Sholas are typified by a three-storey vegetational pattern. The shrub storey is a closed canopy tightly bound locally by creepers into a solid mass of foliage. The under storey is composed of a variety of trees ranging from 4 to 9 m in height. The upper storey is an irregular layer of trees attaining a height of 12 to 21 m. My study of the Nilgiri langur showed that this vegetation was used for a variety of purposes (Table 3.1).

TABLE 3.1
Vegetational Usage by Nilgiri Langurs

Storey	Representative flora*	Hourly usage	Pattern of use
Upper storey 12-21 m	*Syzygium, Euodia, Acacia, Eucalyptus*, etc.	1700-1800	sleeping, sunning, feeding
Under storey 4-9 m	*Cinnamomum, Glochidion, Litsaea, Strobilanthes, Mahonia*	0900-1700	feeding, resting, movement
Shrub storey 1-3 m	*Berberis, Meliosma, Rubiaceae, Acanthaceae*	0900-1100, 1300-1500	feeding
Forest floor	*Ulex, Cytisus, Urticacease, Labiae* grasses	rarely used	feeding, fighting, escape, play

*For fuller identification of these plants, see Appendix above, pp.63-78.

Although superficially similar, sholas vary considerably in height, growth, accessibility and floristic composition depending upon their situation. Certain trees or shrubs such as *Symplocos spicata* (179) and *Eugenia* spp. (71) are found almost everywhere. Others like *Hydnocarpus alpina* (84) grow principally on the eastern plateau and *Eurya* (74) is most abundant on the western plateau. Sholas at lower elevations exhibit better growth and gradually merge with the evergreen forests below. Sholas occur on a variety of soils and are exacting only in soil moisture requirements. They avoid swamps and commonly occur in areas sheltered from high winds. The percentage of shola to natural grassland decreases from the eastern to the western half of the plateau.

Sholas are somewhat similar to the moist evergreen tract 2 forests. However, being situated at a higher altitude the trees are of different genera and species. Trees rarely attain a height beyond 21 m. Sholas are typically evergreen: Myrtaceae, Lauraceae and Symplocaceae are the orders commonly represented by the trees. The undergrowth is characterized by rubiaceous shrubs and *Strobilanthes* (171-4). Ferns and orchids abound, among the former *Alsophila latebrosa* (10), a tree fern, is common. There is one species of reed bamboo (*Arundinaria wightiana*, 19) and some shrubby balsams and begonias. Sholas which humans have altered contain at present substantial amounts of *Acacia* (1) and *Eucalyptus* (70), both introduced from Australia.

Tract 2 has a full complement of jungle fauna including elephants, tigers, leopards, snakes and numerous avian forms. Three species of monkeys may be found in some tract 2 forest: *M. radiata*, *M. silenus* and *P. entellus*. In contrast, tract 3 has a somewhat restricted faunal complement. This is suggested in Table 3.2.

TABLE 3.2

Faunal Complement of Tract 3

most abundant	birds and insects
	humans and domesticated animals
	ungulates and ruminants
	Nilgiri langurs and occasional bonnet macaques
	lagomorphs
least abundant	carnivores

A comparison of the major conditions in tracts 2 and 3 is given in Table 3.3. More extensive botanical and ecological details are given in the preceding article by von Lengerke and Blasco.

TABLE 3.3

Comparison of some Major Traits of Tracts 2 and 3

	Tract 2 (Moist evergreen forests of the slopes	*Tract 3* (Sholas)
Elevation	900-1050 m	1650-2250 m
Rainfall average 1962-65	2100 mm	1500 mm
Main monsoon	southwest	southwest
Air humidity	40-91 per cent	41-93 per cent
Wind	....	inhibited
Vegetation	three storeys	three storeys
Human population	increasing	increasing
Nonhuman primates	bonnet and lion-tailed macaque, Nilgiri langur	Nilgiri langur, occasional bonnet macaque
Faunal complement	full complement of jungle fauna	limited
Primate predators	humans and some carnivores	humans
Comments on ecology	malarious jungle; some areas border plantations; heavy destruction in some areas	some areas border cultivated plots; heavy destruction in many areas

The Nonhuman Primates

Madras Tree Shrew: The tree shrew has only rarely been observed and it is not certain whether it currently exists in the Nilgiris. However, the existence of a Tamil name, *mara mūñcuṟu*, suggests that the animal exists in the Tamil-speaking region to the east. The following information is compiled from Roonwal and Mohnot (1977). The Madras tree shrew inhabits southern India and has been informally reported in the Eastern Ghats and the Shevaroy Hills. This arboreal form lives in tropical rain forests and thorny jungles; it does not seem to frequent areas of human habitation. Its nocturnal life-style probably helps account for the fact that little is known of its habits. Its major food is insects, an ancient primate diet, and perhaps some fruits. Verma (1965) made some observations on *Anathana ellioti wroughtoni* in Madhya Pradesh.

Slender Loris: The slender loris belongs to the family Lorisidae which is restricted in its range to Africa and South and Southeast Asia. Of the five loris genera, two occur in South Asia. Two subspecies of *Loris*, *Loris t. lydekkerianus* and *L. t. malabaricus* are found in South India. The Tamil name for the form in South India is *kāṭṭuppāppā*, *Kāṭṭuppulai*, or *tēyvāṅgu* (Roonwal and Mohnot 1977). Of the two subspecies, *L. t. lydekkerianus* is more likely to occur in the Nilgiris. Its range has been reported to be from the Eastern Ghats west to Mangalore (Roonwal and Mohnot 1977).

As with the tree shrew, practically nothing is known about the naturalistic behaviour or habitat of the slender loris. Most published information on *Loris tardigradus* comes from captive studies. This is a result of many factors, not the least of which is the fact that it is noctural. *L. tardigradus* is found in tropical rainforests, more open woodland and swampy coastal forests. It has been reported to inhabit evergreen forests at about 600 m elevation in South Tirunelveli (Webb-Peploe 1947) and the High Wavy Mountains in Madurai District (Hutton 1949).

During the day *Loris* sleeps in shady, inaccessible places in tree hollows or at the ends of branches. Its favourite trees in South India are the pipal (*Ficus religiosa*, 77), banyan (*Ficus bengalensis*, 76), mango (*Mangifera indica*, 108) and tamarind (*Tamarindus indica*, 183). *Loris* is primarily an insectivorous form. However, it is also reported to feed on shoots, young leaves, bird eggs, small birds and lizards. Most dietary observations come from laboratory colonies and the extent of its natural diet is unknown.

Little is known either about the extent of the home range or the existence of territoriality among *L. tardigradus*. Ilse (1955) reports olfactory marking of territories among two young captive males. There are several reports of loris breeding behaviour. There is considerable difference of opinion concerning the seasonality of breeding; however, Rao and Narayana (1927, 1932) report that *L. tardigradus* breeds twice yearly, April through May and October through November.

The slender loris is primarily a solitary animal or lives in mated pairs. It is aggressive, and when caged together individuals are involved in considerable fighting behaviour.

Lion-tailed macaque: One of three species of macaques in India and one of two species inhabiting South India, the lion-tailed macaque is still relatively unknown. Recent studies by Green and his colleagues have added considerably to our knowledge of its behaviour and habitat. The lion-tailed macaque is rare across its very restricted range and is now extinct in many areas where it once flourished. Its range is principally the Western Ghats; the animals have been observed at 800-1300 m in the Nilgiri, Anaimalai and Cardamom Hills and in the vicinity of Lake Periyar, Kerala (Poirier 1967; Roonwal and Mohnot 1977; Sugiyama 1968). They are reported with the greatest frequency in Kerala (Krishnan 1971) and in scattered areas in the hills of southern Tamil Nadu. Green and Minkowski (1977) note that their stronghold has always been in the extensive *Cullenia*-dominated forests south of 11°30' N. In the Nilgiris this includes a tract of 90 sq. km in the Silent Valley-Bhavani Valley areas on the western face.

Poirier (1967) notes that lion-tails are found in many areas containing Nilgiri langurs, only lion-tailed macaques inhabit the lower elevations. (Local people seldom differentiate between the two forms, although they are clearly different morphologically. In Tamil they are both referred to as *karu kuraṅgu*, the black monkey.) Apparently, lion-tailed macaques are not differentiated into subspecies. Hutton (1949) reports the occurrence of two forms, a large form (11.3-15.9 kg) on the Wynaad Plateau and the High Wavy Mountains, and a smaller form (6.8 – 9.1 kg) in the Anaimalai Hills.

Macaca silenus' survival potential is precarious. According to Poirier (1967) they are being affected by the same destructive forces affecting the Nilgiri langurs. Habitat destruction has been rampant over most of their range. Coffee, tea and cardamom plantations have destroyed many portions of their range. Hunting pressures have also destroyed many of these rare and beautiful animals. Poirier (1971) reports that once the noisy and constant hooting of these animals, and the Nilgiri langur, a ringing 'hoo-ha hoo-ha hoo', was a familiar sound in native forests. Green and Minkowski (1977) note that the best prospects for their preservation are in the Ashambu Hills.

As one wanders through local markets one encounters merchants selling indigenous medicines. Both the Nilgiri langur and the lion-tailed macaque are hunted primarily for their flesh and 'glands' which are used to prepare such Ayurvedic (indigenous) health tonics (Poirier 1971). The medicament made from the 'glands' is called *karu kuraṅgu rasāyanam*, Black Monkey Medicine, and is rather widely distributed in South India, even though obtaining the prime ingredients means killing protected species. Black Monkey Medicine can also be obtained in drug stores. Around 1967 local English and Tamil language papers advertized the use of this medicine as a 'blood builder' and good for building strength in young and old. Each purveyor of this medicine seems to have his own variant of the recipe which he is unwilling to share. Some feel that any part of the animal is useful while others state that only the glands (unspecified) are suitable.

While the medicine is advertized as a general cure-all, it is considered to be particularly valuable as a cure for lung ailments such as whooping cough. Supposedly its medicinal value is derived from the fact that the black monkeys feed on many leaves and shoots which do, in fact, have a medicinal value when commercially processed. Hockings (1980b) has pointed to the symbolic link between the whooping cough and the animals' cry. Most of my informants praised the mixture if either they or one of their relatives had taken the medicine. I received highly laudatory testimonials as to the efficacy of this medicine from a western-trained nurse and doctor. Green and Minkowski (1977) feel however that there is no scientific basis for ascribing any special properties to monkey flesh. Those who eat the flesh often exist on a protein- and calorie-deficient diet. Thus they may in fact experience improved strength after eating the flesh and the belief then becomes validated.

Many Nilgiri langurs and lion-tailed macaques are also destroyed because of their beautiful skins. Although I seldom saw a skin used as a rug or wall decoration, numerous people told me that the skins were highly prized for decorative purposes. In tea and cardamom areas monkeys are hunted for pleasure by estate people who are allowed by law to shoot any monkeys raiding their gardens or crops, something Nilgiri langurs do quite well. Some estate managers supplement their incomes by killing the monkeys and selling the flesh and other wanted parts to local therapists who practise the Hindu science of health and medicine.

Additionally, the blood of the Nilgiri langur (and perhaps the lion-tailed macaque) is supposed to have rejuvenatory powers, but I never witnessed any rejuvenation nor did I meet anyone admitting to having consumed the blood. Life is allegedly prolonged if one drinks the blood of a freshly killed langur. They say that after drinking the fresh blood an individual must begin to run three miles before three minutes have elapsed. This latter precaution is taken because the langur's blood is said to coagulate rapidly and supposedly, if one doesn't run three miles, the blood clots causing death. I have also heard it said that the langur's quietness reflects the fact that it is a gentle, religious animal. This assumption perhaps gives rise to the belief that one can gain a religious disposition by eating the monkey or drinking its blood.

A final reason given for the destruction of the black monkeys is that their skins make good drum heads.

Lion-tailed macaques inhabit dense evergreen or semi-evergreen hilly forests of high trees (over 20 m) in secluded and unfrequented areas between about 600 and 1050 m in elevation (Blanford 1891; Daniel and Kannan 1967; Roonwal and Mohnot 1977; Webb-Peploe 1947). Lion-tailed macaques also inhabit stands of relatively intact undisturbed forests in the midst of extensive tea and cardamom plantations (Karr 1973; Poirier 1967). They overlap the ranges of the Nilgiri langur and bonnet macaque. Thus they occupy a niche, at least perhaps in the Nilgiris, that is seemingly secondarily preferred by Nilgiri langurs and bonnet macaques. Sugiyama (1968) reports that lion-tailed macaques are obligate rain-forest dwellers which are only rarely terrestrial. They occupy a unique position as the only true arboreal macaque (Southwick and Siddiqi 1970).

Lion-tailed macaques usually feed on the treetops, although Krishnan (1971) reports they occasionally feed on the ground. They are primarily frugivores and omnivores according to Green and Minkowski (1977). The chief vegetable foods include the fruits of *Cullenia* (45) and *Artocarpus* (18), and parts of 81 other species of higher plants together with bracket fungi and mushrooms.

Insects also form part of the diet. Karr (1973) reports that an animal will occasionally break off dead branches at the end of tall trees in search of insect food. Green and Minkowski (1977) note that the most striking animal items ingested are agamid lizards, tree frogs and giant walking-stick insects. Fast-moving lizards, insects and frogs are 'hunted' both arboreally and terrestrially by stalking and flushing before a pounce and grab. Slow-moving grubs, etc. are foraged by exposing and picking. Green and Minkowski note that 'tools' are prepared from leaves and twigs and then used for handling and rendering edible offensive stinging caterpillars.

The lion-tails' diet is thus rather eclectic, and Green and Minkowski (1977) note that to fulfill their dietary requirements they utilize every stratum of the forest. The same authors also make a very important point regarding conservation: this eclectic diet may be obtained only from undisturbed mature rain-forests. Such forests are at a growing premium in South India generally and in the Nilgiris specifically.

As with most nonhuman primates, the lion-tailed macaque's diet is seasonal. *Cullenia* and *Artocarpus* are the most important year-round foods. To obtain whatever fruit which forms the bulk of the diet, in any particular month, the monkeys must roam widely. Again, the rapidly decreasing extent of the forest is a major threat to the animal's survival.

In the south, major potential food competitors for the lion-tailed macaque are the Nilgiri langur and the giant Malabar squirrel (*Ratufa indica*). In the northern part of its range, the common grey langur (*Presbytis entellus*) replaces the Nilgiri langur, and the bonnet macaque may occasionally overlap the lion-tails' range. The langur's diet is primarily folivorous (leafy), minimizing potential competition with the omnivorous lion-tailed macaque. Poirier (1970b) and Horwich (1972) discuss the various dietary patterns. Giant squirrels have some seasonal dietary overlap with lion-tails when *Cullenia* is fruiting. Green and Minkowski (1977) feel there is no evidence that arboreal mammals are competing with the lion-tailed macaque for a food supply.

Bonnet macaques usually live in the drier deciduous forests and scrub jungles at lower elevations than those preferred by the lion-tailed macaque. However, bonnet macaques may move seasonally into the wet forests inhabited by lion-tailed macaques. All observers who have seen the animals together report an absence of competition. However, Green and Minkowski (1977) note the possibility that inter-specific competition with bonnet macaques may have prevented lion-tailed macaques from occupying lower elevation forests successfully.

Very little is published so far on the social behaviour of these elusive and rapidly disappearing animals. This situation will be rectified by publications of Green and his co-workers. The lion-tailed macaque lives in small groups of 4 to over 30 animals, with a group mean that varies between 10 and 20 animals and with 1-3 adult males per group (Blanford 1891; Webb-Peploe 1947). Sugiyama (1968) reported that two groups in the Cardamom Hills were composed of 16 and 22 animals respectively. Green and Minkowski's (1977) two main study groups contained 12 and 34-37 animals. In Sugiyama's study area the socionomic sex ratio (the number of adult females to adult males) was 7:2 in a group of 16 animals (with 7 young) and 10:2 in a group of 22 animals (with 10 young). In Green and Minkowski's two groups the ratio was 3:1 in a group of 13 and 9-12:3 in the larger group. Insofar as adult females predominate over adult males, lion-tailed macaques follow the basic nonhuman primate pattern.

Interestingly, the social organization of the lion-tailed macaque has a number of similarities with that of the Nilgiri langur which inhabits some of the same forests. Thus it might be suggested, from this parallel that the habitat is a strong influence on the social organization. Green and Minkowski (1977) report that subgroups have fragmented from the main body of the large group. This may be a first step in the process of fissioning from the larger group. Such fissioning has been reported in a number of nonhuman primates and seems to be related to food supplies and inter-animal relationships. These fragmenting lion-tailed groups contained adult females. In the smaller group Green and Minkowski (1977) report that the adult male and older subadult male may move alone for more than a day at a time. Also one adult female and her infant occasionally move alone. The latter is an unusual occurrence in terrestrial macaques and perhaps the lion-tailed situation reflects a lessening of predator pressure in the arboreal habitat, a point Poirier (1970b) suggests to account for a similar occurrence in the arboreal Nilgiri langur.

Unlike many other nonhuman primates, the lion-tailed macaque does not seem to exhibit birth seasonality. This appears rather strange in the light of other nonhuman primate data, and may be the result of inadequate study. Sugiyama (1968) observed the sexual behaviour of estrous females on numerous occasions from mid-January to late February in the Periyar Lake region to the south. Green and Minkowski (1977), who have the most extensive data, do not report any sharp seasonality in births. Juveniles could not be grouped into distinct year-classes. This is in sharp contrast to the sympatric Nilgiri langur. Green and Minkowski (1977) note that *M. silenus* bear their first infants at about five years, somewhat later than many macaques, and may have only one or two young in their lifetime. This again is in sharp contrast to other macaques and may be attributable to their short life-span. Green and Minkowski (1977) report that the oldest observed female was nine years old.

Other macaques are estimated to live to twenty or more years.

If Green and Minkowski's estimates are accurate then the lion-tailed macaque has added problems of survival. Given low birth rates and a short life-span, these animals must have relatively large deme sizes for survival. Green and Minkowski (1977) report that a viable deme of *M. silenus* must contain ca. 500-2000 animals. Given the many adverse factors affecting their survival these high figures augur doom.

The communication system of the lion-tailed macaque is at present virtually unknown, although Green and his co-workers are attempting to remedy this situation. Sugiyama noted a call system of 10 vocalizations, one of which is the loud 'hoo', a vocalization somewhat reminiscent of the whooping call of the Nilgiri langur and of langurs generally. Among langurs this vocalization is a spacing mechanism.

As is the case with many nonhuman primates, the inter-troop relationship varies from peaceful intermingling to rapid directional reversals by both troops to avoid encounters. Sugiyama (1968) reports that troops may avoid one another by loud vocalizations emitted by the males. Such calls are heard widely in the forested regions and, if I may insert a personal note, are quite fascinating to hear. While Nilgiri langurs produce similar whoops, Sugiyama reports that lion-tailed macaques do not respond to these.

The relationship which lion-tailed macaques have with other nonhuman primates seems to vary. Bonnet macaque troops have been seen in the same area as lion-tailed macaques. Green and Minkowski (1977) report that a bonnet macaque subadult male joined and moved with one of their lion-tailed groups for a period. Poirier has suggested that these two macaque species may be vertically stratified. However, Green and Minkowski express doubt and their opinion should be considered more accurate.

Hutton (1949:69) notes that when *M. silenus* and *P. johnii* appear in each other's territories 'some terrific battles take place in the trees, and there are casualties on both sides.' *M. radiata* may be dominant to *M. silenus*, even given the latter's larger size. Webb-Peploe (1947), on the other hand, states that both the Nilgiri langur and bonnet macaque shy away from the lion-tailed macaque.

The most pressing problem facing the lion-tailed macaque is its survival. Rapidly facing extinction, this primate may unfortunately go the way of many of the world's treasured fauna. Green and his colleagues are making valiant efforts to preserve this species and happily the Tamil Nadu government and other state governments, as well as the central government, seem to be concerned with their survival. For further information on the conservation of lion-tailed macaques one should consult Green and Minkowski (1977).

Bonnet macaques: Most bonnet macaque field studies have occurred in Karnataka. Although also found in the Nilgiris, bonnet macaques have not been studied here. Bonnet macaques range throughout peninsular India north to Satara and the Godavari River. They seldom range above 2,100 m in elevation

(Krishnan 1971; McCann 1933; Roonwal and Mohnot 1977; Simonds 1965). Krishnan (1971) reports that a miniature bonnet macaque, hardly half the size of the common species, occurs in the Wynaad District of Kerala.

The following information is based on studies done outside the Nilgiris. It can be expected, however, that the behavioural descriptions of animals outside the Nilgiris are broadly representative of the species across its range. The bonnet macaque is not an obligate forest dweller. In fact, it is more common in rural suburban regions than in the forest's interior. In the Nilgiris the animals are found at an elevation of 2,100 m. They inhabit a variety of climatic zones from the semidesert of the central Deccan Plateau to the Kerala rain forests. In cultivated areas they inhabit large *Ficus* (76-7) trees. The large banyan trees, up to 30 m tall, apparently provide an optimal environment. Simonds (1965) feels the existence of the banyan tree is necessary for the bonnet macaque's survival. Bonnet macaques are seldom found in forests with low trees and bushes, where *P. entellus* may be common.

The bonnet macaque is more arboreal than the north Indian rhesus macaque. However, it spends considerable periods of time terrestrially, much more so than the lion-tailed macaque. The amount of time spent in the trees or on the ground is influenced by the habitat: in heavily cultivated areas, such as where Simonds worked, monkeys spend considerable time on the ground; in the Nilgiris, I seldom saw the monkeys on the ground except in areas of human settlement. Here they forage for food in the fields and along the roadsides.

Bonnet macaques are omnivorous. They feed on leaves, young shoots, flowers, insects, cultivated grains, birds' eggs and other items. With their eclectic dietary pattern, there are few environments into which they cannot move and survive, even if only temporarily.

Simonds (1965) reports the home range to be fairly well defined and about 5 km^2 in size. The group covers about one-third to one-half of its range daily. The home range includes a core area; one stand of sleeping trees was used 90 per cent of the time. The home ranges of bonnet macaques and common langurs may overlap, as in the Mudumalai Reserve, for example. In this case the smaller macaque is dominant to the larger langur.

There is considerable disagreement over whether bonnet macaques defend their home ranges against incursion from other bonnet macaque troops. Whatever the case, intertroop fights are rare. If a fight occurs, the larger group displaces the smaller one, which quietly retreats.

Group sizes vary from 7 to 76 individuals. There appear to be major size differences between groups inhabiting the forests and those living near human habitations: the latter groups are usually much larger. Adult females usually outnumber adult males in a group; however, bonnet macaques have the lowest socionomic sex ratio of all the macaques (Roonwal and Mohnot 1977; Sugiyama 1971). In contrast to loosely knit Nilgiri langur groups, bonnet macaque groups are tightly knit social organizations.

TABLE 3.4
Comparison of some Major Traits of Lion-Tailed and Bonnet Macaques

	Lion-tailed Macaque	*Bonnet Macaque*
prevalence	highly restricted	common
preferred habitat	*Cullenia* forests, dense evergreen or semi-evergreen forests; avoids human habitations; restricted range	seems to prefer human habitation areas, drier deciduous forests and scrub jungles; wide range
overlap with	Nilgiri langurs	common grey (Hanuman) langurs
arboreality	mostly arboreal	considerably less arboreal, considerable time terrestrial
diet	frugivorous/omnivorous	omnivorous—seemingly more commonly so than the lion-tailed macaque
group size	4-30 with means of 10-20; 1-3 adult males per group	7-76 animals; lowest socionomic sex ratio of macaques
birth seasonality	none reported	occurs seasonally
predation	humans	carnivores, village dogs; humans will trap and harass

Young males seldom leave the bonnet macaque troop. However, those instances where so-called peripheral or isolated males have been reported (i.e. Simonds 1972; Sugiyama 1971) indicate that under certain situations this process does occur. Peripheralization is common among rhesus and Japanese macaques. Occasionally, isolated males form their own social group which moves in a small portion of the home range of the main group from which they are derived. Whether these males eventually rejoin their natal group or whether they later join another group has not been determined.

Females show subtle signs of estrous, making it difficult for the observer to distinguish between estrous and nonestrous females. Copulations occur throughout the year, with a peak in September-November. Males are the aggressors during the mating season; they actively examine the females daily. Dominance seems to have little influence upon sexual activity. Consort relationships, a pairing of a male and a female, are unusual.

Simonds reports that in southern Karnataka births occur from January to early May, partly coincident with the dry season when the deciduous forests are comparatively bare. However, this season is followed shortly by the April rains which rejuvenate the deciduous forests.

In contrast to the langur situation, bonnet macaque mothers do not allow other females to hold their infants (Simonds 1965). Even subordinate mothers will not allow their infants to be taken by more dominant females. Mothers with young infants form their own subgroups and sometimes handle each other's infants. By the end of the second month a mother leaves her infant alone while she is feeding.

Bonnet macaques are socially gregarious animals. Play and grooming behaviour are common. A complex dominance hierarchy exists among adult and subadult males. Dominance among females is weaker and less defined (Simonds 1965). Because of the presence of a clearly defined hierarchy, most dominance behaviours are accomplished without aggression. Highly dominant males are found within a central hierarchy in the group. These males are active in social interactions such as directing movement and controlling intragroup aggression by their mere presence.

The bonnet macaque's range overlaps that of other monkeys. In the Dharwar forests, for example, they live in the same habitat as the common langur. There does not seem to be any marked difference in food habitats here. Sometimes the two species follow the same activity patterns for days and may travel together in the same trees (Sugiyama 1967, 1971). The bonnet macaque is dominant over the langur. Parthasarathy (1972, 1975) notes that when langurs and bonnet macaques meet, the former slowly drift away. On those occasions when I witnessed bonnet macaques and Nilgiri langurs in the same area there were no aggressive interactions. *M. radiata* and *M. silenus* may overlap in the high forests. Sugiyama (1968) notes that in any aggressive encounter, *M. radiata* is dominant; Webb-Peploe (1947) states however that *M. radiata* avoids *M. silenus*.

Nilgiri langur: This langur inhabits South India in the Western Ghats south of Coorg, and the Nilgiri, Anaimalai, Brahmagiri and Palni Hills, usually at a height not below 900 m. To date, the most extensive study of Nilgiri langurs is that by Poirier during 1965-6 (Poirier, 1968a-c; 1969a-c; 1970a-c; 1971; 1972a, b; 1974; 1975a, b; 1977).

The Nilgiri langur commonly inhabits the sholas (for a fuller discussion of the botanical details of the shola, other than that provided previously, see Blasco 1971, and Krishnan 1971). To a lesser extent these langurs occupy the evergreen forests at about 900-1200 m. Although generally arboreal, they will come to the ground to cross from one shola to another, during territorial encounters, and to raid cultivated plots. The Nilgiri langur is considerably more arboreal than the common grey langur *P. entellus*. In the Palni Hills the Nilgiri langur does not descend below about 900 m altitude; however, it has been reported in the foothills of the Tirunelveli Hills (McCann 1933). In the High Wavy Mountains of Madurai District Hutton (1949) found that it prefers the more sheltered valleys and seldom ventures to the tops of the higher ridges.

During my studies in the mid-1960s I reported that the Nilgiri langur was threatened by habitat destruction and hunting. However, Krishnan (1971) feels its position has improved recently. Kurup (1975) reports that in the Western Ghats its distribution seems to be progressing towards being continuous along the unbroken monsoon forest belt; quite unlike *M. silenus* which inhabits the same forests.

The Nilgiri langur's diet is principally folivorous. However, it may occasionally eat insects and earth (Horwich 1972; Krishnan 1971; Poirier 1968a, 1970b, 1971; Prater 1965). Poirier (1970b, 1971, 1977) also reports on its habit of raiding cultivations in the Ootacamund area. Here it seems to be extending its diet—at least the youngsters seem to be doing so—by including leaves of the *Acacia* and *Eucalyptus* trees which were introduced by the Forest Department. Nilgiri langurs feed mainly in the trees; the group is dispersed while feeding. A considerable part of the day, as much as 7 – 8 hours, is spent in feeding. Most foods are conveyed by hand to the mouth. A complete listing of the animal's diet in the Ootacamund area appears in Poirier (1967, 1970b). A total of 52 floral specimens were consumed; most tasted distinctly bitter or sour to me. Preferred food plants change seasonally and there are local preferences. Thus there are minor dietary differences in widely dispersed groups.

The home ranges of troops in the Nilgiris District cover an area of 0.6 – 2.6 km^2. Home range sizes vary according to group sizes and the availability of suitable vegetation for food and sleeping (Poirier 1968a,b,c, 1969c, 1970b). Groups inhabiting areas of high population density have smaller home ranges than groups in low density areas. Within the home range certain areas are preferred over others, and activity is concentrated in these core areas. A group may spend as much as 70 per cent of the day within a small area of its home range. Sleeping trees, resting areas and preferred food sources are usually concentrated within the core areas. Travel routes within the core area form a dense and tangled web. Although neighbouring groups may have considerable overlap in their home ranges, core areas do not overlap.

Home ranges in the Periyar Sanctuary of Kerala are smaller than those reported by myself. Horwich (1972) reports home range sizes of 5.6 – 8.3 ha, with 0.11 – 0.56 individuals per ha.

Adult males roam freely about the home range and will move into the home ranges of neighbouring groups. Adult females, however, rarely leave the home range and generally confine their activities to the core areas. I have suggested (for example in Poirier 1968b, 1969a, 1972a, 1973, 1977) that this provides the female's infant with better protection from predators since the mother stays within a small area with which she is thoroughly familiar. Adult males are more apt to leave the home range than are adult females. This is particularly true when a home range is destroyed; see Poirier (1968a, 1969c) for a discussion of this occurrence. Even adult males, however, completely abandon a home range with seeming reluctance.

Group sizes vary. Webb-Peploe (1947) reports a group range of 20 – 30 animals, whereas Krishnan (1971) reports a range of 6 – 12 animals. My studies have shown bisexual groups to be the most common. These range in size from 3 to 25 individuals, with an average of 8 – 9 animals. Most groups contain only one adult male; however, larger groups may contain as many as four adult males. In retrospect, it appears that these larger groups eventually fission to form groups of one adult male with multiple adult females.

All male or bachelor groups often range adjacent to bisexual groups. As has been reported for the common langur (*P. entellus*), all-male groups may eventually join a bisexual group. Among Nilgiri langurs this is a relatively peaceful occurrence (Poirier 1968b, 1969a). Among *P. entellus*, on the other hand, such an occurrence often leads to violence, the killing of all infants in the bisexual group, the killing or driving out of the adult and subadult males and a sudden increase in sexual behaviour between the new male leader and the adult females in the group. There are many interpretations for these phenomena (Poirier 1974).

What appears to be one social group is actually a combination of a number of loosely knit subgroups of animals of like age and sex. This results in an overall weakening of the group's structure and may help explain the rather high incidence of group change among the animals I studied (Poirier 1969a). Infants and juveniles have little opportunity to interact with adult males. Adult males and adult females rarely interact with one another, especially in the one-adult-male bisexual groups. In such groups the adult male normally remains socially and physically on the group's periphery.

As noted above, structural fluidity is common. One group underwent six changes in a seven month period (Poirier 1969a). There is periodic addition to and departure from a group. This helps prevent any substantial inbreeding and allows gene flow between groups spread throughout the sholas. Although there is no quantified data on the amount of intergroup flow, physical similarities of animals in the far-flung sholas of the plateau suggest that there is considerable gene flow among disjointed populations. The current agricultural expansion, with its concomitant destruction of the sholas, will affect this pattern. The Nilgiri langurs of the plateau region now present some intriguing possibilities for genetic investigation (Poirier 1977).

Adult males play a relatively minor role in the defence of the group. This contrasts with many nonhuman primates and probably results from a combination of the lack of predation pressure and the presence of the arboreal niche whereby each animal flees from danger on its own. The adult male also plays a relatively minor role in group social interactions. His major role seems to be in fathering the next generation and territorial battles to maintain the group's integrity. Nilgiri langur groups are female-focal social organizations: females form the group's social core.

Nilgiri langurs in the plateau region show a depressed level of social activities compared to most nonhuman primates. This may result in the structural fluidity mentioned earlier. There is a lack of play and grooming behaviour, two important integrating behaviours for most nonhuman primates. Even grooming and play between a mother and her infant are curiously rare.

Mating behaviour was never seen during my study. Females show no physical or behavioural signs of estrous. In the plateau region the animals exhibit two major birth peaks. The main birth peak occurs in May and June and corresponds to the southwest monsoons. A second birth peak of lesser intensity occurs in November and probably extends to February. This coincides with the north-east monsoons. Webb-Peploe (1947) observed quite small infants in June but a few births occur year-round. Krishnan (1971) observed infants in March in the Nilgiri Hills and in April in Kerala. In the plateau the birth rate is rather high: 21 of 30 females in my study area were associated with infants under one year of age. This is a reproductive rate of 70 per cent per annum.

Poirier (1968a, 1970b) made a rather intensive study of the mother-infant dyad. Females have the major burden of caring for and socializing the young. Adult males and infants generally avoid one another. This mutual avoidance lasts until 12 – 15 months of age. Older juveniles will approach adult males and sit nearby. Very occasionally an adult male will assume a protective role towards an infant.

As is the case with most colobines (leaf-eaters), Nilgiri langur mothers soon allow other females access to their infants. When the infant is about 10 days of age the mother allows other females to handle it or take it from her. Infant transfers usually occur in one of two ways: a mother may simply deposit her infant on a branch next to another female, or another female may approach a mother and eventually leave with her infant. Such transfers are most frequent at 3 weeks of age and seem to cease at about 7 weeks. Any female may take another's infant; the females' relative dominance does not seem to affect infant transfers. Not uncommonly, a female is found with two or even three infants in her care while their mothers feed nearby.

Nilgiri langur mothers seem to be less solicitous and protective of their infants than most nonhuman primate mothers. I have suggested that this may be a function of the lack of predators in the arboreal habitat. Occasionally infants are seen wandering alone. Mothers often leave their infants precariously dangling from a branch while they go to feed. By 14 weeks the mother continually leaves the infant, and the latter is forced to find its way through the trees to her. Thus the infant learns the proper routes through the trees and begins practising, at any early age, those locomotor skills on which its life ultimately depends.

Weaning is generally completed by one year. The weaning process is traumatic for the infant. It constantly attempts to obtain its mother's nipple and sit or sleep by her. Mothers often physically rebuke the infant during the latter stages of weaning. If a female does not bear another infant, its yearling may be allowed to continue its close association with her, maintaining mouth-nipple contact during resting and sleeping. Even after the birth of another infant, a juvenile may be allowed to sit beside its mother and new infant. While in one respect the Nilgiri langur mother-infant relationship is lax, in another sense it seems to be relatively long-lasting. At least this seems to be so for mothers and their daughters.

In contrast to terrestrial species, Nilgiri langurs do not have a wide range of facial signals. However, they have a rich system of postural, gestural and vocal signals. From an observer's perspective, the most impressive communicative signal is the male whoop display. In this display, a vocalization, loud whooping, is coupled with wild and frantic movement through the trees. This display is restricted to territorial encounters and is exchanged solely between adult males of interacting groups (Poirier 1968b, 1969b, 1970c). While Tanaka (1965) recognized only 4 vocal patterns, Poirier (1970b, c) distinguished 19 different vocalizations.

Nilgiri langur intergroup interactions include peaceful feeding in proximity, peaceful withdrawal of one group, and most frequently an exchange of visual and vocal signals by the males with occasional chasing (Poirier 1968a,b, 1969b, 1970b). The main intertroop spacing mechanism is the whoop vocalization. While the Nilgiri langurs can easily avoid intergroup conflict, adult males sometimes seem to go out of their way to encounter males of other groups to engage them in these fascinating territorial encounters. Such encounters rarely include physical contact between the interacting males. The adult females and young animals are never harmed. In fact, the latter seem to largely ignore these male 'games' and continue their leisurely feeding, resting, or other concerns.

From my observations, Nilgiri langurs interact peacefully with other nonhuman primates whose range they occasionally cross. Their relations with ungulates and ruminants are peaceful and occasionally mutually beneficial, each being alerted to the presence of an enemy by the sudden flight of the other. Deer are often found beneath the trees of feeding langurs, eating discarded fruits. The major predators are human beings and an occasional leopard.

Hanuman or common grey langur: Sixteen subspecies of *Presbytis entellus* are recognized. There are slight differences in size, colour and tail carriage across its range. For details of the Hanuman langur's habitat in peninsular India consult Krishnan (1971). The subspecies most likely to be inhabiting the Nilgiri region is *P. e. priam* (Blyth 1843).

Behavioural differences have been noted in various regions, some attributable to habitat. However, the daily rhythm is essentially similar in all locations. A considerable number of studies have been conducted on *P. entellus* throughout its range, the Nilgiris being one exception. What is now said is gleaned from the various studies in South India and can probably be generalized to those populations inhabiting the Nilgiris.

Hanuman langurs are considerably less arboreal than the Nilgiri langurs. In various parts of the Hanuman's range it may spend as much as 80 per cent of the day on the ground. However, it is as graceful in the trees and seemingly as accomplished an arborealist as the more arboreal Nilgiri langur (personal observations). Like Nilgiri langurs, the common grey langur is a vegetarian, primarily a folivore. Cultivated crops will be taken when these are found within the range. Krishnan (1971) reported over 30 species of food plants used in peninsular India. Additionally, *P. entellus* regularly licks stones and hard earth from termite mounds, or eats pieces of dirt (Roonwal and Mohnot

1977). I have seen this numerous times in the Mudumalai Sanctuary, where I also observed animals scraping bark from a tree with their incisors. Whether this was to obtain insects, the bark, or simply to sharpen the teeth was not determined.

Home range sizes and the existence of territories vary with the locale. Home ranges of bisexual groups are considerably larger than those of all-male, bachelor, groups. Home ranges in open habitats are more extensive than those in the forests. This is probably due to the existence and concentration of various types of food source. As might be expected, larger groups generally have larger home ranges than smaller groups.

Group size varies across the langur's range. In the Dharwar area of peninsular India, Sugiyama (1964) found differences in group size to be dependent upon the habitat. In relatively open areas the mean group size was 17.1 while in forested areas the group mean was 14.4 animals. The average group size for both areas was 15.1 with 8 females and 5.3 infants and juveniles. Although the proportion of all-male to bisexual groups varies with the location, the latter type usually predominates. In peninsular India, Sugiyama (1964, 1965b) found that unimale groups were more common than multiple-adult male groups. Unimale groups composed 56 – 95 per cent of the sample, depending upon where the sample was taken. Solitary males are occasionally encountered, but perhaps are less frequent than is true for Nilgiri langurs.

TABLE 3.5
Comparison of some Major Traits of Nilgiri and Hanuman Langurs

	Nilgiri Langur	*Hanuman Langur*
prevalence	increasing	common
preferred habitat	shola forests, evergreen forests	varied, drier deciduous forests and shrub jungle
overlap with	bonnet and lion-tailed macaques	bonnet and possibly lion-tailed macaques
arboreal/terrestrial	mostly arboreal	up to 80 per cent terrestrial
diet	folivorous	folivorous
group size	3-30, average 8-9; one-adult male groups; all-male groups	17.1 in open areas, 14.4 in forested areas; one-adult male and all-male groups
territoriality	yes	yes
social activities	depressed	more frequent
group structure	fluid, no infanticide	fluid, infanticide
infant transfers	yes	yes
predators	humans, carnivores	carnivores, little human interference

Social change both within groups and between groups is rather frequent and is usually associated with attacks by the males of other groups. Groups are continually fissioning and being rearranged. During a two year study, Sugiyama (1965b) and Yoshiba (1968) observed more than ten major social changes. In Rajasthan, Mohnot (1968, 1971) observed at least four such changes in a three year period. Changes are always caused by contact between all-male groups and bisexual groups. The dominant male of the attacking all-male group assumes leadership of the bisexual group. Changes are accompanied by increased sexual behaviour and often by infanticide. Changes can also result from intragroup tension. Hrdy, following her (1974) study of the langurs of Mount Abu in Rajasthan, visualized three hypothetical stages in the occurrence of group change.

Various theories have been suggested to explain langur infanticide (e.g. Demarest 1977; Poirier 1974). Among the most widely accepted explanations is one that infanticide prevents incestuous matings between a male and his female offspring and between siblings (Itani 1972; Sugiyama 1965b, 1967). In southern India, among the animals studied by Sugiyama, infanticide occurred with regularity about once every four years. At this interval, excess males living either alone or in all-male groups approach and attempt to join a bisexual group. One outcome of this attempted joining is a high rise in aggressive behaviour. The intruding males kill all the infants in the bisexual group (those mortally wounded are left to die by their mothers), and the intruders drive out all the adult males. This harsh action brings the adult females in the group to sexual receptivity and the new male leader copulates with the females. In this way the new male ensures that all the subsequent infants are his own and the adult females ensure that their infants will receive the benefits of the new male leader (Hrdy 1977).

What makes infanticide so very interesting is the fact that it has not been observed among all langur groups, a phenomenon that is not yet adequately explained. The periodicity of the behaviour, its occurrence being approximately once every four years, is an effective means of preventing incest. Since it takes langurs approximately four years to develop to sexual maturity, infanticide stringently limits the possibility of the male leader's (the father of all youngsters in a group) mating with his newly maturing daughters and of newly maturing brothers and sisters mating with one another. Because of this, infanticide is often seen as an adaptive reproductive strategy.

It might be worthwhile to note that infanticide by adult males also occurs among the suborder Prosimii and among members of the New World primates, as well as among humans. The closest parallel to this behaviour among non-primates occurs among lions. Considerable discussion of male infanticide is found in Hrdy's (1977) book *The Langurs of Abu* and the reader interested in this topic could profitably consult it.

An estrous *P. entellus* female may initiate copulation. Generally only the most dominant male copulates, forming a temporary consort relationship lasting from 2 to 24 hours. There are indications of birth seasonality. In peninsular India newborns are seen throughout the year but most are born between December and April, with mating probably occurring between May and October (Parthasarathy 1972; Sugiyama 1967).

As among Nilgiri langurs, *P. entellus* females allow other females to handle their infants soon after birth. Sugiyama (1967) observed that in peninsular

India mothers allow young infants to be handled not only by females of their own group but sometimes by females from other groups. Older siblings may monopolize and temporarily care for new infants. In this process they learn the mothering role (Poirier 1968b, 1972a, 1973, 1975b, 1977). Males generally do not interact with youngsters. An infant stops receiving preferential adult treatment when it is about 14 months of age. By 15 months the youngster is independent of the mother.

The behaviour of male and female juveniles differs. I have suggested in a number of places (i.e. 1972a, 1973, 1975b, 1977) that this is in preparation for adult roles. Male social relationships are primarily with age-mates. Females, on the other hand, spend a considerable part of the day associated with adult females and infants. Females are generally less active and aggressive than the males.

In contrast to macaques, langurs are not necessarily characterized by a strong and obvious dominance hierarchy. In some regions, however, a male dominance order does exist. In contrast to macaques, langurs are generally peaceful animals, a trait that is reflected in local folklore.

TABLE 3.6
Comparison of Colobine Group Structures

	P. entellus (Dharwar)	*P. entellus* (N. India)	*P. johnii* (Nilgiris)	*P. johnii* (Thekady)
adults	66%	51%	60%	56%
adults/sub-adults		60%	74%	
subadults		9%	14%	
juveniles		15%	10%	3%
juveniles/infants	20%	41%	27%	7%
infants		26%	17%	
no. of groups	8	4	16	5
sex ratio	4.7:1	2.1:1	1.3:1	8:1
investigator	Sugiyama	Jay	Poirier	Tanaka

Adult females spend much of their day mutually grooming, and ties among such females seem to be particularly strong. As among Nilgiri langurs, common grey langurs are characterized by female-focal social groups. The dominant male in bisexual multiple male groups is often the centre of a grooming cluster of a number of females. Such males however rarely groom other animals. Grooming is rare in all-male groups.

The existence of territoriality appears to be dependent upon where behaviour is sampled. In one of the first studies of common langurs, Jay (1965a,b, 1968) reported the lack of territoriality. In southern India, on the other hand, Sugiyama (1964, 1965a,b, 1967), Sugiyama *et al.* (1965) and Yoshiba (1968) reported that group encounters occur daily and whoop vocalizations are used as displays against other bisexual groups or all-male groups. Males of different groups may contest one another: however, fights are not severe and usually are characterized by an exchange of displays. If actual fighting occurs, it is usually limited to the adult males. The pattern in South India is quite similar to what I have reported for Nilgiri langurs.

In South India the common grey langur overlaps the range of the bonnet macaque. Bonnet macaque groups are larger and their home ranges are usually at least double the size of those of the common grey langur. However, the population density of the bonnet macaque is lower. Mixed groups often occur without any aggression. My observations in Mudumalai suggest however that langurs generally avoid bonnet macaques and move to avoid contact. If aggressive contact occurs, the bonnet macaque dominates.

For many reasons, including their place in Hindu religious lore, common grey langurs are generally not bothered by human populations. There is nothing of the persecution experienced by Nilgiri langurs and lion-tailed macaques. The principal predators of the Hanuman langur appear to be carnivores such as the leopard and, near human habitations, the pariah dog.

The Hanuman langur has a close relationship with the cheetal or spotted deer. We noted this many times in Mudumalai. The deer's olfactory sense coupled with the langur's visual acuity seem to be mutually beneficial. They both flee on each other's alarm call. Cheetal are often found beneath a tree where langurs are feeding, and the cheetal feed on the fruit which the langurs drop.

Conclusion

The need for strong conservation practices: The Nilgiris present a fascinating floral and faunal array. However, much of the district has experienced rapid and traumatic change through human intervention. Such change has had a strikingly negative impact upon natural forests and their animal inhabitants. Unless this trend is reversed, human populations may also experience negative effects. Pressures for removing the forests and utilizing their products are unremitting. In the Ootacamund area where I did my research, many magnificent sholas were cut and burnt, to be replanted with *Acacia* and *Eucalyptus* trees. Economic reasons notwithstanding, such trees offer little food and protection for the native animals. The list of rapidly disappearing floral and faunal species in this region due to this forest programme is distressing.

Forest planners are looking for new ways to utilize evergreen tree timber and to make foresting in remote areas economical. Underplanting sholas with cardamom for its lease revenue has been widespread with little thought given to the effects of this practice on the ecological balance. Tea plantations have required the vast destruction of many once beautiful forest tracts. Extensive tracts now given up to cultivation were once covered with thick growth; the stony ridges and heights excluded. The numerous shola trees which stand singly or in groups in the middle of farm lands lend credence to this statement. The frequent suffix–*kāḍ*, which denotes jungle or forest, in the names of localities where now hardly a tree stands is a further sad evidence of this total destruction.

Where reforestation has occurred it has often been accomplished with exotic species which do not serve the same functions as the shola forests in conserving rainfall and preventing floods and erosion. Changing weather patterns in parts of South India may have been induced by human environmental alteration. Deforestation can have drastic effects on rainfall patterns. Replacement of shola vegetation with species like eucalyptus and wattle, or other economically valuable but non-evergreen vegetation like teak or sal, can severely alter rainfall amounts. Non-evergreens are incapable of high transpiration rates with the first rains; they cannot adequately maintain atmospheric humidity to the same degree as evergreens.

What remains of the Nilgiris' flora and fauna can be saved if strict conservation efforts are employed. These efforts should be undertaken by the indigenous populations and not necessarily by specialists from outside. Local populations should appreciate the unique and valuable resource they have in their midst. In the schools more emphasis could be placed on conservation and on the special character of the indigenous wildlife. Perhaps the education programme formulated by the Wildlife Clubs of Kenya can serve as a model. If local inhabitants were made aware of the richness of their environment, destruction could perhaps be slowed.

If these forests are to be exploited economically this can be done without their destruction. For example, tourism can be promoted. The sholas of the Nilgiris would be a source of interest for tourists interested in wildlife and faunal diversity. Comparable situations have been tapped in Africa and South America. India's sholas have as much, if not more, to offer. The Nilgiris need not only attract tourists from outside India; India's indigenous populations would also benefit from the proximity of these forests.

Beside an emphasis on more conservation, more studies are needed. There is woefully inadequate data on most of the Nilgiris' nonhuman primate species. The best known form, the Nilgiri langur, has been the subject of only one intensive study, that by myself. Comparative work is lacking. Green and Minkowski's work will greatly clarify the situation on the lion-tailed macaque. The loris and tree shrew are yet to be studied, and the common grey langur and bonnet macaque are best known through studies undertaken outside the Nilgiris.

NOTES

1. The bonnet macaque is well known in regions below the Nilgiris; little is known however about its habitat and behaviour within the Nilgiris proper.
2. Two sources were particularly valuable in writing this account: Green and Minkowski's (1977) recommendations for conserving the lion-tailed macaque, and Roonwal and Mohnot's (1977) book on South Asian nonhuman primates. I wish to thank Paul Hockings for his care-

ful reading of the manuscript, and François Blasco for correcting some botanical identifications.

3. There is continuing debate over the primate status of the tree shrew. Although the tree shrew may eventually be deemed a non-primate by most investigators, it is worthwhile discussing it in this report. Furthermore, there are so few sightings of this nocturnal creature that it is possible that it may no longer—or perhaps never did—include the Nilgiris as a part of its habitat.

Nilgiri Prehistoric Remains

WILLIAM A. NOBLE

4

Of late years, the Neilgherries have been so exposed to the pick-axes of indefatigable archaeologists, that their huge store of curiosities has been almost exhausted. Little now remains but the fixtures. In many parts almost every hill is crowned by single & double cairns, enclosing open areas, which, when opened, were found to contain numerous pottery figures of men & animals. There are some remarkable remains.....; all, however, have been rifled of the funeral urns & the other relics which they contained. Vases holding burnt bones & charcoal, brass vessels, spear heads, clay images of female warriors on horseback, stone pestles, pots & covers ornamented with human figures & curious animals, have been taken from the barrows that abound in different parts of the Neilgherries.

—Sir Richard Burton (1851:313-14)

Megalithic Cults

It is remarkable that people in very different parts of the world and in millennia far apart have exhibited a tendency toward the use of stone or earth to fashion circles on the ground. Paralleling this tendency was another in which dolmens were formed with orthostatic stones emplaced vertically and capped with large horizontal stones (Fergusson 1872). The term *megalithic* ('large stone') has been applied to the cults which were related to stone use in earthen or stone circles and dolmens. Related to these features too are others not so simply classified.

The inventive genius responsible for the varied megalithic sites in western Europe was linked to funerary practices, worship, and astronomical observations. In Great Britain and Ireland there is an amazing series of earthen and/or stone circles, dating between 3200 and 1500 BC and including the famous Stonehenge IIIC—completed by about 1500 BC (Burl 1980: 39, 46, 54). The early Neolithic circle builders there used no metals, but in the three centuries following 2600 BC the immigrant Beaker people introduced the use of copper and gold (Burl 1980: 43-4). By 1200 BC, when the circle cult had ended, the Late Bronze Age had started. In far western Europe some simple—but in some instances large and massive—dolmens were constructed (Daniel 1980:

80-1, 84-5; for many well-illustrated examples in northeastern Spain, see Pericoty Garcia 1950). These were either free-standing or covered with earth. From southern Sweden to Spain and Sardinia, there are early passage or gallery tombs in which the dolmen principle and the earliest known examples of corbelling were used (Renfrew 1979: 122-9). Some of these megalithic tombs in Brittany, France, were built prior to 4000 BC. The tombs were generally used for collective burials by Neolithic people, and only in the late ones is there evidence for the use of copper. In Malta, within the western Mediterranean basin southeast of Sardinia and not far from Sicily, the Neolithic members of a distinctive megalithic cult employed massive stones to build the world's first free-standing stone temples before 3000 BC (Renfrew 1979: 147-9, 152).

Megalithic cults in India[1] have been related to the construction of features similar to those in western Europe. Many earthen and/or stone circles and dolmens are present in parts of southern Asia, mostly south of the Godavari River in the southern peninsula (Leshnik 1974: maps on 228-31; Gururaja Rao 1972: Fig. 2). In contrast to its use in the massive passage and gallery tombs of western Europe, however, the dolmen principle in India was used to create many funerary cists which are enclosed vertically by orthostats. The capstones of some cists are below the surface, but in other cists the capstones are located either at ground level or at varying heights above the ground (mostly no higher than 2 m). The two most common means to inter the remains of the dead and related artifacts were to use either a capped cist or a pit within a circle. Capstones were not infrequently emplaced over burial pits. In my terminology for discussing India, I define a *dolmen* as an above-ground rock structure with partially enclosing orthostats and a capstone or capstones; thus, at least one opening is always present. The Indian earthen and/or stone circles, stone cists, and burial pits were generally related to corpse exposure followed by bone interment in graves not infrequently reopened (Leshnik 1974: 226). Most dolmens cannot be related to the depositing of human remains, but some are indisputably associated with either the ritual for or memorialization of the dead.

By contrast with western Europe, Indian megalithic sites generally date back no further than 800 BC (Gururaja Rao 1972: 326). They are therefore post-Neolithic, and relate to the spread in use of iron. Black and Red Ware pottery is characteristically found within the sites, although red or black pottery is also present (Leshnik 1974: 154). Some megalithic sites have remained in use into this century, mainly among tribals. For example, in the 1960s some Pulayans in extreme southern India lived in dolmens (Williams 1969: 607). Outside the southern Indian region, in northeastern India, the Khasis continued a megalithic cult in the honouring of their dead at the family, lineage, and clan levels (Roy 1963). A megalithic cult related to head-hunting among the Nagas, who also live in northeastern India, was reported in the 1920s (Hutton 1926).

Probably because of much cultural and cultic variation in both time and place, there has been in India an exuberant variation of megalithic features. Apart from features discussed above, and because of the warm tropical conditions and resultant laterite of the southwestern Malabar coast, such unique features as rock-cut caves, hood stones, and hat stones occur in northern Kerala. In extreme southern India, in both Kerala and Tamil Nadu, large funerary urns were commonly used to inter the remains of the dead. Stone circles often identified the urn sites, but many marker stones have since been gathered for other purposes. Particularly within the region that is in and near Madras City, terracotta sarcophagi—typically supported by feet—were used to inter collected bones within stone circles. As indicated previously, both urns and sarcophagi were buried either within cists or burial pits, and those within pits sometimes had capstones placed over them.

Further impressions of the wide array of megalithic features which exist in India can be obtained from these few examples: In his well-known report which stimulated others in megalithic research, Taylor (1865) recorded a large number of dolmens, above-ground cists, and stone circles at Rajankallur[2], Hegaratgi, and Yummuguda in the midst of the Deccan. Some dolmens and cists had side orthostats reaching at least 4.6 m in length and over 2 m in height. The boxed dolmens with one side open were typically empty, whereas the characteristic megalithic pottery and a few urns (one over 0.9m in height) were found in the enclosed cists housing the remains of the dead. Many cists had portholes ranging from 26 to 61 cm in diameter in one side and thus exhibited another feature found in both western Europe and southern India (Childe 1948: map on 4). At Yummuguda there were double stone circles around each of four cists. At an additional place called Jiwarji, Taylor counted no less than 268 stone circles. Within these, earth and stones were used to form mounds covering over the centred burial cists. In a much more recent report, Murthy (1976) has provided details for two above-ground cists surrounded by stone circles at a place called Kadiriraya Cheruvu, in the Chittoor District of Andhra Pradesh. The main box cist has four orthostats arranged in a swastika (not uncommon), and these enclose a rectangle approximating 2.5 x 1.5 m. The overlying capstone measures approximately 3.2 x 2.8 m. This cist without a porthole is surrounded by six stone circles of decreasing height outward (from about 3.4 m high to about 0.45 m high). Around the outer circle there is a dry stone wall about 0.35 m in height. An adjacent above-ground cist with eastward-facing porthole (the most common orientation) is surrounded by only four stone circles of outward decreasing height. King (1877) reported on one of the most intriguing of all megalithic sites in India. At Mangahpett, in the Warangal District and close to the Godavari River in Andhra Pradesh, he found some above-ground cists with closely fitting orthostats, rectangular capstones, rectangular portholes cut into the tops of eastern-facing orthostats, and elongated depressions cut into the underlying stones—presumably to hold the remains of the dead. The cists were centred in stone circles formed from stones carefully sculptured and fitted together, and stone crosses stood nearby.

The Nilgiris and Wynaad

Some aspects of the prehistoric remains (i.e. predating the coming of the English in the early nineteenth century) of the Nilgiri region reveal affinities with others in southern India. Attention in the remainder of this chapter will focus upon selected prehistoric remains within the Nilgiri region. In the Wynaad, and primarily in the Malabar portion of Kerala State, funerary urns have been found at a number of sites (Cammiade 1930). Graveyards with urns form urnfields. No accompanying stone circles were discovered there. All the urns at one site were protected above by stone slabs, the general size of which approximated 1.4 m x 1 m x 13 cm. Some urns were covered by fashioned stone caps, and various types of smaller pots were used to cover others. The largest recorded urn was 92 cm high x 76 cm in diameter. Within the urns, non-cremated bone remnants were accompanied by a characteristic array of smaller Black and Red Ware, black and red pots. Memorial stones, some having vertical sculptured panels of a characteristic type occurring widely into Madhya Pradesh, far to the north, stand within the Wynaad and in the low-lying Mysore Ditch just north of the Nilgiri massif. A well-preserved example, dating perhaps to the thirteenth or fourteenth century AD, is incorporated into a veranda wall in the manager's bungalow at Pambra Estate, in the Malabar Wynaad. The bottom panel shows the hero meeting his death in combat with another warrior (both hold short swords). In the middle panel the warrior is being lifted upward by two female celestials. In the top panel the hero (presumably, in the afterlife) is worshipping a *linga*, a priest holds an incense burner, and a sacred humped bull (*Nandi*), a sun and a moon are also represented.

Our focus again shifts, and is concentrated hereafter only upon the megalithic remains of the upper Nilgiris. These fall into the broad categories of stone and earthen circles, a burial rectangle, above and below surface cists, and dolmens. They generally conform to the southern Indian megalithic criteria, and a classificatory system developed specifically for the upper Nilgiris will provide a reference frame for the broader spectrum of features there. Interments demarcated by the circles are typically associated with filled-in burial pits topped by stone slabs. Unique features will also be revealed. For example, in all of India, it is only in the upper Nilgiris that there are stone circles with built-up walls having both inner and outer perpendicular sides. Similar circles occur in Egypt, but these pre-date the Nilgiri ones by well over 1000 years (Lal 1963). Despite a similar time gap, ring stands and Black and Red Ware, black and red pots are characteristic of both southern Indian and Egyptian sites. By contrast, the upper Nilgiri pottery is generally of a completely different tradition, and ring stands have not been excavated from the Nilgiri sites. Whereas there were whole bone fractional burials in adjacent areas, including the Wynaad, bone remnants from cremations were buried in the upper Nilgiris. Out of the few areas with bronze ware in southern India, sites in the Nilgiris contained one of the richest assemblages of bronze ware. The Nilgiri sites held unusual pottery types, including a unique one in which each pot appears stylistically to contain superimposed pot forms. Artifacts in no other area of India are comparable to the charming upper Nilgiri series of human, animal, and bird pottery forms either separate or attached to lids. From all that can be gleaned by me, it appears that no other area in India has such a concentration of sculptured orthostats within dolmens. It seems likely, too, that for several centuries some of these orthostats provided surfaces for the sculptured commemoration of periodic tragic events.

This effort contains the results of a preliminary survey. Because many prehistoric sites in the upper Nilgiris are located on the tops of peaks, the climbing of most peaks is essential to the completion of an adequate survey. My field survey was, of necessity, limited. As no excavation was possible, archaeological evidence must come from published sources. The outstanding contribution by Breeks (1873) continues to be the foundation. Only one scientifically acceptable excavation has been reported—that conducted by the editor of this book (Hockings 1976).

Site Characteristics and a Classificatory System

Brief coverage of some selected prehistoric sites not only reveals their characteristics, but also provides some insight into complications stemming from more recent tamperings and human use after the initial construction. As an aid to classification, broad categories are systematically employed. However, because features at a site may vary, most variants not in the category being treated are covered as well. In this way, the reader becomes aware that there is a mixing of types at some sites. This survey is keyed to Fig. 4.1 which shows archaeological sites in the upper Nilgiris.

Walled Circles: The characteristic Nilgiri circles with walls of piled stone—typically rising to about one metre in height—were normally on summits in grass, forming balds. Such stone circles and other archaeological sites were made more conspicuous by rhododendron trees, the leading natural colonizers (Pl. 13, A; Site X in Fig. 4.1). It has become increasingly difficult to find these sites, for man-created acacia and eucalyptus forests now shroud many Nilgiri peaks. Although some walled circles with inner and outer perpendicular sides are well preserved, others exhibit partial collapsing inward or outward. Post-construction religious use, probably dating primarily to historic times, tended to be stimulated by the presence of walled circles.

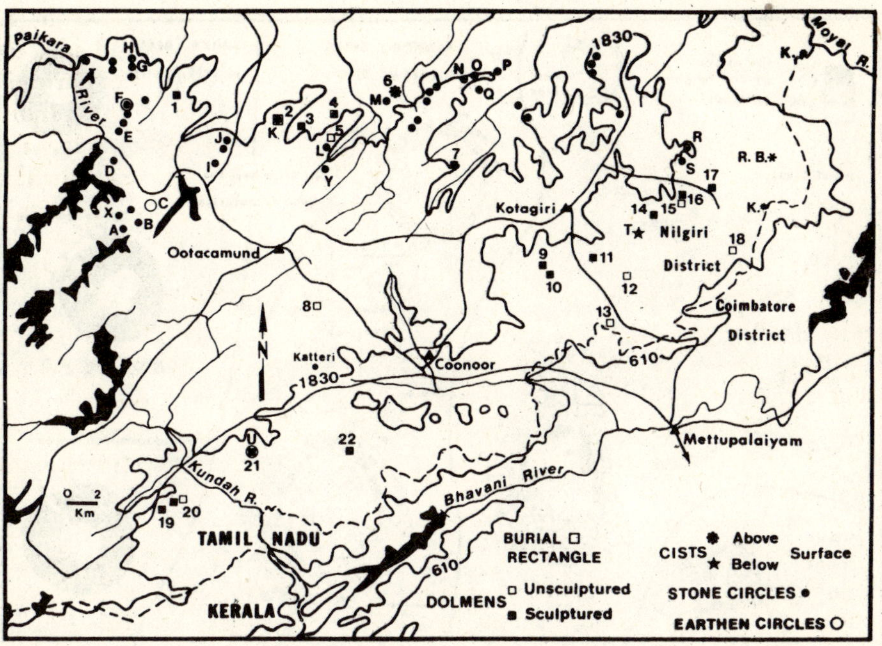

FIG. 4.1 Nilgiri Archaeological Sites.

Site A, on top of a peak, has two walled circles. Excavators probably stood capstones against the interior wall faces. Inside one circle, five upright stones and a flat stone placed before the largest central stone later served as an altar. Nearby Site B has four walled circles strung over the summit on Anikal Betta (Fig. 4.2; Pl. 13,B). The three lower wall sections are attributable to excavators. The southernmost low circle of piled stones, of a different type, probably once had upright stones standing around the interior. A supplementary dolmen shrine with a Ganesh image proves additional religious use of the site. Evidence for another former dolmen shrine also exists. When I climbed to the top of this peak in 1978, so thick were the eucalyptus trees that not one archaeological site could be seen.

Site I, next to a ghat road and within easy access of Ootacamund, was well known to English pot-hunters. Seven walled circles straddle the summit. Three are well-preserved, but the four others show advanced slumping or infilling with stones. The westernmost one was probably used as a source for stones to construct a shallow piled stone circle with entrance just to the west. Between the easternmost two there is a different small circle with some remaining stones on end, probably indicating the manner in which the entire circle was constructed. The rest of the circle now has stones lying on their sides. Associated religious structures at this site include a dolmen shrine with three rocks for worship on the inside, a *linga* over one metre high and a nearby

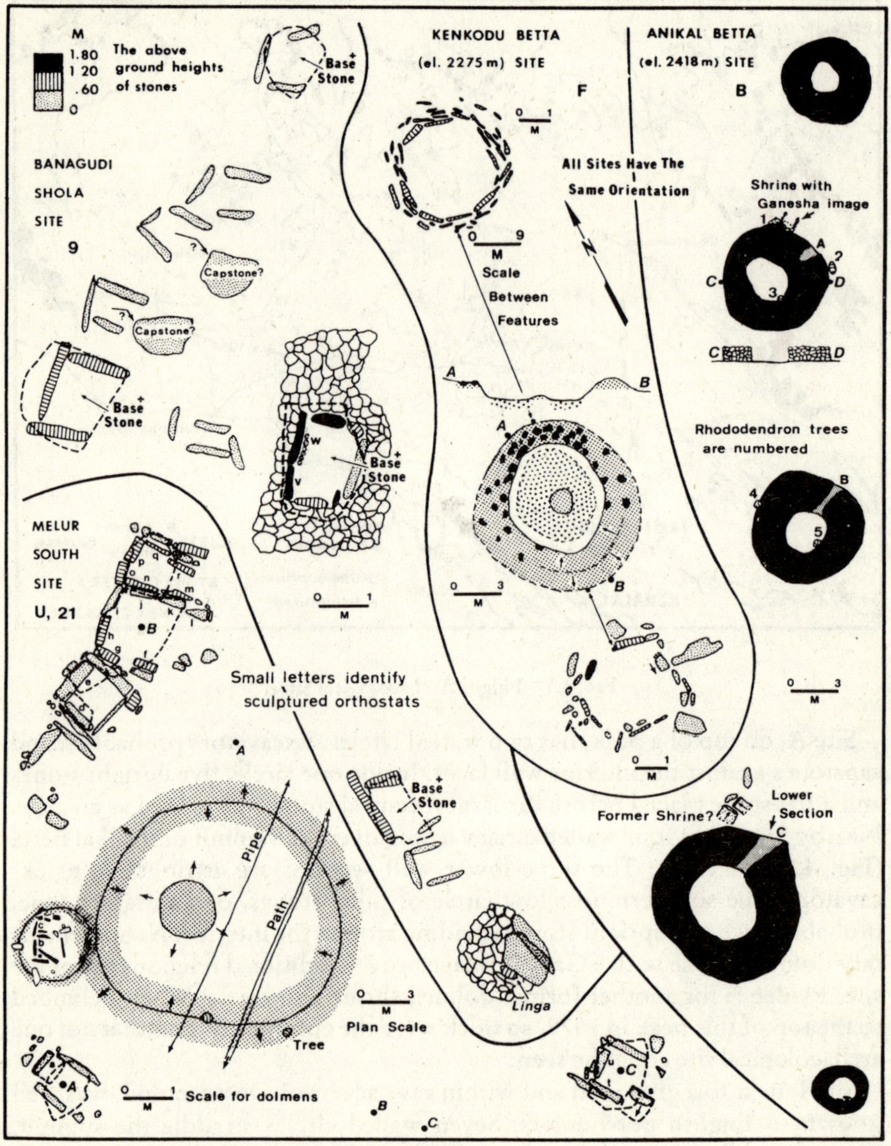

FIG. 4.2 Selected Megalithic Sites.

shrine covered with small stones, one small free-standing uncovered shrine, and three small uncovered shrines abutting the well-preserved perpendicular side of a walled circle. Farther to the northeast, at Site J, there is but one walled stone circle standing on an outstanding peak with a view down into the Mysore Ditch. Two blackened stones close to a metre high lean against the inner perpendicular side. Again, despite the fact that this walled circle stands well-preserved, both the stones could have been capstones. A pile of stones rising about 60 cm above the top level of the circle (about 1 m high) covers over a side chamber within which there were several worship stones, a pottery incense burner, two small pottery lamps, and paraphernalia for making sandalwood paste.

Site Q has a single walled circle, slumped outward in places. Pottery fragments indicate past excavation, and two vertical stones (1.22 m and 1.83 m high) standing against the well-preserved inner perpendicular wall could have been capstones. With the flat stone placed at their base, an altar now exists. A stone pile (about 55 cm high) stands on one side of the circle (about 1 m high). At Site R there are seven walled circles strung along a prominent ridge behind the easternmost Kota village. Seven Fort Hill, so-called after the seven circles, is the ideal site with walled circles. At perhaps no other Nilgiri site were walls piled so well and so high. Most walls rise above 1 m and one measured outer wall height of 1.37 m was recorded. In a published depiction of this site (Noble 1976: Fig. 1 on 95 and Pl. 1, b), it was easy to reconstruct the original configuration of the circles. It should, however, be pointed out that one side of each circle was dismantled outward by excavators.

Other Piled Stone (or Cairn) Circles: Obviously having an affinity with the walled circles are those circles with a band of stones piled in a single layer, or the low-lying circles with relatively level tops and stones piled somewhat higher (tending towards 50 cm in height, or less). There are also circles with well-constructed inner perpendicular sides reaching to a height of even a metre or more, and with outer sides sloping to ground level. Other piled stone circles have sloping inner and outer sides. In some more complex circles, one or two inner circles of vertically emplaced stones are followed by piled stone sides, either outward-sloping and/or in a horizontal layer.

Site G illustrates how varied the Nilgiri prehistoric remains can be (Fig. 4.3). The northernmost piled stone circle has stones reaching a centred height of from 30 to 45 cm, and there is inward and outward sloping from this crest. The next circle southward, which is walled, generally reaches no higher than 84 cm. One can but speculate that this circle may once have had the function of isolating that which was sacred from the more secular, in that it appears to have two entrances originally incorporated, and because it has dimensions suited to the enclosure of a Toda conical temple. The next circle is one of the most interesting of all Nilgiri remains. This relatively small circle of piled stones rises no higher than 45 cm. Its centre is entirely covered by a large capstone of sixteen or more cm thickness. This offers proof that a centred

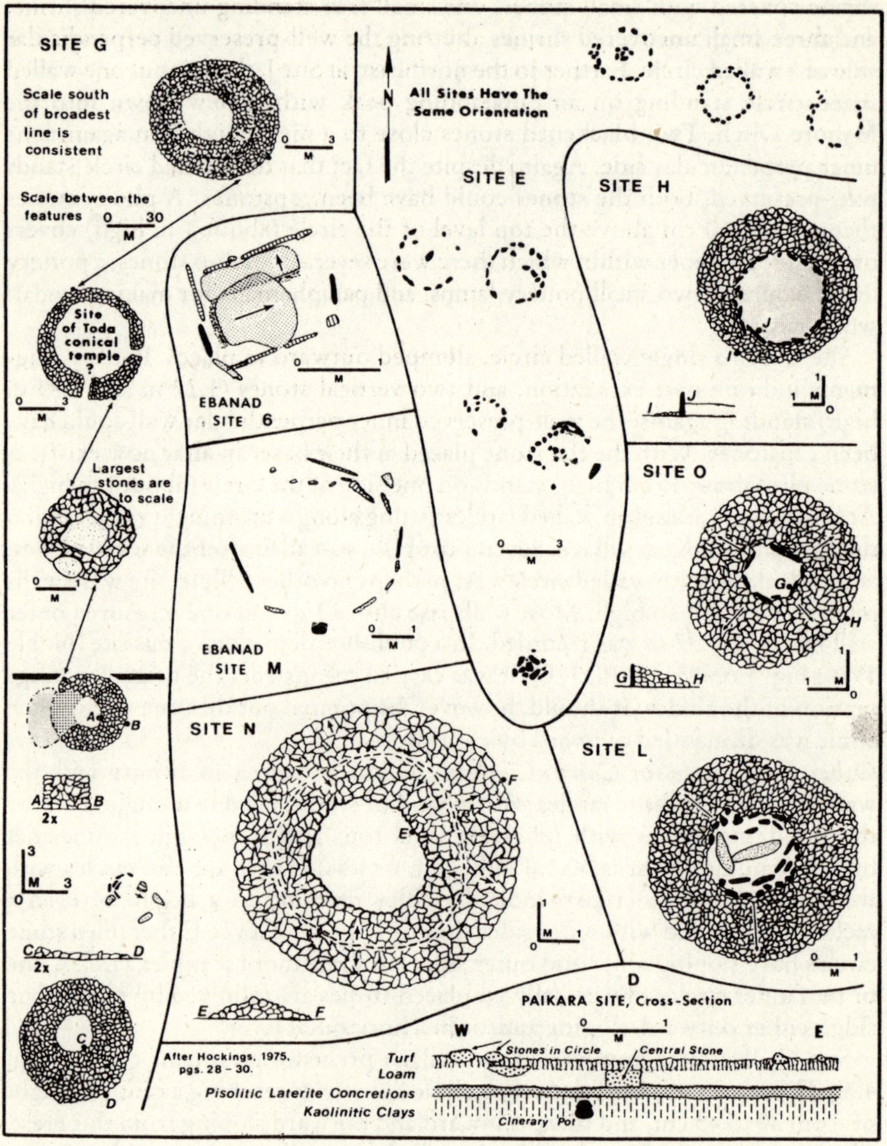

FIG. 4.3 Further Megalithic Sites.

capstone covering a burial pit could simply be laid at or above ground level, without being covered over with earth. The next circle is a walled one of the characteristic Nilgiri type. It rises to the typical height of about a metre (was there some measured unit used to establish height, or was height set by seemingly sensible visual parameters, or, considering the size of stones used, was height controlled by an optimum beyond which it became disadvantageous to pile stones any higher?).[3] With this feature reconstruction was avoided, so that the reader may grasp the effect of outward demolition. The next remains, of a different category, must once have formed a small circle of erect stones (highest measured one was 70 cm tall). Some constructional stones lie flattened nearby. The southernmost stone circle is largely single-layered, but there was some additional piling near the open centre. Farther north at Site H (Fig. 4.3), a single circle with erect stones (highest rising to about 59 cm) is surrounded with piled stones forming either a single layer or, with some additional piling, a downward sloping gradient from the interior. Earth fill brought the central enclosure to about 30 cm above ground level. Did this result from earth brought and deposited in the last stages of construction, or was earth deposited after each in a series of interments at the one site?

At Site K, 2, there is a piled stone circle with well-constructed inner perpendicular side estimated to have been originally over one metre high. The other side slopes outward to an edge which blends with the surrounding ground. To those who might speculate that this type of feature was formed when a walled circle disintegrated outward, this site offers convincing proof to the contrary. The width of this piled circle, for example, is similar to the piled stone widths of some walled circles. Nearby there is a burial rectangle which will be treated later. Site L (Fig.4.3) has one complex circle with two coarse inner and adjacent circles of erect stones followed by piled stones forming a downward slope to the outer edge. The two fallen stones indicate the size of the erect inner stones and, if re-emplaced in a reconstruction (done in Noble 1976: 95), would complete the two inner erect stone circles. Site N (Fig. 4.3) has one simpler circle with stones piled upward to a central crest about 60 cm in height. From the crest there is a downward sloping to both edges. Site O (Fig. 4.3) has one complex circle with only one inner circle of erect stones followed by piled stones with downward slope to the outer edge. Site P, shown already in another publication (Noble 1976: 95), is another ideal Nilgiri site. The four stone circles at this site represent an affirmation of the four piled stone circle types identified at sites H, K, N, and O. At Site S, below Seven Fort Hill and on a southern slope toward the easternmost Kota village, there are three small and shallow piled stone circles (at the highest point, about 64 cm tall) with relatively level tops.

Non-piled Stone Circles: Stones in these circles were originally laid upon the ground or embedded in a vertical position. Apart from the somewhat haphazard use of individual stones to form configurations approaching circles, more clearly defined circles were constructed. By laying stones close to each other, a covered circular feature could be formed. This might then be surrounded with separated stones forming a circle. Or a simple circle might be formed by laying or vertically emplacing stones apart from each other. More complex circles were formed by the emplacement of stones to form two or three adjoining circles of erect stones, or an approximation thereof. No multicircle features formed with laid stones were found, but these might also exist.

Site D is where, in 1963, Paul Hockings excavated a feature approaching a circle (see Hockings 1976: 26-8, and op. Pl.). Thirty-one rough, irregularly shaped stones were here laid on the ground. Hockings determined that the buried contents of two nearby features had been entirely removed. One of these is a piled stone circle with an inner perpendicular side and the opposite side sloping outward. The other is a smaller, low-lying piled stone circle (tallest measured point was about 45 cm high) with a relatively flat top. Site E, across the road and not far from Site D, is another ideal site. Here there are at least eight features. To show these more clearly, the representations of prone stones were blackened (Fig. 4.3). Illustrated at this one site are stones placed next to each other to form a covered-over circular feature, another similar feature surrounded by a circle of separated stones, a well-formed circle of separated stones, two haphazard arrangements of separated stones which form a slight resemblance to circles, and three features with some erratically placed stones accompanying separated stones which generally form three circles. Thus, this spectrum of features seems to offer evidence for divergent features being so created by the original builders.

Site F (Fig. 4.2), spreading downward from the top of Kenkodu Betta, is another ideal site. At the summit there is a complex stone circle with vertically emplaced and separated stones forming, for the most part, at least three somewhat coarse circles (true, one can argue that in a part of the complex there are no more than two circles). Farther south and downslope is an earthen circle, discussed on p. 113. Next are the remains of a partially ruined complex stone circle which may once have had, for the most part, two adjacent circles of erect stones. Site M (Fig. 4.3) on the village green at Ebanad, has the most impressive simple circle of erect stones in the Nilgiris (also see Noble 1976: Pl. 1, c). On a conspicuous summit, just south of the Badaga village of Anikorai, are the ruins of a similar circle (Pl. 13,C; Site Y in Fig. 4.1).

Earthen Circles: Because I have identified only four earthen circles in the upper Nilgiris, all are covered in this section. One lies not far below a summit, but the other three are on two lower, rounded ridges. Each was formed by the piling of earth excavated from its interior, and each offers evidence of the piling of earth over a centred burial pit.

In 1978 I stumbled into an earthen circle at Site C. Covered by grass, it blended imperceptibly into the surrounding grassland. The interior depression, however, was clearly discernible. The centralized raised area was inconspicuous. These site characteristics perhaps illustrate why so few earthen circles were recorded: often, for one to become known, a horse-rider or a hiker must literally ride or walk into such a feature. Todas, Kotas, and Badagas, thoroughly familiar with the locations of many prehistoric remains, are the finest source for more rapid discovery. The ideal Site F (Fig. 4.2), with outstanding examples of erect stone circles, also has the most informative earthen circle. Here the builders cut into the slope and piled earth into the downslope portions of a circle. The most northerly portion of the circle, lying flush with the surface, was outlined with the aid of stones placed on a level which perhaps was cut slightly below the surface (or did stones originally placed at the surface work themselves down into the turf?). With gradient downslope, the function of outlining stones became less important and earth as an outliner came to dominate entirely. The inner low-lying depression was levelled at a depth of a metre or more below the crest of the surrounding circle. The centred higher portion probably covers the burial pit.

The other two earthen circles, at Site U, 21, Melur South (Fig. 4.2) are located at a leading Nilgiri religious centre. We may postulate that the earthen circles were the original features constructed, and that they came to be related to worship at this site. Always afterward, then, this site remained a religious centre. This came to be the only place where circles and the larger prehistoric, megalithic cult dolmens are found together. A group of dolmens was actually centred in the small earthen circle, and other dolmens stand near the larger one. The only Badaga round temple that we have a record of (Breeks 1873: Pl. 75) was once to be found at Melur South. Both Badaga Gaudas and Toreyas now have Mahalinga temples, and these are located on opposite sides of the earthen circles and dolmens. A small dolmen and fire-walking pit are located close to the Gauda Mahalinga Temple, and it is to this temple that a Kuṟumba comes each year to perform pre-planting ritual which starts off the annual agricultural cycle. The two earthen circles lie on the village green. Two trees grow out of the larger one, and a pipe for water and a path cross it (Pl. 13, D). The large one, which is probably the largest prehistoric circle in the Nilgiris, was made from earth piled up from inside. Its crest does not rise much above 30 cm. By contrast, the top of the mound positioned within and toward one side rises to about 53 cm. The small earthen circle, simply constructed with piled earth from its interior, partially crosses the other one. Due to the presence of the dolmens, the delineation of a mound within has become impossible. It is unfortunate that the dolmens have been tampered with (under supervision by Breeks, to enable the taking of photographs?), for here stands the finest of all Nilgiri sculpturing Pl. 13, E) on the orthostatic stones for dolmens.

Burial Rectangle: At Site K, 2, on Bilikal Betta, there is a unique feature which appears to have once consisted of erect stones forming an enclosure measuring at least 1.25 m x 2.5 m. The largest stone upright is up to 1.4 m high by about 1.2 m wide. The rectangle has been partially destroyed and offers evidence of having been excavated within. Large flattened stones assist in its reconstruction.

There are too few of these to suggest seriously their past use as capstones, even if these are conceived of as being smashed. It also seems highly unlikely that erect stones so thin and of such varied height could have supported capstones which, to bridge the interior, would by necessity have to be large and heavy. It appears, then, that this feature is associated with subsurface burial followed by the erection of a rectangular enclosure open to the sky. It is therefore called a burial rectangle.

Above-Surface Cist: Away from the village green and on the outskirts of Ebanad there is a religious centre with a ritual arch on an upright and circular platform, an erect stone on another upright and circular platform, an erect stone under a large and old tree, and a nearby feature, which appears to have been an above-surface cist. This feature, at Site 6 (Fig. 4.3), is a rectangle almost entirely enclosed by erect stones. Two large stones slanting up from the ground on the inside may have been capstones, so it seems possible that this feature may have been an above-ground cist. If the inner stones had once stood erect in the enclosing rectangle, there is the likelihood that this feature is another burial rectangle instead.

Below-Surface Cists: At Site T there was once, within an area covered by savanna vegetation, a series of stone circles delineated by individual stones set apart. Some stones were laid on the surface, but others were erected. In contrast to the dominant Nilgiri system in which burial pits within circles were covered over with capstones, here the builders constructed below-surface cists to house the remains of the dead. The capstone of each cist lay at ground level. Although we cannot be fully certain, because the necessary pottery remains have been lost, this may also have been the only Nilgiri site with typical South Indian megalithic pottery—the Black and Red Ware, black and red pots.

As Congreve was the first and last to report on intact cists, the related specifics are now outlined. In looking at two illustrations of one, entitled 'Closed Cromlech at Bellike, Neilgherries, Front View' and 'The Same Side View' (Congreve 1847: Pl. 9 op. 123), a person might come to the conclusion that this was an above-surface cist. However, the potential misunderstanding is resolved when one reads, 'The *Kistvaens* were nearly buried in the vegetable soil, a fact, considering their height (five feet) [1.52 m] that sufficiently attests the high antiquity which must be assigned to them' (Congreve 1847: 123). Congreve reconstructed how the cist once looked above the surface, for he believed that as time passed soil had built up and surrounded the cist. His excavation report provides additional clarifications:

> After removing a large slab five feet [1.52 m] long, three [.91 m] broad, and one [31 cm] thick, which served as the roof of one of the Cromlechs, I proceeded to excavate the earth that had fallen inside, and reached the floor, another large flag eight feet [2.43 m] long by six [1.83 m] broad. Here I found fragments of clay vessels, probably remains of funerary urns. The chamber being cleared presented four walls, each consisting of an entire

> stone, and was seven feet [2.13 m] long by five [1.52 m] broad. The Monolith constituting the eastern wall was pierced by a circular aperture about nine inches [23 cm] in diameter (Congreve 1847: 123; my emendations).

Congreve's measurements for the capstone may have been incorrect. If they were not, the capstone would most probably have been laid over the interior of a subsurface cist having earth piled in over the interred human remains and grave goods.

In his report twenty-six years later, Breeks (1873: 106) wrote:

> They [the kists] are all so much alike in construction, that a description of the one photographed will be sufficient. It was surrounded by a circle of single stones 18 feet [5.5 m] in diameter. Four large slabs were standing edgewise in the ground (natural soil, not vegetable earth or refuse), their tops just level with the surface of the ground; another large slab lay at the bottom. The covering slab had been removed and was lying outside the circle.
>
> The earth inside the kist was mixed with charcoal, and was loose vegetable soil; the outside undisturbed, as if a square hole had been dug and the slabs put in.
>
> The kist measured 3 ft. 6 ins. [1.06 m] from E. to W., and 2 ft. 6 in. [76 cm] from N. to S. In the middle of the eastern slab was a round hole, varying in different instances from 12 to 15 in. [31 to 38 cm] in diameter. The kistvaens had all been rifled; we dug round some of them, inside the stone circle, and found a broken dagger, CCCXIX., and some fragments of pottery, thick and highly glazed, quite different from that of the cairns. (my emendations).

To obtain the only photograph there is of a Nilgiri porthole (Breeks 1873: Pl. 76), most of the cist was smashed apart and destroyed. The recorded artifacts were not preserved.

When I visited this site in 1963, only two stone circles and the partial remains of another were left. One oblate circle measured 5.5 m x 4.3 m, and the two other circles had diameters of approximately 4.6 m x 3 m. Only a portion of one cist was left (Noble 1976: 95). Excavated pits at the three circles indicated past cist depths of no greater than 91 cm. Because retaining walls were being constructed in this area during 1963, it is probable that all the prehistoric remains have by now been destroyed.

Dolmens: In contrast to the stone and earthen circles, which for the most part stand on summits or ridges at higher elevation, the Nilgiri dolmens generally lie within valleys or on nearby slopes at lower elevation. The locational differences add support to the concept of the circles being related to persons with a herding tradition, whereas the dolmens are most easily related to farmers. Whereas the circles identify actual funerary sites, there is no evidence to identify dolmens as being permanent depositories for the remains of the dead. They can, in a few instances, be linked with funeral ritual. They have served primarily for the memorialization of the dead, or for worship. The ideal Nilgiri dolmen has three vertical orthostats at an angle of approximately 90°

to each other, and there is thus an opening on one side (Pl. 13, F). Its capstone stands at a height of close to a metre above the surface, and seldom is there an underlying stone in a dolmen. At some sites there is only one dolmen, but others have aligned dolmens. The most complex dolmen, at Doddamanaihatti (Site 16 in Fig. 4.1), has ten orthostats and three capstones. Dolmens may or may not have sculpturing on the interior faces of orthostats. Although several dolmens stand at Kavilorai (Site 3), here there is also a series of free-standing erect stones aligned next to each other. There are more sculptured stones here than at any other Nilgiri site, and only a few of these stones are within dolmens.

For the reader desiring a more detailed grasp of the variations related to Nilgiri dolmens and post-construction dolmen use, visual imagery in photographs and the site plans of dolmens at 1, 3-4, 7-8, 10-14, 16, 18-20 in Fig. 4.1 have already appeared in published form (Noble 1976: 97, 110, 119, and Pls. 1, 2 and 3). The burial rectangle at Site 2 and the above ground cist (?) at Site 6 have some constructional affinity with the dolmens. The feature at Site 5 is a dubious one. A stone of about 1.37 m x 0.85 m x 8 cm in thickness rests on a rocky surface to one side and is supported at no higher than 20 cm by two rocks on the other side. This sitting place (and temporary altar?) is probably a recent construction. At Site 15, in Tuneri, there are two aligned and unsculptured dolmens. Choking vegetation proved their disuse in 1963. Local inhabitants claim that they were used for worship prior to the construction of the nearby Mahalinga Temple. Worship stones in the back interiors substantiate the claim. At Site 17, in Sholurmattam, there is a single dolmen with a sculptured back orthostat. It has been incorporated into a wall next to the road and surrounding a school compound. With an altar formed by a stone slab placed in its back and another stone for breaking coconuts, the shrine so formed is still used.

Site 22 is near the location of Tudurmattam, which, according to persisting tradition, was totally abandoned by Adikaris as a result of Kuṟumba magic and sorcery (Francis 1908: 316). Tea in Woodland Tea and Coffee Estate now covers the area. The so-called round temple at Site 22 is actually composed of seven sculptured and five unsculptured orthostats emplaced in a rough circle. Two capstones partially cover the structure. Close by are another two sculptured and four unsculptured erect stones which were probably once incorporated into dolmens. That there were once aligned dolmens here seems undisputable, but the account of them attributable to Commissioner and Mrs Breeks (1873: 104-5) is probably flawed:

> ...there are carved cromlechs [dolmens] in Major Sweet's plantation beyond Kārtēri [Katteri in Fig. 4.1]. ...The sculptures, besides the usual *Basavas [Nandis]*, sun and moon, man spearing animal [pig], woman holding fan, &c., have a horse caparisoned, but not mounted, and a figure on horseback with an umbrella held over him. These cromlechs yielded, besides *Deva-kotta-kallu* [river-worn memorial stones associated with the Kuṟumbas, and of a type also stored in Irula memorial temples], a number of iron and bronze armlets, sickles, rings, two small iron hatchet heads, and

> a small rough common chatty [pot], but no bones or charcoal. The iron was much less rusted than in the cairns (my emendations).

Further in the book (Breeks 1873:137), in a listing of artifacts found, there is a glaring discrepancy: '12. Bits of bone' are listed under the headings of 'Found in Cromlechs' and 'Major Sweet's plantation'. In the account above is the specific statement, 'no bones or charcoal'. We may suspect that Mrs Breeks, who edited her deceased husband's manuscript, mixed data pertaining to two sites. The first part of the data above is basically correct. If bronze and iron ware and a pot had been found in the dolmens, this would offer the most substantive proof of the circle builders also being dolmen builders. In the separated listing, 1-10, artifacts referred to in the account are called bracelets, small sickles, iron and bronze rings, and small hoes (instead of hatchet heads). I believe that all the artifacts listed, 1-10, and 12-14, came from a stone circle excavated on his estate by Major Sweet, probably in 1869. The artifacts may well have formed the topic of discussion at the February, 1870, meeting by members of the Asiatic Society, Calcutta. The artifacts discussed were 'quite recently dug out from a cromlech on the estate of Major Sweet in the South side of the Nilgherry plateau' (Saxton 1870: 52). Because artifacts within dolmens could easily be removed, they would not be 'dug out'. The English were constantly confused over the use of terms, and the writer believes that the incorrect application of the term cromlech in this instance may have caused Mrs Breeks to blend data pertaining to two sites on Major Sweet's estate: one with dolmens, and the other with at least one stone circle. Another question is fitting at this point. In the article by Walhouse, to be considered later, did the artifacts now in the British Museum come from a stone circle, and perhaps the same stone circle, on Major Sweet's estate? The iron artifacts described by Walhouse are probably the best preserved prehistoric ones found to date in the upper Nilgiris.

The partial plan of the Melur South Site U, 21, and a plan of the Banagudi Shola, Site 9, hitherto unpublished, can now be used to demonstrate some more specific site and dolmen characteristics (also see Noble 1976: 102; Pl. 1, e, f, g, and h). As mentioned previously, Melur South is a leading Nilgiri religious centre. *Bana* (secluded)-*guḍi* (temple) *Shola* (montane forest remnant), true to its name, is an isolated worship centre within a forest. Melur South has a combination of single and aligned dolmens (Fig. 4.2). All but one of these have sculpturing on at least the back orthostats, and some side orthostats are also sculptured. The dolmens at Banagudi Shola represent a range of size in dolmens (Fig. 4.2). The largest is the most unusual one in the upper Nilgiris (Pl. 13, G). It departs from the norm in several ways: 1) it is the largest, and 2) the tallest, and has 3) a stone base, and 4) many stones piled next to it. Not unusual are the two sculptured back orthostats, sacred ash placed on the foreheads of each sculptured human representation, and a set of seven worship stones at the base of a sculptured orthostat. A *linga* held upright by stones at the ruins of another dolmen is also ritualistically significant. Some nearby Badagas and others come to worship here. There is no sculpturing in any of the other smaller dolmens, or their ruins. Three of these do, however, have underlying stone bases.

In summarizing the classificatory system which has been used, and to move toward the provision for a distinctive system for the upper Nilgiris, we may tentatively divide the prehistoric megalithic remains into six categories (A through F): Most funerary sites are identified with the emplacement of the cremated remains of the dead in burial pits filled with earth and capped with stones. Their locations were then demarcated by A) stone or B) earthen circles. There is some possibility of there being two types of funerary sites not demarcated by circles: C) burial rectangles and D) above-surface cists. Dependent upon proof only possible through excavation, there may only have been burial rectangles and no above-surface cists. Each funerary feature at only one site known thus far was demarcated by a stone circle, but the presence of E) some below-surface cists was by far the most distinctive facet. Because of the advanced degree of destruction, we will probably not be able to determine if these features—by contrast with the majority—were associated with the widespread southern Indian practice of fractional burial and use of the characteristic Black and Red Ware. There are also F) dolmens, and some of these have interior sculpturing on at least the back orthostat.

Because there is such variation among the stone circles, we may divide them into two series: one with stones piled into circles, and a second with individual stones forming circles.

In Series One there are circles with

1. walls having inner and outer perpendicular sides rising to about one metre or even more in height (type sites: B, R, and X),

2. a band of stones piled in a single layer, or with stones piled somewhat higher and with relatively level tops (type sites: G and S),

3. inner perpendicular sides and outer sides sloping to ground level (type sites: K, 2, and P),

4. inner and outer sloping sides (type sites: N and P),

5. an inner circle of upright stones followed by a band of stones piled in a single layer and/or a piled stone wall sloping outward to ground level (type sites: O and P),

6. a profile similar to 5, but with an accumulation of earth in the centre (type sites: H and P),

7. a profile similar to 5, but with at least two inner circles of upright stones (type site: L).

Stone circles in Series Two were formed by

1. stones laid close to each other, to cover each circle (type site: E), or

2. either lying or vertically emplacing separated stones in a single circular alignment (type sites: E, M, and Y), or

3. combining each covered circle with a surrounding circle of separated laid stones (type site: E), or

4. emplacing two or more adjacent circles of upright stones, with the uprights in each circle somewhat paralleling each other (type site: F).

The Typical Archaeological Assemblage (or Aggregate?) and Artifacts

Apart from the atypical below-surface cists surrounded by stone circles, for which some information has already been provided, site reports by several individuals enable us to visualize the typical archaeological assemblage (or aggregate?) and artifacts found within earthen circles and the remaining stone circles.

Apart from his fanciful illustrations, Congreve's description of an excavation in a walled circle clearly indicates the possibility of artifacts forming an aggregate:

> After clearing away the trees and brushwood that overgrew the interior I excavated the soil in the middle of the cairn [walled circle] to a depth of two feet [61 cm] and alighted upon two large stones; these being removed I found two circular urns [the cinerary urns] and some articles in brass [bronze] and iron, ..., in holes about three feet [92 cm] below the stones. The urns contained charcoal and bones. Surrounding the interior [farther out from the centre] of the cairn numerous urns of the description figured hereafter appeared a few inches [probably 8 cm or more] below the surface, some standing upright with lids on, others thrown upon their sides with the lids beside them, many broken into pieces and some of their fragments contained in entire urns: this arrangement of the vessels manifested that they had been disturbed at different periods in order to make room for the burial of other urns. Figures of animals and human beings were dispersed in every direction (Congreve 1847: 88-9; with my emendations).

What is reported was basically repeated so often at other sites that serious attention must be paid to the possibility of artifacts having been periodically deposited at close to the surface levels within some funerary circles. Because Congreve probably excavated more sites than any other person, and at a critical time, we must hope that his fascinating collection of artifacts will be discovered.

This final resumé by Breeks, following his brief notes for over 35 excavated earthen and stone circles, shows that the basic pattern described by Congreve was generally followed:

> The above extracts show that the general features of the cairns [stone circles] and barrows [earthen circles] vary little. Above and between the slabs, ..., round the circle [within the interior] near the surface, lie the rough pots ...., large deep narrow vessels, pointed at the bottom, so that they cannot stand upright, with rough figures of men and animals on the lid[s], and

> empty, or containing only earth, as far as their almost invariably broken state allows us to judge. The number of these is surprising. Baskets full of heads, horns, and tails of buffaloes and other figures may be carried away from some cairns; but in most cases they lie so near the surface penetrated by the roots of trees and bushes that nothing but fragments can be recovered. Below, at depths varying from one to four feet [31 cm to 1.22 m], are the cinerary urns, superior in quality and make.
>
> There does not seem to be any rule as to the arrangement of the actual interments. Sometimes the bones are at the bottom of the urn [cinerary pot], sometimes in a bronze vase contained in it, sometimes under the inverted bronze. Often the bronze is not in or near the urn. Some of the urns do not contain bones, but only implements and ornaments, and some only earth. Sometimes the number of interments corresponds with that of the slabs; but this does not occur often enough to prove design [Breeks 1873: 93; my emendations].

From a note covering one excavation (Breeks 1873: 75), we have proof that there is sometimes more than one layer of capstones: 'The second layer of slabs under the first seemed to me strange, but Mr Metz said he had found it so in several cases'.

Hockings (1976: 40), whose report on the Paikara Stone Circle (at D in map) is essential reading, agrees with Breeks: 'We can only concur with his general description of the cairns and barrows, a description which fits our own case too.' The most unusual feature of this excavation was the discovery that the single capstone did not cover the single cinerary pot or any other artifact (Fig. 4.3).

To summarize, with an alternative viewpoint providing an assemblage in each case, this basic constructional sequence is suggested: First, there was at each funerary site the digging of a pit which narrowed downward. The niche for a cinerary pot, or niches for several pots, deposited at the lowest depth seem in some instances to have involved the final digging of holes barely wider than the pot diameters. Although cinerary pots tend to be similar in shape, the fact that their contents vary as noted by Breeks is understandable. A particularly hot cremation will leave scant human remains, and these may further disintegrate through time. Thus a cinerary pot might eventually appear to have no remains from cremation. Although bronze, iron, and even gold artifacts tended to be placed in each cinerary pot and/or nearby, the nature of the artifacts used must have depended upon each bereaved family's possessions and the final decision-making processes. There is, for example, much variation among the bronze vessels. There seems also to have been a propensity toward placing cremated remains in or under a bronze vessel and, as Breeks pointed out, such a vessel was not always placed in a cinerary pot. Once each cinerary pot was emplaced, the fill-in would start. A spectrum of artifacts tended to be deposited at higher levels, most probably by relatives of the deceased. After a large portion of the pit had been filled in, consensus among those present would result in the decision to emplace a capstone, or capstones, or even two layers of capstones. These would be positioned over level fill. They would then serve as guide markers for a final climactic rite. Those attending each funeral, both from near and far, would bring their terracotta effigies or pots with effigy lids to the funerary site. These would be emplaced upward in the loose soil surrounding a capstone or capstones. If there were many mourners, as there may have been at the funeral of a particularly revered elder, it seems likely that some grave gifts would have become smashed at the stage of initial deposition. After gifts of the climactic rite had been emplaced, it is easy to envision donors walking away from a funerary site without looking back. Some attendants would at the end cover over the effigy artifacts and, in many instances, the capstone or capstones. Each demarcating circle may have been constructed in the days following by males having specific, well-defined kinship relationships to the person or persons who had died.

Brief coverage of characteristic artifacts, all of which were collected by Breeks, is now offered. In Plate 14 the illustrated artifacts 1 to 4 and 22 were photographed by me in the Tamil Nadu State Museum, Madras.[4] Artifact illustrations 5 to 17, 19, and 25 to 39 were extracted from photographs in Pls. 36 to 41 and 43 at the end of the volume by Breeks (1873), and these provide most of the substance in this section. Artifact illustrations 18, 20 and 21, 23 and 24, numbered 582, 537, 538, 820, and 822 by Foote, were extracted from Pls. III, IV, X, and XIII in Foote's catalogue (1901). For those wishing to see more, there is Naik's unpublished catalogue (1966) covering artifacts mainly collected by Breeks and now housed in the British Museum, London. This will eventually be supplanted by a more comprehensive catalogue of Nilgiri artifacts in the British Museum, now being prepared by J. R. Knox, Assistant Keeper of Oriental Antiquities. For the first time since the 1930s, there is also a display of Nilgiri artifacts in the British Museum.

Illustrations 1 to 4 show effigy lids which once covered pots made from light red clay, typically emplaced close to the surface. Perhaps a hyena, leopard, buffalo, and peacock (highly stylized) are represented. The effigies were moulded by hand and then decorated with different devices. Rounded indentations were made with round blunt or hollowed objects (porcupine quills?). Styluses were used to produce rows of depressions or indented lines.

Illustrations 5 to 9 are of effigy pots, so distinctive of the Nilgiris. One can see how effigy lids were fitted onto the pots, but the most significant feature is the varying and distinctive manner in which the decorative device of a pot over a pot was utilized. Light red clay, of a type still used by Kotas, was used to fashion the effigy pots on rotated potter's wheels. Paddles and fingers were probably used to close the gap at the bottom of each pot when removed from a wheel. Pot 5, with it sword-bearing mounted male on a fanciful horse effigy, indicates the past presence of horse riders able to engage in combat. Both the effigies in 20 (probably off a lid) and 21 (probably free standing originally) are of mounted males, and both have heads fashioned in the same manner as the male's head in 5. Is the male in 21 depicted as riding on a donkey? In 19 (probably once attached to a lid), a sheep effigy with a bell hanging from its neck is depicted. The female effigy in 6 has an interesting pointed cap. Effigy 7 illustrates how fanciful some fashioned objects were. Effigy 9 is of a humped bull or cow (The buffalo in 8 is enlarged in 3).

Effigy pots 10 and 11 are of another distinctive type with flaring mouth. Is a stylized peacock shown in 10? It appears that there is a fanciful flightless bird effigy on the lid of 11. In pots 12 and 13, with effigy lids missing, the flaring mouths are clearly discernible.

Cinerary pots, typically placed in the lowest portions of burial pits, are represented by pots 14 to 16. Despite their variation in shape and size, all are similar to a degree. As shown in 15, there was often the modelling of closely fitting lids to enclose the contents of these pots. Pots 17 and 18 are utilitarian ware of unusual design. Pot 17 has a spout and 18, with its portrayal of breasts, reveals the maker's creative departure from the norm.

Illustrations 22 to 27 show both the variation and sophistication of Nilgiri bronzes. Due to a high tin content (29.89 per cent in one analysis), it would appear that the resultant brittle and non-malleable bronze caused metal-workers to depend primarily upon casting for the basic shaping of all the Nilgiri vessels (Naik 1966:217). Two views of the best-known Nilgiri bowl are shown in 22 and 23. It demonstrates an initial casting of parts (the base was later attached to the bowl) followed by decorative flutings done with the aid of a steel scriber and a steel punch struck with a hammer. The interior lotus motif was employed in other Nilgiri bronzes. The two parts of round bowl 24 may have been turned on a lathe before being put together. The horizontal decorative lines in the smaller bowl 25 were probably fashioned as this bowl was being turned. Both the smaller and shallower bowls 26 and 27 have a centred upraised projection in their interiors.

Illustration 28 to 39 are of artifacts mainly made of iron. The potentially combative nature of their former users is revealed in artifacts 28 to 32: a spearhead, two arrowheads, and two different short-swords. Artifacts 33 to 36 could have been used by either farmers or herders. A sickle (33) may be for cutting forage grasses, and billhooks (34 through 36) are useful in the removal of shrubby growth or in firewood collection. Artifact 37 was a shaving razor, and it is suggested that artifact 38 was a razor for cutting leather. Illustration 39 is of a lamp.

The diversity in prehistoric Nilgiri weaponry is specifically demonstrated by the iron artifacts removed from One Stone Circle. According to Walhouse (1873: 276-7), the stone circle having a diameter of about 1.83 m was made inconspicuous by 'stones of moderate size, only just appearing above the ground'. Instead of being on a prominent peak, this circle was on a slope between Coonoor and Katteri (Fig. 4.1). That there was a unique departure from the norm in funerary practices is indicated by the iron weapons having been remarkably preserved by a carbon layer, apparently resulting from a flaming pyre within the stone circle itself. Apart from the additional 'two

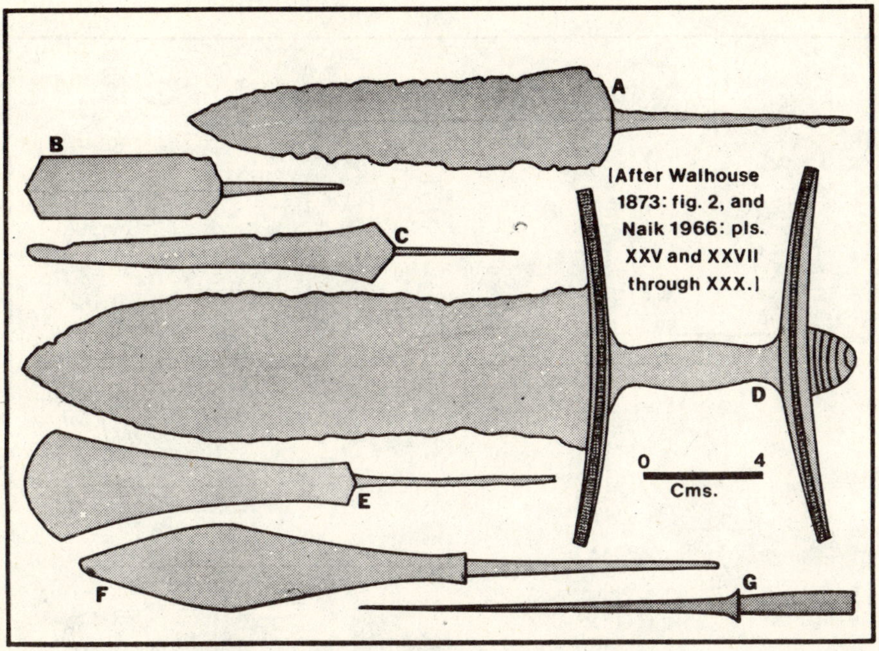

FIG. 4.4 Iron Artifacts from One Stone Circle.

pairs of bronze or copper bangles...and several other less noteworthy weapons and objects' from the layer, other artifacts from this layer and lower levels may have reached other individuals (see previous references). In his selected spectrum of artifacts (they and others are in the British Museum), Walhouse (Fig. 4.4) illustrated a short-sword (D) and a series of projectile points (A-C, E-F) which might have been attached to spears or javelins. A four-sided point (G) may have formed the base end of a javelin (therefore, two different points on the same shaft).

Sample Sculptures in Dolmens

To enable the reader to obtain some understanding of the sculptures within dolmens, some specific examples of them—all shown in Fig. 4.5—are now covered. These fall within a spectrum ranging from the crudest to the finest.

At Site U, 21, Melur South, there are crudely sculptured figures of two men on a single side panel (Fig. 4.2, orthostat d). Each man holds a short sword or dagger in the left hand. While one holds a spear in the right hand, the other holds a sword. Both are depicted as wearing waist garments, necklaces, and no shirts or upper covering. The sculpturings are typical of those honouring male heroes only. By contrast, there is at the same site the finest sculpturing (previously mentioned). Photographic coverage does more justice (Pl. 13, E),

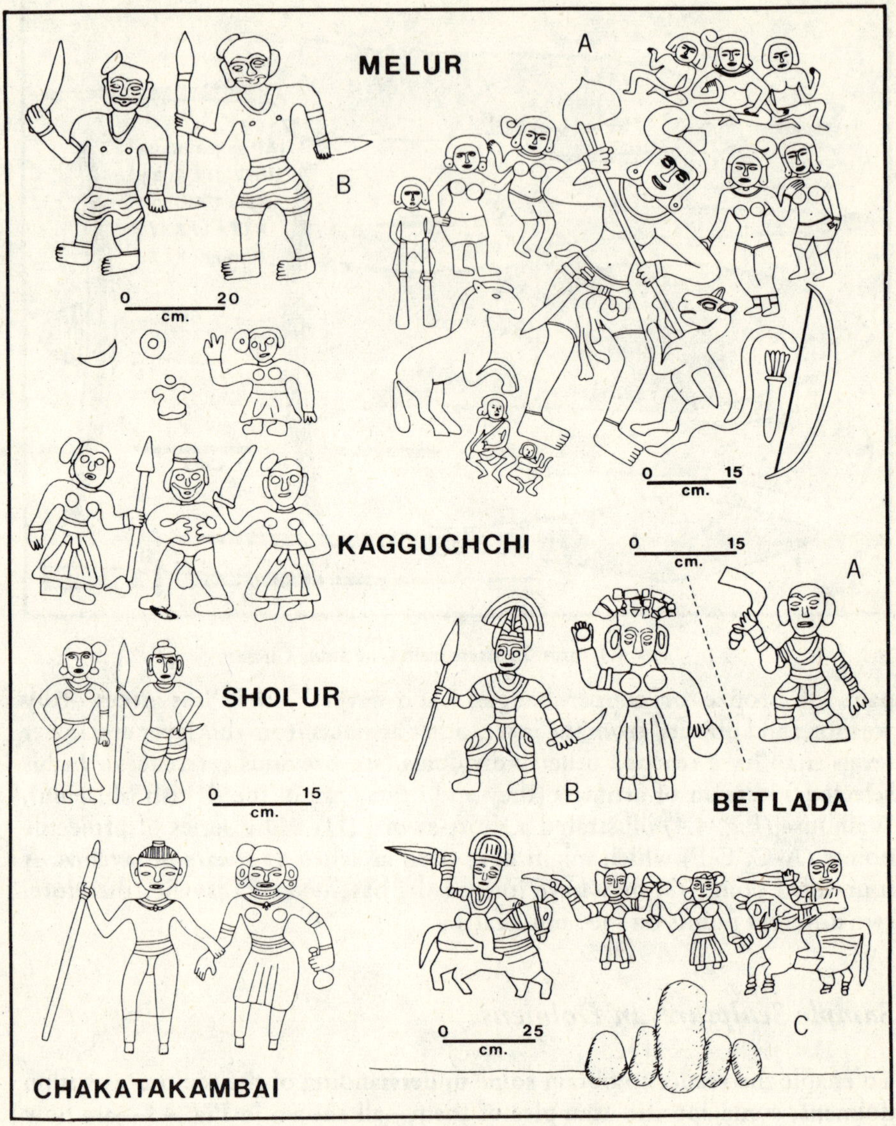

FIG. 4.5 Carvings on Orthostats.

but it was impossible to photograph the whole panel (Fig. 4.2 orthostat c) without shifting site features; therefore a line drawing is used. Clearly, this carved panel was related to the development of a single theme. Through the techniques of centring and sculpting this figure larger than any other, attention is drawn to the main character being honoured. His headgear, the emblem of authority to the right of his head, sword attached to waist, a bow and quiver with arrows, and a smaller horse assist in conveying the message that this was an important personage—perhaps a member of royalty and/or an army commander. This warrior was probably killed in the mêlée after piercing a predator (a leopard?) with his spear. As one customary device to honour a *sati* was to depict her as a female with lime in palm, it may be conjectured that the female to the hero's left was cremated on her husband's funeral pyre. A second wife, farther to the warrier's left, may have died in the same way. Were the hero's three children depicted to his right? Could they have died before the hero died, and are their spirit selves depicted above? And did the hero leave two living younger children, depicted below? As the rare exception in the Nilgiris, there are on the rock face with this single panel some coarse Tamilian letters not permitting the reconstruction of words. These were in 1962 dated to the fifteenth century AD by members of the Government of India Epigrapher's Office, then located at Ootacamund. Earlier, Chakravarti (1936: 97) had dated the letters to the seventeenth or eighteenth centuries AD. In a nearby dolmen, on a back panel (Fig. 4.2, orthostat e; Noble 1976: 102), there are other letters respectively dated to the twelfth and thirteenth centuries AD by the same authorities. Despite the discrepancies in dating, the letters tend to support continuing dolmen construction and use over some centuries.

The crude figures at Kagguchchi were sculpted on a side orthostat (Noble 1976: 97, orthostat *B*). Did a Badaga, with tongue in cheek, poke fun at the ancients who had sculpted other orthostats? A female with spear and another female with sword to each side of an unarmed male provide a seemingly hilarious touch. However, the above woman with upraised arm and the sun and moon are seriously emblematic of a *sati*. Therefore, in reconsidering the lower depiction, it is suggested that two women may indeed have fought to protect a man's life. Could the man have been ill at the time an enemy attacked? A wife (also depicted in ghost form in the after-life?) may have become a *sati* after the man died.

The sculptures at Sholur and Chakatakambai typify the majority, in which males and females are depicted together. The male figure at Sholur is representative of a common pose. This figure is apart from but close to a female. The Chakatakambai figures illustrate female and male body contact, not infrequent among the Nilgiri sculptures. If a male figure with spear in right hand does not have the left hand on his waist, the hand is often depicted as grasping any part of a female's arm—or, *vice versa*, the female may be depicted as grasping any part of the man's left arm. The bulbous container in the female's left hand is a rather common *sati* symbolization in the Nilgiris. It represents a container with oil which was poured over a woman just before she became a *sati*. The whole unique series of male with female sculptures in Nilgiri dolmens is indicative of centuries in which male heroes and related *satis* were memorialized.

At Site 11, Betlada, there is the most outstanding series of sculptured males bearing arms in the Nilgiris. Events related to armed garrisons in a nearby fort, now in advanced deterioration, may have provided memorable episodes to be commemorated in stone. Here there is also evidence that sculptures honouring those who died at different times were systematically incorporated into sculptured rows on the faces of three orthostats, now standing in two shrines (corresponding with two original dolmens) within a shed constructed by Badagas (Noble 1976: 110, *A*, *B*, and *C* stones). The use of one stone face to memorialize a series of deaths separated in time is one of the most remarkable, and perhaps unique, characteristics in Nilgiri sculpture. One sculptured warrior (on *B* stone) bears a boomerang, and thus offers proof of boomerang use in southern Indian combat. Does another sculptured warrior (on *A* stone) holding a spear and sword prove that cotton-quilted armour was used? Next to this figure is the representation of a *sati*, well symbolized with lime in upraised right hand palm, the characteristic pot held downward in the left hand—and perhaps even a crude, stylized representation of flames rising above her head? In a bottom row close to the floor (on *B* stone) and five upright river-worn worship stones there are the crudely sculptured figures of two warriors on horseback, each carrying a sword. Each horse appears to be covered with trappings—clothes for a triumphal parade perhaps? A woman might be related to each of the two warriors on horseback. Although the crudity of sculpture prohibits positive identification, each may be holding a pot symbolic of a *sati*.

In a first attempt to establish clearly the use of one orthostat to memorialize deaths separated in time, a photographic series in Pl. 15 is used. Photograph 15, A shows one sculptured orthostat at Site 1, Sholur (Noble 1976: 110, orthostat *B*). Photographs 15, B to F are of individually sculptured panels on that orthostat. The varying angles at which these panels lie, the separateness of each panel, the stylistic differences, and some reflection of each sculptor's need to adjust his efforts to confined space tend to prove the use of the one orthostat to memorialize those who had died at five different times. Generally, it may be argued that the men who died and their related *sati* wives are honoured in every panel.

The topmost panel (15, B) honours two warriors with spears. Notice how one of the spears blends with the panel edge, and how the adjacent panel is cut deeper into the rock on the other side. Is one of the warriors holding the right arm of a female, and is that female holding a *sati*-related pot? The next panel down (15, C) is more easily interpreted. The female figure holding a pot in the right hand and with left arm upraised in a blessing pose represents a *sati* who perished on her husband's funeral pyre. The sculptor depicted him as a warrior with sword in upraised right arm. The next panel down (15, D) has a similarly sculptured representation of another *sati*. However, is she and the sculptured representation of her husband clasping the hands of their upraised arms? Or is she holding an upraised lion-mane whisk close to the upraised arm of her husband? If so, and in other distinct examples of such which are *sati*-related, the lion-mane whisk symbolizes waiting upon her lord, the husband. It may even be suggested that she and her husband uphold a fire-brand with their arms, and this would then be a *sati*-related symbol. In the next panel down (15, E), the *sati*-related pot is held in the right hand of the female. Are the female and male jointly holding onto the trunk of a tree of life? If so, this would be emblematic of continuing connubial relationships in the after-life. The lowest panel (15, F) has a male with both arms raised and right hand holding a sword. The female's right arm is upraised in the blessing of a *sati* but again it is impossible to determine what the sculptor intended in the portion involving the overlapping of upraised male and female arms.

The last two photographs are of sculptured orthostats at Kavilorai (Noble 1976: 97, orthostats *M* and *S*). In 15, G there are two sculptured rows and, before the stone was weathered away, there was probably enough space for the sculpture of another couple. The figures are conventional, with most females holding the characteristic *sati*-related pots. The bottom centred group in which a man holds hands with two women probably does commemorate an event in which a man's two wives became *satis* through cremation on his funeral pyre. In 15, H there are three separate panels, and enough space remains for the sculpture of another panel. Each of the two lower couples could have been sculpted at different times. Three of the four women bear *sati*-related pots. One man holds onto a spear, and it is difficult to ascertain what the three other men are holding. Might they be boomerangs?

Dating and Identity

The evidence necessary to date the Nilgiri prehistoric remains adequately is at present lacking. We can, therefore, only tentatively reach some understanding of when these remains were constructed. For features apart from the dolmens, these are our most positive clues:

1) On a potsherd discovered by Das (1957: 147-8) there are Brahmi characters which epigraphists of the Government of India Epigrapher's Office have dated to the first century AD.

2) A gold coin of the Roman Byzantine Series was excavated from a stone circle in Ootacamund, and this has been dated (Srinivasan and Bannerjee 1953: 112) to the fourth century AD.

3) From sample GIF-2345, with carbon content from pottery excavated on the Nilgiris by Das (Hockings 1976: 45), a radiocarbon date of 910 ± 90 years

BP was obtained. This we can approximately date to AD 1100.

Based on these few clues, it may be suggested that there is no evidence for people having lived on the upper Nilgiris in BC times. Most prehistoric remains were probably constructed between AD 100 and AD 1100.

In the case of the dolmens, from what was previously written, we know that none can be dated to earlier than AD 1200. Kuṟumbas presently construct and use dolmens to house the waterworn stones (called *Deva-kotta-kallu* by Breeks) related to spirits of the departed, or to house offerings to these spirits, or for the performance of ritual. It may, therefore, be suggested that the construction and use of dolmens have continued to the present, for a period of over 700 years.

Three recent attempts at tentative dating basically support the above suggestions. Leshnik (1970 and 1974: Appendix A, 255-67) compares the stylistic evidence of Nilgiri artifacts with other evidence available in South Asia and elsewhere. On this basis, the Nilgiri evidence suggested to him a clustering in the third, fourth, and fifth centuries AD. These centuries embrace the ascendancy and flourishing of the Gupta Period. While at the University of London, Naik studied Nilgiri artifacts in the British Museum. She concluded that 'the analysis of the metal and terracotta objects favours a period from AD 700-1100' (Naik 1966: 143). Hockings was affected by Naik's conclusions and by the radio-carbon date. He has, in comparison with Leshnik, opted for a later period of settlement. He goes a step further than the others, and suggests an historic sequence leading to the settlement and establishment of those who constructed the Paikara and other remains around the eighth century AD:

> There is an epigraph dating to AD 769-770 which actually refers to a defeat by the Pāṇḍyan ruler Jaṭila-Parāntaka of Kurumbas [or Pallavas] a few years before, in the district of Nāṭṭukurumbu, west of Erode (Sewell 1932: 30).
>
> It would seem that we have in fact some support for the following hypothesis. The Töwfiḷy section of the Todas were descended from Kurumbas living in the Mysore plains, who fled to the hills in the eighth century on the occasion of the Pāṇḍyan attack to which we have just referred. They settled in a fairly restricted area beside other Todas already there, who were the ancestors of the superior To:røas̱ moiety (Hockings 1976: 47).

And this brings us to the inevitable question, what relationships may ancestors of the traditional Nilgiri tribal people have had to the prehistoric remains? The Todas have generally occupied hamlets located close to prehistoric earthen and stone circles. As I once offered an argument for ancestral Todas being the actual domesticators of buffaloes (Noble 1968: 55-9), it is understandable that I find it impossible to accept Todas in one moiety as being descended from Pallava Kuṟumbas—even if these were distant from those who ruled at Conjeeveram. While further considering the Todas, who were living in the Nilgiris by AD 1117 (Rice 1898: 16, 10, inscription 83), I argued against them being descendants of the circle builders (Noble 1976: 108-9) and then offered these alternative speculations:

1) The basic concept of Rivers (1906: 686-91) and Hockings is followed, but in reverse. Those who came later were ancestors of those in the To·ṟøaṣ moiety, and they established a superior position over descendants of the earthen and stone circle builders who came to be associated with the Töwfiḷy moiety.

2) The Todas dwelt in the Nilgiris before the circle builders came, and then preserved their cultural identity while the circle builders were neighbours.

3) Todas filled an environmental niche in which a cultural void existed. Withdrawal long after a successful colonization or drastic population decline through epidemics may have ended a cultural milieu dominated by the circle builders.

Without a series of radiocarbon and/or other dates covering the prehistoric remains and the duration of Toda settlements as well, the speculations will remain.

As there have historically been no more than seven Kota villages, it seems hardly likely that Kota ancestors could have constructed the majority of the earthen and stone circles. Of all the traditional tribal people, however, it is the Kotas who could most logically have provided the herder/farmer-artisan symbiotic relationship. If ancestral Kotas had lived in hamlets close to herders who were circle builders, they would most probably have supplied grain, iron and pottery artifacts, and music to the latter. If basic cast bronze vessels were brought in from the plains, the Kotas could have completed the final decorations on these vessels. There is clearly a need for dates indicating the antiquity of Kota settlements. Has any Indian archaeologist seriously considered excavating in a Kota village?

Elsewhere (Noble 1976) I have already offered evidence supporting the two hypotheses that 1) Kuṟumbas probably erected most dolmens, and that 2) Saivite Badagas, with women who became *satis*, played a major role in the sculpture of orthostats within dolmens. Tied to the two hypotheses is the concept of the Kuṟumbas being a refugee group descended from, perhaps, the earliest inhabitants of the Nilgiris. These people were refugees in the sense that they retreated into wilder areas after lands they had once used were first appropriated by Badagas, and later by the British as well—who primarily needed land for plantation agriculture. Although Kuṟumbas were shifting (swidden) farmers when the British first came, it seems entirely possible that their earlier ancestors depended primarily upon foraging.

Because of its potential to improve insight, I can offer fresh evidence of a cultural practice leading to dolmen construction away from the Nilgiris. Most memorial stones (called *naḍukals*) in other parts of Tamil Nadu stand there by themselves. However, through boxing by the vertical emplacement of two additional side orthostats and then the placing of a capstone, a dolmen was formed. In this way the structure became at least a shrine, or even a temple honouring the person or persons (*satis* included) memorialized in the sculpture (Nagaswamy 1973: 268-9 and Fig. 5). At Vadivullamangalam in the Coimbatore District, next to the Nilgiri District, there is such a dolmen dating to the reign of Rajendra Chola. Apart from the sculpture on its back hero stone, one of the side orthostats has an inscription. On a trip from Mettupalaiyam to Kurnool, now in Andhra Pradesh, and at a place called Udenhalli, Congreve (1878: 162-3 and illus.) recorded two sculptures of warriors on a hero stone within a dolmen. He reported that natives used the term 'house of a hero' for such a dolmen.

Knowing of these cultural traditions enables us to better understand why the Nilgiri Kuṟumbas have used dolmens for essentially religious reasons. As I found no positive proof that the Kuṟumbas themselves have ever sculptured orthostats in dolmens, the possibility remains that Badagas and others were responsible for the construction of and sculpture within at least some dolmens. Did Kuṟumbas then adapt dolmens without sculpture for their own religious uses? At Kavilorai, the only place in the upper Nilgiris with free-standing sculptured stones, there is a clear indication that the remains may represent a cultural shift through time (centuries?) from a period in which only free-standing memorial stones were erected, into a transitional period in which some dolmens were erected, to a period in which only sculptured dolmens honouring male heroes and *satis* and unsculptured dolmens were erected, or *vice versa*.

Conclusions

In their structural and substantive parallels, Nilgiri prehistoric remains show affinities with prehistoric remains elsewhere in India. Yet the upper Nilgiri prehistoric remains are in some ways different. For example, in the upper Nilgiris there are 1) walled stone circles demarcating some of the burial sites, 2) a unique series of pottery effigies both separate from and attached to lids, 3) variant pot forms with either flaring mouths or what I have called a pot-over-a pot decorative device, 4) characteristic cinerary pots buried at the lowest levels, 5) comparatively rich offerings of bronze-ware left in the graves, 6) a concentration of sculptured dolmens, and 7) dolmen orthostats which were periodically used to sculpture memorializations of deceased men and *satis*. In this area there is also an absence of Black and Red Ware pots, of pottery ring stands, of large funerary urns, and of memorial stones with vertical panels related to theme development.

Based on a survey of the features at over forty sites, a classificatory system can be established just for the Nilgiris. This covers stone circles in two series, earthen circles, a burial rectangle, above-and below-surface kists, and sculptured or unsculptured dolmens. Further archaeological work, however, may prove that there were no above-surface kists. A Nilgiri prehistoric site with cremated remains typically has at least one cinerary pot at the lowest level, either one or more capstones covering the burial pit, and effigy figures close to the surface. Many of these figures are attached to the lids of pots, but some are separate. The question as to whether an assemblage or an aggregate exists in any site with cremated remains is unanswered, but I have suggested a reconstruction which would make each such site an assemblage. A spectrum of characteristic Nilgiri pottery, bronze, and iron artifacts collected by Breeks is demonstrated with thirty-nine photographic samples. To show the diversity in prehistoric Nilgiri weaponry, iron artifacts obtained from one site by Walhouse are illustrated. A spectrum of the coarsest to finest sculpture in dolmens is also demonstrated with selected examples. In a photographic series, a first attempt is made to establish clearly the use of single orthostats to memorialize deaths separated in time. According to the present state of our knowledge, most Nilgiri funerary structures were probably constructed between AD 100 and 1100. None of the dolmens can be dated earlier than AD 1200, and dolmen construction and use has continued to the present. There is some likelihood that ancestral Badagas, Kotas, Kuṟumbas, and Todas were in varying ways related to the Nilgiri prehistoric remains or to the sculptures within dolmens.

Nilgiri archaeology is in a state of infancy. There is much left to be done. A comprehensive survey and detailed mapping of the prehistoric site features should be completed.[5] The suspected presence of burial rectangles and of no above-surface cists can be verified or disproved through excavation. Only careful excavation and the collection of uncontaminated samples will yield the series of needed dates. Such efforts might also clarify areal variations and those through time. An effort by archaeologists to establish the time span of carefully selected Badaga, Kota, Kuṟumba, and Toda sites might be most revealing. This approach raises the possibility, for example, of solving the vexing problem of whether the Kuṟumbas have descended from ancient Nilgiri ancestors who were primarily gatherers, or from far more recent farming immigrants. Excavation within rock-shelters might be helpful here: such an endeavour within a photographed Kuṟumba rock-shelter shown by Thurston and Rangachari (1909, IV: 169) is recommended. It may also be discovered that within Nilgiri rock-shelters there is representative Neolithic art (Pl. 13, H and I). Congreve (1847: 138-41 and illus.) reported on two rock-shelters in the Bellike Valley on the eastern Nilgiri slopes. Breeks (1873: 105 and Pls. 80 and 82) obtained two photographs of the rock-shelter art there.

NOTES AND REFERENCES

1. The basic fieldwork was supported in 1962-3 by the Foreign Field Research Program, sponsored by the Geography Branch, United States Office of Naval Research. For her contributive efforts in that period, the writer owes a debt of gratitude to Louisa B. Noble. Other

financial assistance from the Faculty Research Council, University of Missouri—Columbia, ensured the completion of the line drawings and photographic illustrations.

2. Throughout this chapter out-dated spellings of some place-names have been changed to more recent ones.
3. Paul Hockings points out that in the Nilgiris there was an indigenous unit of linear measurement, the *ma:*, roughly equivalent to the cubit or 50 cm. It is the distance from the elbow to the fingertips. *2 ma:=1 jagatu*, the pace, which is perhaps the measurement referred to here.
4. The writer thanks the authorities at the Tamil Nadu State Museum for assistance and for permission to use these photographs.
5. After this chapter was drafted, Dieter B. Kapp published his valuable survey, 'The Kuṟumbas' Relationship to the "Megalithic" Cult of the Nilgiri Hills (South India)' (1985).

The Languages of the Nilgiris

M. B. EMENEAU

5

The language of the Bergies is a dialect of the Canarese; that of the Todevies and Koties is supposed to be a dialect of the Tamil; but it is a singular fact that the Todevies cannot speak the language of the Koties, or the Koties that of the Todevies, and that the language of both these classes is equally unintelligible to the Bergies.

—Anon. (Sullivan?) 1819: lv

Unique as are the Nilgiri communities in many respects in India, they are unique too linguistically. They are described elsewhere in this book as consisting of a nucleus which is a mini caste-system, surrounded by a number of peripheral jungle communities which also display caste-like features, e.g. in the economic relationships of some of them with the nuclear communities. It will be remembered that communities in Indian caste complexes tend to insist on their differences, both for mutual differentiation and as symbols of intra-community solidarity.[1] Such differences are to be seen in the linguistic sphere since castes or clusters of castes within the area of one language often speak differentiating dialects of that language. There are also numerous instances throughout the subcontinent of immigrant castes, occupational or of other characters, retaining as home language the different language that was brought from the original homeland; e.g. the Saurashtrian weaver caste in Tamil Nadu still has its original Indo-Aryan language as its home language, the Aiyar Brahmans in Kerala still speak Tamil at home, the blacksmiths in Hunza still speak the Indo-Aryan language Dumaki in this Burushaski-speaking country. But the uniqueness of the Nilgiris lies in the extreme to which this tendency is carried. The three most conspicuous communities, Toda, Kota, Badaga, speak three mutually unintelligible languages, all Dravidian to be sure, but still very different. For the other, peripheral communities the matter has not yet been sufficiently researched to admit of clearcut statement. So far it is clear that these communities do not speak the languages of the nuclear communities. It is also clear that several languages or dialects are to be recognized; preliminary reports have claimed separate language status for Irula and the various Kuṟumba languages. We have no certain reports yet on such matters as mutual intelligibility between these last speech forms and between them and the sur-

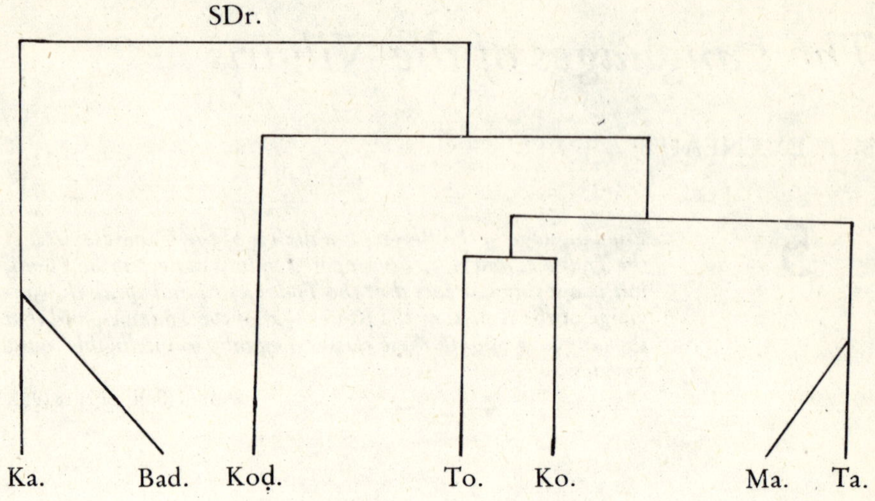

FIG. 5.1 Stemma for South Dravidian.

rounding major languages (especially Tamil and Kannaḍa), except that Irula is reported (Zvelebil 1973: 3) to be an independent language from Tamil on the ground of mutual unintelligibility.

These diverse languages all belong to the Dravidian family. Moreover, historical analysis of the family has set up a genetically closely related South Dravidian sub-family, to which all the Nilgiri languages belong. The well-known members of the South Dravidian sub-family are Tamil and the very closely related Malayalam, the Nilgiri languages Toda and Kota, Koḍagu of Coorg, the somewhat divergent Kannaḍa, and (probably) the even more divergent Tulu.

The position of Toda and Kota within the South Dravidian sub-family and their relationship to one another have been investigated in Emeneau 1956, and the conclusions reached there and sharpened somewhat in Emeneau 1967b[2] still stand. Many features of Toda and Kota make it clear that they belong to the South Dravidian sub-family; detailed examination of the formation of the past tense stem of the verb (Emeneau 1967b) shows that their nearest relatives are Tamil/Malayalam and Koḍagu, Kannaḍa having diverged in this feature from the rest of the South Dravidian sub-family. Subrahmanyam (1971: 517) has taken Kota/Toda to be a subgroup distinct from Tamil/Malayalam/Koḍagu, stating several important innovations which establish both the distinction from the latter group and the unity of Kota and Toda as a group. These innovations include most importantly, the use of the inherited past tense stem of the verb as the basis also for the present tense and some other formations; several other innovations are of a more complex nature and need not be stated here.

In my publication (1956: 35-46) I stated that the Toda past paradigms have a sibilant suffix as the past morpheme and, corresponding to this, Kota has a past unreal (irrealis; e.g. 'I would have done, if so-and-so had been the case') with suffix-c-; these suffixes are related to a past suffix *-c-[3] that is reconstructable for proto-Dravidian (Subrahmanyam 1971: 221-5), but that in the South Dravidian sub-family appears as a distinguishing morpheme of a paradigm only in the Nilgiri languages and in Old Tamil. Consequently, it is argued that Toda/Kota are closely related to Old Tamil.

Since a conspicuous Tamil/Malayalam innovation, palatalization of initial *k- before front vowels, which had taken place before the Tamil record started, does not appear in the Nilgiri languages, we must say that the Nilgiri languages split off from pre-Tamil before the palatalization reached the territory from which the Nilgiri aborigines migrated into the hills. In (1967b: 367) I drew the inference that the split must have taken place 'probably at the period round about the beginning of the Tamil recorded texts'; this of course is a *terminus ante quem*, and the period earlier than this at which the split might have taken place is quite unknown.

The innovation of the Nilgiri languages' extension of the original past tense stem to other paradigms besides the past may be taken as evidence for the languages of the Todas and Kotas having been unified at the time of the migration and for an indeterminate period thereafter. Many innovations in the languages of the two communities, and especially the very complex and numerous phonological innovations on the Toda side, have produced the two very different languages of the present day.

Badaga is obviously closely related to Kannaḍa. For example, in the formation of the past tense of the verb it is in most details like Kannaḍa rather than the rest of the South Dravidian sub-family; moreover, it has the unique Kannaḍa change of original initial *p- to h-. However, even with our present imperfect knowledge of Badaga, it is possible to identify several features in which it agrees with the languages of the Nilgiri subgroup (Kota and Toda) rather than with Kannaḍa.[4] Such traits are to be regarded as the result of diffusion from the Nilgiri subgroup.

When Badaga becomes better known than it is now, it may turn out that it is to be classified as having merely dialectal status, on all fours with the other already well-recognized local dialects of Kannaḍa. Most references to Badaga have treated it thus. The major differences from Kannaḍa that have already been recorded give one pause, but on the question of mutual intelligibility with Kannaḍa Hockings [1980a: 11] writes that 'it [Badaga] and the Kannaḍa of Mysore are, with some effort, mutually intelligible'.

Irula, i.e. the Irula dialects (including Kasaba, Urali, and possibly Sholega), is closely related to Tamil, but the published accounts (Ẓvelebil 1973, 1979) treat it as a separate language rather than merely a dialect of Tamil, and it and Tamil are mutually unintelligible.

Material on the languages of the various Kuṟumba tribes has so far only been published in a preliminary way. The language of the Pālu Kuṟumbas (those best known to the Todas and Kotas; their chief village is that known as Toḍiki or Bāni, the latter being the Toda Po·ny) is classed by Kapp (1978a: 512) with Tamil/Malayalam, but is said to present 'more or less strong Kannaḍa influence'. The Ālu Kuṟumba language is said (ibid.: 513) to be much more Kannaḍa-like, and judgment is reserved on its exact classification. Such names as Jēnu Kuṟumba, Beṭṭa Kuṟumba, Muḷḷu Kuṟumba are in the record, but too little has yet been published for anything to be said. The same must be said for Paṇiyan, Yerukala/Korava, Wainad Chetti, etc.; they are still only names (Diffloth 1968).

The polyglot structure of the people of the Nilgiris must obviously have presented difficulties of communication at all periods. In the present period, i.e. since penetration by outsiders from the beginning of the nineteenth century, some proficiency in Tamil has been the major addition for all who have had contacts with the outsiders; English was a bad second language during the British régime.

Internal communications, however, which presented the major problem during the many centuries before outside contacts became pressing, are probably of greater interest. At the same time the subject presents great difficulty because of the almost total lack of direct evidence. The most informative account is that of Rivers (1906: 15-16); it is worth reproducing in part:

> ...the chief source of error in previous accounts of the Todas. In their extensive intercourse with the Badagas, the Todas use the language of this people, with which they appear to be perfectly familiar. The Toda language is very difficult to understand, and the literature shows that from the first, most of those who have investigated Toda customs have used the Badaga language or Tamil as their means of communication. Every Toda village, every Toda institution or office, and nearly every object used by the Todas has its Badaga name as well as its proper Toda name

and in consequence the early accounts of the Todas and of Nilgiri toponymy by Western observers almost always were presented with Badaga words rather than Toda hence the need for Rivers (1906: 734-7) to provide a list of Toda hamlets with both Toda and Badaga names. This is good evidence that communication between Todas and Badagas involved bilingualism for the Todas rather than for the Badagas. That many of the Todas were (many still are) somewhat, or even very, proficient in the Badaga language is clear from the infiltration of Badaga words into the Toda language, whether of everyday use or of song. Examples of the latter are given in Emeneau 1979b: 43-4, such as use of the doublets püṣy (Toda) and üly (Badaga) for 'tiger', po·b (Toda) and o·f (Badaga) for 'snake', in which the Badaga equivalents huli and hāvu have initial h- corresponding to Toda initial p- but Toda pronunciation deletes h-. In general Toda usage (prose or song), oky 'bedding, sleeping-place' must be a borrowing from a Badaga word (hagalu; Old Kannaḍa paṟke, pakke, modern Kannaḍa hakke; *DEDR* 4407)[5]. That Toda pronunciation of

Badaga is idiosyncratic is obvious, but the Toda 'accent' was (and is) undoubtedly not too difficult for Badagas to comprehend. How many Badagas reciprocated by learning Toda is unknown, but there must have been some, to make Badaga participation in the Toda council effective on the occasions when the debate concerned relations between Todas and Badagas (Rivers 1906: 550)—unless indeed on such occasions Badaga was the language used. During the 1930s I often saw a Badaga at Toda ceremonies and was informed that he spent most of his time with the Todas; unfortunately, I failed to inquire whether he understood Toda, but imagine that he did. My impression is that Badaga knowledge of the Toda language was rare in the early part of this century and that the Todas were the bilinguals in Toda/Badaga communication situations.

That the situation was not always thus is however to be inferred from the phenomena of convergence that have already been identified.[6] Badaga, although basically a Kannaḍa dialect, and in spite of the paucity of Badaga data so far available, has already been shown to have some features that it can only have acquired from the Nilgiri languages. This must mean that at some period Badagas, in some numbers, were fluently bilingual in Toda or Kota. Considering the realities of the caste situation and of the economic situation as regards Badaga acquisition of land from the Todas, Toda is more likely than Kota to have been the second language. The period of such bilingualism must have been the beginning of Badaga settlement when they were comparatively few in numbers and were much more dependent on Toda favour than they became in recent centuries. Whether the Badagas would ever have deigned to speak in the low-caste language of the Kotas, when they were dickering for their menial services as blacksmiths, potters, or musicians, is probably unlikely.

There is much less known or to be inferred about Kota. Nothing, so far as I know, has been recorded about the language used between Todas and Kotas on the fairly numerous occasions when economic interchanges (funeral music and disposal of slaughtered buffaloes at Toda funerals, acquiring of iron tools and pottery from the Kotas, etc.) were taking place. These communications were fairly rigorously channelled, and it can be imagined that utterances were rather formulaic, but there is no direct evidence. Considering the low status of the Kotas, I would guess that the Todas spoke in Toda, and the Kotas had to do the hard work of understanding—in other words, such bilingualism as there was devolved on the Kotas.

Similarly, in the situation involving Badagas and Kotas, it seems probable that the Kotas were forced to be bilingual (as was implied above). Certainly, Kota like Toda shows the influence of Badaga in its lexicon. An example is the verb or-(oṯ-) 'to undertake (vow, responsibility), obey' (alongside the inherited por- [poṯ-] 'to bear (child in arms), obey the words of', *DEDR*: 4565). An even more weighty piece of evidence is the Kota verb o·g- (o·y-, o·n-) 'to go' (*DEDR*: 4572), which should have initial p- but shows the influence of the

Badaga verb ho·g- (Old Kannaḍa pōg-), the initial h- of Badaga being deleted in Kota borrowing as it is in Toda (Emeneau 1965: 20-2; 1980: 80-1).

On the situation as it involves Kuṟumbas, apparently in the main the Pālu Kuṟumbas of Bāni *vis-à-vis* Todas and Kotas, and Ālu Kuṟumbas *vis-à-vis* Badagas (Kapp 1978a: 512), we have practically no evidence. However, since Kapp reports that the language of the Ālu Kuṟumbas is strongly influenced by Kannaḍa (? Badaga), it is possible to infer that these Kuṟumbas have had to be largely bilingual in Badaga.

Of the situation as regards the Irulas, Zvelebil (1979: viii) reports that in addition to their own language 'they know mostly Badaga, the lingua franca of the Nilgiris, and Tamil (the Kasavas [alias Northern Irulas] tend to know rather Kannaḍa than Tamil)'.

Taken altogether then, in the more recent aboriginal times, Badaga, the language of the most numerous community, became the second language of the bilinguals in practically all the rest of the Nilgiri communities. There is however evidence within the Badaga language itself that at an earlier time, probably a period of still undetermined length (decades or centuries?) immediately after the arrival of the first Badagas in the Nilgiris, the Badagas were the bilinguals, being forced by the nature of their situation to learn the language of the earlier inhabitants with whom they had to deal for favours, i.e. the Todas. That it was a period of intense bilingualism seems clear from the nature of the structural traits in which Badaga shows Nilgiri influences as discussed below.

Recognition of this extensive polyglottism and bilingualism/multilingualism leads one to attempt consideration of the Nilgiris as a 'linguistic area'. Linguistic area studies in recent time have usually focused on areas 'which include languages belonging to more than one family but showing traits in common which are found not to belong to the other members of (at least) one of the families'.[7] However, there has been discussion as to application of the theory and methodology to situations in which languages of the same family or dialects of the same language are involved (Emeneau 1965: 27; 1980: 127). 'Convergence area' and 'diffusion area' were suggested as alternative terms, and in the present situation 'multilingual convergence area' would be very specifically suitable. The term 'microarea' has also been used (Zvelebil 1980: 4, n. 15, following Diffloth 1975: 55, n. 5), and applied especially to those smaller areas within India in which the effects of convergence can be identified. The languages may belong to different families; e.g. the representation of palatals by *ts* and *dz* affricates is a trait that identifies a microarea including Indo-Aryan Marathi and southern Oriya, Dravidian Telugu, northern Kannaḍa, and some of the tribal Dravidian languages of central India, and Munda Korku (Emeneau 1956: 7-8; 1980: 111); the languages of old Bastar State, Indo-Aryan Halbi, Dravidian Parji and various Gondi dialects, have many lexical items that are found nowhere else (communication from T. Burrow). Or the languages may belong to one family or even to one sub-branch of a family, and in this sense the Nilgiris can be treated as a linguistic microarea.

A methodological point in area studies must be recalled. The traits that are of greatest interest in the identification of a linguistic area are structural and typological rather than lexical (Emeneau 1980: 2-3). Lexical items may easily diffuse from any one language to any other without the question of a linguistic area arising; e.g. Chinese words for *tea* appear in all the languages of Europe. Consequently, areal studies have concentrated on structural traits, and lexical data have been used in a subsidiary way as secondary evidence to be adduced once an area has been identified on the basis of structural convergence (Emeneau 1971a: 46; 1980: 179).

Nilgiri areal study has been begun by Zvelebil (1980), and already shows interesting results. Identification of structural traits has so far been difficult, since few of the languages have been studied in sufficient depth to allow such traits to be identified in a broad range of languages. One trait of the verb morphology, though only found so far in three languages, is clearly a feature of the Nilgiri area, since it is only found there. Distinction in the 1st. personal plural between exclusive (speaker + referee, e.g. 'I and he') and inclusive (speaker + addressee ± referee, e.g. 'I and you') is a Dravidian trait both in pronouns and in the pronominal reference of verbs (for this and the following details, Subrahmanyam 1971: 406-9). In the South Dravidian sub-family there has been a loss of the distinction in the verb inflection in Tamil/Malayalam, Koḍagu, and Kannaḍa, including Badaga. Some of the Tamil grammarians give -am/-ām for inclusive and -em/-ēm for exclusive; modern Tamil has -ōm for first plural without distinction. Comparative evidence (from Telugu) leads to reconstruction of *-am/*-ām as the suffix of the inclusive. Several of the Nilgiri languages have the distinction in the verb, the inclusive form being: Toda -um, Kota -o·m, Irula -o/-om. Reconstruction on the basis of Toda and Kota is *-om/*-ōm (Emeneau 1979a: 227-8), and the Irula evidence agrees perfectly with this. It has been suggested tentatively (Zvelebil 1973: 25) that the origin of the Irula suffix was an original, but unrecorded, use of the Tamil suffix -ōm for inclusive rather than for general 1st. plural. However, at the moment it seems easier to regard the use of the suffix in the Nilgiri languages as an innovation, probably originating in Toda/Kota. Whatever the final solution may be, it is an areal trait of a structural kind. (Kuṟumba evidence, as supplied in a preliminary way by Kapp, places Ālu Kuṟumba with the generality of the South Dravidian sub-family. Pālu Kuṟumba has inclusive -a·mu, exclusive -e·mu [like the statements of the Tamil grammarians!] the significance of this is still uncertain.)

That Badaga shares some areal features with Toda and Kota, having been influenced by them, has been demonstrated for several features. Badaga has numerous pairs of intransitive-transitive verbs in which the contrast is carried by contrasting past tense suffixes. This is a feature of the South Dravidian languages Tamil/Malayalam, Koḍagu, and Toda/Kota, but not of Kannaḍa. It has been argued in detail (Emeneau 1967b: 366, §1.5, n. 2; 392-4, §5.4) that the stimulus to building up this morphological feature in Badaga came from Toda/Kota but that the Badaga details were developed to some degree independently. No description including this trait has yet emerged for Irula. It is stated that the trait is found in Ālu Kuṟumba and Pālu Kuṟumba; historical interpretation is not yet at hand.

Similarly, the South Dravidian languages Malayalam, Koḍagu, and also Tamil as described in the old Tolkāppiyam grammar have two verbs meaning 'to give', *taru when the indirect object is 1st. or 2nd. person, *koṭu when it is 3rd. person. Toda and Kota also have representatives of these two verbs. Kannada does not, but Badaga does, and it is clear (Emeneau 1967b: 391, §5.2) that Badaga owes this trait (structural or lexical?) to the influence of the old Nilgiri languages. It is recorded by Kapp that these two verbs are 'common in Ālu Kuṟumba and Pālu Kuṟumba'. The matter of usage is not yet clear in Irula, although both verbs have been recorded.

In phonology, Toda and Kota, Irula, Pālu Kuṟumba (Kapp 1978a: 513), Beṭṭa Kuṟumba, and Paṉiyan have preserved, in the main without change, contrast between dental, alveolar, and retroflex stops and, in the case of the former two positions, trills (Kota has lost this latter contrast but Ālu Kuṟumba preserves it) (Zvelebil 1980: 18-19). Since this proto-Dravidian contrast is preserved in Malayalam, Old Tamil, and in the matter of the trills in several other Dravidian languages, and is to be inferred as a whole in the prestages of some other Dravidian languages, this can only be a retention in the Nilgiris area—but it is a retention that is significant once the area has been defined otherwise. Old Kannaḍa had lost the triple contrast in the stops, the alveolar coalescing with the dental, but retained contrast in the trills; modern Kannaḍa has lost the latter contrast also. Badaga agrees with modern Kannaḍa, and the areal contrasts do not appear in Badaga.

Once this linguistic area (or microarea) has been established by the identification of structural traits—and I think it is now established, at least in a preliminary way—it may be bolstered by examination of lexical items. *DEDR* had included numerous lexical items from Toda and Kota which were found in these two languages and had no cognates elsewhere, as well as others which, though related to words in other Dravidian languages, showed peculiar developments of form or meaning in these two languages only; occasionally Badaga material was available to add to Toda and Kota. Zvelebil has now added to numerous such instances items from others of the Nilgiri languages, using his Irula material and Kapp's Kuṟumba material; Kapp has added more Kuṟumba items. Some examples illustrate this type of enquiry.

A fairly large group of etymologies identifies items of the Nilgiri flora. For Toda and Kota there are about a dozen etymologies for which we do not yet have counterparts in any other languages; e.g. Kota kavḷ, Toda kafïḷ 'bulb (edible) of *Ceropegia pusilla*' (*DEDR* 1343); Kota kiḏ, Toda kïḏ '*Olea robusta*' (*DEDR* 1602); Kota polg '*Elaeagnus latifolia*', Toda pïs̱x '*Rhododendron arboreum*', pum bïs̱x '*E. latifolia*' (*DEDR* 3719). Of wider areal interest are the Toda/Kota/Irula/Badaga etyma for the *Strobilanthes* ('conehead'), whose profuse blooming (in cycles ranging from six to eighteen years depending on the species) was observed by the Todas and remembered by the individual as a measure of his age (Rivers 1906: 415-16); outside the Nilgiris the common word is Tamil kuṟiñci or the like (*DEDR* 1849). The Nilgiri words are (*DEDR* 1154): Toda/Kota kaṭ, Badaga (recorded by Hockings) kaṭṭe, Ālu Kuṟumba kaṭṭe, Irula ebbukaṭṭe; for Ālu Kuṟumba ebbukaṭṭe is recorded as 'a profusely flowering kind of *Strobilanthes*'; in the Irula and Ālu Kuṟumba words ebbu is to be related to Kota peb 'a profuse flowering of *Strobilanthes*' (*DEDR* 4411), another Nilgiri areal etymology. Another areal etymology (*DEDR* 3377) for an item of flora concerns the tree *Meliosma pungens* (also *M. Wightii*) used ritually by the Todas (Rivers 1906: 433-5) and the Ālu Kuṟumbas (Kapp 1978b: 177): Ālu Kuṟumba tūḍe, Kota tu·ṛ, Toda tï·ṛ, Badaga tu:ḍe (as given by Hockings; Lushington's toḍe is incorrect).

Among other nouns than names of flora is the name of the bamboo flute (*DEDR* 4239): Kota bugi·r, Toda puxury, Badaga buguri, Irula bugari, bugiriya, Ālu Kuṟumba buguri, Pālu Kuṟumba bugiri.

A Nilgiri specialization of meaning is seen in words meaning 'waterfall' (*DEDR* 999): Kota oyḷ, Badaga ole Toda was̱y, Irula uli, Ālu Kuṟumba and Pālu Kuṟumba oli; the related words in other languages mean 'to flow' or 'river'.

In several etymologies Nilgiri languages agree on derivational material which is different from that found in other languages (*DEDR* 546) Kota i·ruv, Toda ü·ruf, Irula (one dialect) i·rvo, Ālu Kuṟumba (one dialect) i·ruvu 'liver', with *-v-, whereas Tamil/Malayalam and another Irula dialect have *īral (*DEDR* 533) Kota i·p, Toda i·py, Irula i·ppi, Pālu Kuṟumba and Ālu Kuṟumba i·pi 'a fly' have *-ppi, while the other languages have only *ī or this plus other suffixal material (e.g. Telugu īga, Brahui (h)īl̤h). In Kota vayr, Toda pary, Irula bäri 'roof', Ālu Kuṟumba bari 'thatched roof' (*DEDR* 5264), both meaning and formation show special development, since the basic verb *vari- means 'to bind, tie' and Tamil uses the verb also in the meaning 'to fix as the reepers of a tiled roof' and has for 'reeper of roof' the derived nouns variccal, variccu.

That the foregoing account fails to mention dialects of the languages is because so little is known about the matter. Even though the Toda community is so small, the two moieties (*jātis*) which make up this 'caste' community were said by my informants to have slightly different dialects. A very small amount of data was collected bearing on the dialect of the Töwfiḷy moiety—untrustworthy data since my chief informants came from the other moiety, and the Töwfiḷy moiety, and their dialect, are regarded as 'inferior'[8] for this linguistic ranking of the two Toda sub-castes; cf. my remarks on p. 133 and also Rivers 1906: 681, 687).

The Kota community also, small as it is, probably has more than one dialect. My informant belonged to Kolme·l village, but he gave me a very few differing items from other villages; G. Subbiah's informants belonged to Aga·l village (Kotagiri) and he recorded for the verb somewhat different forms from those recorded by me.

The large Badaga community has dialectal differences, differentiated probably by locality and subcaste, but there is little certain information at hand. Hockings (1980a: 11) suggests that there are six dialects.

Our best account of dialects so far is that given for Irula by Zvelebil (1973: 1979). His numerous data allowed him to state phonological differences, and his vocabularies record numerous dialectal forms.

The most interesting matter that has emerged so far is the nature of the Toda song language. It is different in many respects from everyday language. Many differences of detail are merely 'metrical'. However, the verb structure is really different (Emeneau 1979b). It has been established that in the song language the verb has a tenseless paradigm which is 'archaic' and very close in detail in part to the reconstructed past tense of the South Dravidian sub-family (as seen, e.g. in the Tamil past tense and the Kota present-future tense). Verb forms of this type are also put into the mouths of the gods in some old myths and legends and are used in the solemn language of assembly decisions. The song language also has two types of verbal adjective (unlike those of prose), which correspond in formation to Tamil and reconstructed Dravidian forms; these also have prose use in myths and in prayers. This is a fairly typical case of 'diglossia', in which different language forms are dictated by different 'social' contexts.

Linguistic recording and study began in the Nilgiris almost as soon as there was English penetration. A listing of the early attempts at vocabulary gathering is of no great interest and need not be given here. Henry Harkness in 1832, Bernhard Schmid in 1837, and Friedrich Metz in 1856-7 published remarks on the Toda language and vocabularies, especially the last named. Robert Caldwell in *A Comparative Grammar of the Dravidian or South-Indian Family of Languages*[9] included both Toda and Kota in the Dravidian family, using Metz's material for both languages. Caldwell also, in his second edition, made use of G. U. Pope's reworking of Metz's material, in the form of an outline grammar which was included in Marshall's volume on the Todas (1873). Relying on Pope, Caldwell judged Toda 'to have been originally old Canarese [Kannaḍa]'; on the basis of Metz's Kota material he judged Kota also to be 'more closely allied to the Canarese than to any other Dravidian idiom'. Badaga was early judged to be a dialect of Kannaḍa; e.g., Caldwell called it 'an ancient Canarese dialect'.

The Linguistic Survey of India Dravidian (and Munda) volume of 1906 (Grierson and Konow 1967) is no advance as regards the languages of the Nilgiris. The *Survey* did not collect material in the Madras Presidency. The editor contented himself, for the languages of the Nilgiris, with quoting from Caldwell; for Badaga he added a little material and a bibliography. More searching work began in the 1930s. My own fieldwork in 1935-8 was concentrated on Toda and Kota, with some slight work on Badaga[10].

It is to be noted that the work of these five decades has changed the historical picture, and it is no longer possible to accept Caldwell's grouping of Toda and Kota with Kannaḍa.[11].

NOTES AND REFERENCES

1. For more elaborate statement, Emeneau (1974a: 114-15; 1980: 222).
2. Especially p. 366.
3. Throughout this chapter an asterisk preceding a linguistic form indicates that it is a reconstructed, hypothetical form of an earlier language stage.
4. These are identified and discussed below on pp. 138-41 where there is treatment of the Nilgiris as a 'linguistic area'.
5. *DEDR*–Burrow & Emeneau, *A Dravidian Etymological Dictionary*, 1984.
6. This has been discussed on pp. 138-41.
7. Definition from Emeneau 1956; 1980: 105-25.
8. See my account in *Toda Grammar and Texts*, Philadelphia; American Philosophical Society (Memoirs 155), 1984.
9. First edition, 1856, second edition, 1875, third edition, 1913.
10. Much publication followed, of which the most important were: *Kota Texts, UCPL* vols. 2 and 3, 1944-6; 'Toda, a Dravidian language', *Transactions of the Philological Society*, 1957; 15-66; *Toda Songs*, 1971; *Toda Grammar and Texts*, 1984; and the inclusion of much lexical material in T. Burrow and M. B. Emeneau, *A Dravidian Etymological Dictionary*, 1961, revised edition 1984; and M. B. Emeneau and T. Burrow *Dravidian Borrowings from Indo-Aryan, UCPL*, vol. 26, 1962. Recent work on these two languages has resulted in S. Sakthivel's *Phonology of Toda with Vocabulary*, Annamalai University, Department of Linguistics Publication 41, 1976, and *A Grammar of the Toda Language*, Annamalai University, Department of Linguistics Publication 49, 1977: and G. Subbiah's Annamalai Ph.D. dissertation on Kota, 1973.
11. There has been important fieldwork in the 1960s and 1970s on the Irula and the Kuṟumba languages. Gérard F. Diffloth's University of California, Los Angeles, Ph.D. dissertation of 1968 on Irula has been followed by Kamil Zvelebil's two monographs (1973; 1979). Dieter Bernd Kapp's very extensive fieldwork on the Kuṟumba languages has been presented so far only in the papers listed in the bibliography, lexical material which he has allowed Zvelebil to publish in his etymological notes on Irula vocabulary items, and further lexical material supplied in the course of the writing of this chapter.

    There has been some incidental fieldwork on Badaga (e.g. by Diffloth and K. V. Sriṇivasan), and a much more searching undertaking by C. Raichoor and P. Hockings (Hockings 1988).

The Kotas in their Social Setting

DAVID G. MANDELBAUM

6

> *...the Kothurs are the artizans of the hills, working in brass, iron, silver, and other metals, as also in wood. They have no objection, occasionally, to bear a gentleman's palankeen, but will not carry burdens, nor do any cooley work, except in building. They cultivate the soil to a great extent, and produce some of the finest crops on the Neilgherries. They are also the musicians of the hills... .*
>
> —James Hough (1829: 101-2)

'Who are you?' A man may be called on to answer this question in a variety of situations, asked in a variety of ways. It may be the usual conversational gambit in the tight proximity of a crowded bus, or the more pointed yet still polite enquiry of a merchant in the bazaar, or the blunt query of a government clerk, pencil poised to fill out a form for whoever appears at the head of the queue.

The kind of answer which a Kota[1], or indeed any person, gives to this question depends on who is asking it. Various people may ask the question, and the answer must be suited to the questioner. If the questioner is clearly a stranger to the Nilgiris, then some broad social grouping must be specified, else the reply would make little sense. Should the question come from a man on his way to market with a basket of produce, obviously a local farmer, then some specification of village, perhaps of clan and family, would be in order.

So the reply which defines 'who we are' must select some definition of 'we'—whole group, village, clan, family, or some other 'we'. To a stranger, a man is apt to give the broad answer, as in that phrase which regularly opens a petition to government authority on behalf of the whole group. 'We, the Kotas of the Nilgiris' runs the pat phrase, but it is more than a formal usage, it does bear a meaning for all who know themselves as Kota.

Kotas of the Nilgiris

On a map 'the Nilgiris' appears as a district—an administrative division—of Tamil Nadu. To the Kota it is that and much more. The natural area of hills and plateau so bounded and labelled is also the homeland of the indigenous peoples. The Kota know that here they have lived from most ancient times, in consort with the Todas and Kurumbas. Badagas came later, but still long enough ago to be considered old inhabitants as against the newcomers who now so greatly outnumber the indigenes. First of these outlanders to settle in the hills were Englishmen and their servants. In their wake have come folk of various castes and creeds and tongues from the lands below the Nilgiris—Tamils, Malayalis, Kannaḍa speakers, Hindus, Muslims, Christians, low caste Parayas and high Brahmins, government officials, traders and labourers looking for work.

A Kota readily knows where 'We, the Kotas' belong in space and time, in area and history. His folk, he knows, are part of the original and still distinctive population of this distinctive environment. The term *authochthones* expresses what the Kota understand themselves to be. Further, they understand that they and the other indigenes (but they above any) are somehow entitled to special protection, if not special privilege, by reason of being prior and rooted in that time and in this place.

Not only prior habitation, in Kota eyes, sets the four indigenous peoples apart; they are also distinctive by virtue of the ancient social system which existed among them, and which still has some consequence. Each of the four peoples performed certain special functions for the three others and each needed the special functions of the other three for vital parts of its own culture.

The Todas were a purely pastoral people, whose whole economic and religious interest was centred on the care and cult of their herds of buffalo. They furnished ghee from their dairies and their presence was considered necessary at certain Kota and Badaga ceremonies. The Kurumbas, living in the jungles on the steep edges of the plateau, were mainly a hunting and gathering people whose chief significance to the others was as sorcerers and whose main utility to them was as counter-sorcerers. The Badagas, by far the most populous of the four, were the cultivators who provided the other three with most of their staple grain. While the Badagas have far outstripped the other Nilgiri folk in population, the others, who were very few in comparison with the Badagas a century ago, are now proportionately even fewer.

The seven villages of Kotas hold 1,400 people (1981 estimate). In the first Indian Census of 1871 which was relatively accurate and complete in the Nilgiris they numbered 1,112, and in succeeding census decades they did not vary widely from that total.

The Todas number an estimated 1600. They had totalled 693 in the 1871 count and, after some decline, their numbers are on the increase. The Kurumbas have never been satisfactorily counted; they are too scattered, isolated and ill-defined. The 1981 figure for them is 4,874.

But the Badaga population has soared from 19,476 in 1871 to perhaps 125,000 in 1981, a rate of natural increase far greater than that for the state or for the whole country during the same period. And while in 1871 the four original Nilgiri peoples comprised some 44 per cent of the population of the Nilgiri district (21,894 in the district population of 49,501), by 1981 the flow of plains people into the Nilgiris had more than offset the rapid Badaga increase and the four hill peoples were about 23 per cent of the district's population (about 146,000 in a total of 630,169) (Grigg 1880; 29-35).

These population trends are only hazily apparent to a man who identifies himself as a 'Kota of the Nilgiris'. He does know that there seem to be more and more plains people about all the time. But they are generally of lesser interest than are the peoples of the original Nilgiri tribes. With these he feels a special relationship, though frequently a contentious one, and when the men cluster of an evening their talk is apt to be about these Nilgiri folk.

As the children nestle close to their fathers and listen to the talk, they hear references now and again to the origins of the Nilgiri peoples and to the beginnings of their mutual relations. When a child comes to remember and repeat the stories about the beginnings of things, the scattered references fall into a recognized format.

Differing accounts are told by Kota of the origins of the Nilgiri folk. But all versions agree that the Toda, Kurumba, and Kota were brought into being simultaneously by a parent creator. Commonly the stories tell of three brothers who either transgressed against the parents or quarrelled among themselves. Their father, a supernatural, then assigns to each a different function in life and ordains that the three shall exchange goods and services. The descendants of the three brothers became the three tribes and thus the three peoples have been bound in a common fate since the beginning of time. Hence a Kota thinks it unlikely that the common fate should be totally unbound in the present midcourse of time.

The explanatory versions also agree that the Badaga came long after the three primeval peoples had been established in the hills, that at first they were refugees who depended on the suffrance of the original inhabitants, and that they were given land and social accommodation by the indigenes.

Those who have arrived only lately, as it were, have very little history in Kota reference. A few scraps of recollection exist about some early English officials. A man reminiscing about his boyhood may tell of planters, soldiers, plantation labourers, a raja or a merchant. But these are incidental characters: they do not appear in the Kota telling as players long and continuously involved in the local scene. It is as though, to the Kota, they are in the Nilgiris but not of it. The original groups are viewed in time depth, usually a legendary time perspective. The others are seen more as being in a contemporaneous interplay than in a durable action.

Whatever the action or the time reference, a Kota sees his own people as a plainly distinctive group among humanity. The limits of the group are clear. There are only seven Kota villages. Moreover there is only one Kota language. All Kotas speak it: non-Kotas do not. Many other signs and symbols of being a Kota can be recognized, but speech is the firm hallmark of belonging to the community.

The earliest English visitors wrote of their distinctiveness: an account of 1812 refers to their 'peculiar ideas'; in 1819 an observer notes 'The Koties in appearance have no resemblance whatever to the Todevies and except that both classes go without covering head or foot, their manners and customs are dissimilar.' The first full dress account of the peoples of the Nilgiris, in Harkness's book of 1832, relates of the Kota that 'They are a strange race, bear no distinction of caste, and differ as much from the other tribes of the mountains as they do from all the other natives of India' (Grigg 1880: xlix, liv; Harkness 1832: 30).

Some of the marks of Kota distinctiveness are beginning to blur a bit, so that during the weekly market in town it is not as easy as once it was to pick out the occasional Kota who wanders through. But a Kota entertains no blurry image of his group identity. To him the Kotas are as clearly distinctive in way of life as they are in language. Despite differences among Kota villages and factions within villages, they see themselves as one people and look outward on their Nilgiri neighbours with a common view.

The Kotas were artisans and musicians for the others, furnishing metal tools, pots, wooden utensils, and other products of their handiwork. They provided the music which was essential for many of the Badaga and Toda ceremonies and also disposed of the carcasses of cattle and buffalo which had died a natural or a sacrificial death.

No stringent territorial division prevailed among the four groups: the seven Kota villages were interspersed among the dozens of Toda settlements and the scores of Badaga villages. Kurumbas were the sole inhabitants of the jungle escarpment but their needs and their services brought them frequently to the plateau villages.

The interchange among the Nilgiri peoples is now attenuated but some goes on. The Todas still are primarily pastoral; they still appear for ceremonies at some Badaga and Kota villages; but they are—after long decades of indifference to the urging of missionaries and government officials— beginning to take some interest in a little farming. The Kurumbas are also taking to cultivation; their magical and ritual services are only occasionally called for among the other Nilgiri folk. Some Kota, as we shall see, still provide music and artifacts to the others on something of the old basis—but the economic mainstay for all Kotas is now agriculture. And the Badagas who supplied all the others with basic foodstuff have mainly turned away from the old relations with their neighbours. Those of the Badaga who still keep up a semblance of the old interchange bonds are only a small minority.

The Outward View: Todas

The Todas are an ornamental part of that view. 'My son, you are very beautiful but you are lazy. You must always watch the buffaloes...Use the milk and manure for your purposes and help your brothers. You are called Toda'. So runs a Kota version of the progenitor's instructions to the Toda ancestor.

The note of admiration of Toda beauty and deprecation of Toda behaviour was also attested by early British visitors to the Nilgiris. One of the very first of these visitors, Lieutenant Evans Macpherson, was very diffident about the adequacy of his 1820 report, both because of the limited opportunity he had for observation and '...because I am but little in the habit of writing on any subject [so that] I must solicit a partial perusal of the foregoing remarks'. But his remarks accurately reflect the general impression given in other reports. 'The Todewar is fair and handsome, with a fine expressive countenance, an intelligent eye, and an acquikine nose; his appearance is manly, being tall, strong-built, and well set up; his limbs muscular and finely proportioned' (Grigg 1880: lviii).

Similar admiration of Toda bearing is expressed in Major Ouchterlony's 1848 account, the first official and systematic survey of the area. Ouchterlony writes of the Toda as tall, well proportioned, and athletic, with bold independent carriage; men whose noble features, full beards and flowing hair present a striking appearance. Even the Toda women, Ouchterlony adds, are 'handsome in feature as well as in person.'

The Major has less enthusiasm for the Toda way of life:

> Anything more utterly useless, or unproductive in the social scale, than the life led by the Todas, it is impossible to conceive. Endowed with great physical strength and capacity to endure fatigue and vicissitudes of weather, and hence eminently fitted for a life of agricultural industry or other active employment, this fine race, instead of legitimately developing the powers which have been given to them, devote their lives to the unprofitable end of herding a number of buffaloes, the only use of which is to produce the small quantity of milk required for the use of the few families which congregate together in each mund [settlement], and to furnish sacrifices to the manes of any one of their male proprietors who dies (Ouchterlony 1848: 51, 53).

This mid-nineteenth century opinion finds echoes in the comments made a century later of some reflective Kotas. They too consider the Toda a remarkably handsome folk, to them also there is something in the way in which a Toda carries himself that betokens self-confidence, self-possession and self-interest. But these are mere decorative qualities, and it is more important to them that the economic assets which used to come from the Toda have dwindled. They well remember how a Toda funeral would bring quantities of food into the village when the sacrificial buffaloes would be butchered and the meat distributed. And when a Toda's buffalo died a natural death, the carcass became the property of his Kota associate. The hide was saleable, the meat edible.

Now Toda herds are smaller, fewer buffalo are sacrificed, not all Kota will take the carcasses, many Kota will not eat such meat and even for those who will, there is apt to be competition for the carcass from such of the plains people as are interested in meat and hides. Still, the old relationship has not entirely lapsed. Some Kota men provide music for Toda funerals and ceremonies; they come back with bundles of meat. A Toda occasionally turns up at a smithy and asks for a knife or an axe from the Kota whose family has been reciprocating with his own. He may be put off for a time, told to come later—after all, he brings no cash or ghee to induce rapid delivery—but not infrequently he may get his tool right away. A Toda still commands Kota respect despite his economic inutility. And Todas still are desired fixtures at Kota ceremonies.

Thus at the major festival of the year, the God Ceremony, Toda representatives should be duly invited and should appear bringing ghee for the ritual. At Kolmel village there were several years when they did not turn up; factional argument in the village made the date uncertain and the proceedings distraught. But they came again for the ceremony when some of the dispute had calmed, and were received with the traditional gestures of respect.

In the more conservative and unified village of Kurgoǰ, the village which maintains closest relations with Toda families, the old ritual relationship is little impaired. Todas attend the main ceremonies and also come bringing ghee for the funeral rites of an important person; when they arrive, Kotas give respect by bowing their heads to the Toda's upraised foot. At harvest, the leading Kurgoǰ families set aside some of the crop for their Toda associates. Why? 'The main thing is for our rituals. Todas were honoured guests at the rituals in the days of our fathers and grandfathers; they should be invited, received and honoured in our time as well.'

The Outward View: Kurumbas

Kurumbas are very differently viewed. They are usually avoided rather than invited, shunned except when they are needed to fulfil a ritual function or to provide magical services. Then they are asked to come. They quietly appear in twos or threes and squat beyond the last house of the village until the man who has called for them comes up to transact his business. They do look a bit crude even now—smaller than the villagers, hair and person even more unkempt than is common sight in the village.

The ancestor of the Kurumbas was told by the progenitor (in the version by the especially articulate teller of tales, Sulli), 'My son you are always a deceiver ("humbug fellow" in Sulli's English), you have no sense at all. You will live in the forests...if you want anything go to your two brothers and they will help you.' They have helped but the motivation has been fear rather than any fraternal feeling.

Fear of the Kurumbas persists, though it is not the same now as it was earlier. They are seen with a dual reference. One is the remembrance of what the Kurumbas were 'in our father's times'; the other is what the Kurumba are now. The two views are not entirely different, but there is a difference of degree—mainly the degree of terror.

To 'our fathers' Kurumbas were dreaded sorcerers. Those Kurumbas who knew malevolent spells and the magical use of plants were always on the prowl to pick off an unguarded Kota. A man was first rendered unconscious by magical manipulation with plant leaf and incantation. Then they zipped open his abdomen, extracted the entrails which they cooked and ate, filled the cavity with dirt and stones, then closed it so that no scar showed. On awakening, the victim remembered nothing of what had happened to him, but within eighteen days he would sicken and die. A woman who became their prey would be made unconscious in the same fell way, but she would not usually be killed, only used sexually by the sorcerers. On regaining consciousness, she too would have no inkling of what had been done to her.

These things the Kurumbas would do for their own evil satisfaction: they could also kill magically at the behest and pay of jealous men of the other Nilgiri peoples. Whenever a Kota fell seriously ill and the symptoms spelled witchcraft only one therapy was indicated. His relatives had to secure the services of some other Kurumba who would turn the mortifying magic out of the sick man and back into the sender of the evil. Should this recourse fail, the relatives— especially the father of a dying child—might take direct action, if not to save the poor bewitched one then to ensure that the origin of the evil was eradicated. Such a father would waylay the suspected Kurumba and knock out his two upper incisors. Then that Kurumba no longer could work his magic, because he could never again pronounce the incantations clearly enough. Or a particularly grieved attacker would simply beat the suspected Kurumba to death.

Badagas and Todas no less than Kotas had this fear of Kurumbas and used similar retaliatory measures. The Gazeteer of the Nilgiris records the killing of Kurumbas by vengeful slayers in 1824, 1835 (58 were massacred then), 1875, 1882, 1891 and 1900 (Francis 1908: 155). There were undoubtedly other such slayings which did not reach police notice. An early display of the might and mystery of the British Raj came when a Badaga was hanged for the killings of 1824. Captain Henry Harkness, reporting the incident in 1832, tells how all the Badagas looked upon this killing 'as an act, fearful indeed, but as one deserving of praise rather than punishment. British jurisprudence, however, could not look on it in this light, and therefore, according to the Burgher [Badaga] account, its sentence gave much encouragement to demonology, and tended greatly to increase the insecurity both of themselves and of their cattle' (Harkness 1832: 84).

A century later, the blindness of such government justice was still being lamented. One Kota offered a story of an encounter with Kurumbas and how one of the village diviners had advised the people. 'If you kill them, government will hang you so it is better to knock out their teeth. If we tell Government what the Kurumbas do, they do not believe it'.

Fear of Kurumbas was all pervasive. Children were thought to be especially vulnerable. Women were more fearful than men, but a man walking alone at night through a field or a wood was apt to walk with pounding pulse and readily startled sensibility. A misbehaving child was told to mend his ways or he would be taken by Kurumbas. Not until a child was well grown— perhaps into his teens—was he allowed to go out at night to a nearby field to defecate. He was too vulnerable to Kurumbas even when accompanied to the field by an adult. Women in the menstrual seclusion house spoke in whispers behind barred doors: even a whisper might be heard by a roving Kurumba and attract his vile notice. Men did not like to go anywhere alone; solitary men could not as easily guard against Kurumba wiles. Even when travelling in pairs, men who sighted a Kurumba—or what looked like one—far up on the road would detour if they could to avoid a passing encounter.

Still, fear could be tempered with common sense. One man, Valmand, told a story of his father and his father's brother, both of whom decided that a certain Kurumba had killed the brother's son. The father proposed that they beat the culprit but warned his brother to join him only if he felt brave enough. 'If you think that the Kurumba will kill you after we kick him, don't go . . . because (and he cited the proverb) to him who fears, something will happen'. The story goes on to relate how the two do go, after stopping for two drinks of *arrak*, 'just enough to make them brave but not enough to make them drunk'.

But there were co-operative relations with Kurumbas as well. Each Kota family had an associated Kurumba family, much as there were Toda associates. Thus when a man died and a tall pole was needed for the funeral rite, a message was sent to the Kurumba associate and he soon turned up with the necessary pole freshly cut from the jungle. At other times he brought canes, fruits, and baskets. In return he received Kota manufactures, knives, axes, hoes, ploughs, pots. At harvest and at the proper ceremonies the Kurumbas came to the village bounds (not being welcome in the village proper), made their presence known by calling for their Kota associate and received their due. The Kota turned to his Kurumba associate when he needed magical protection. Not all Kurumbas, indeed, knew the spells and incantations. But if one's associate did not, he knew which of his tribesmen did.

Kurumbas, like Kotas, played music, and sometimes provided music at Badaga and Toda ceremonies. Kotas were not apprehensive on this score. Kotas alone claimed the carcasses and Kurumbas were not their economic rivals in this. Kota musicians had, and still have, no objection to including a Kurumba or a Paraya in the band. This seems to be a privileged context of co-operation.

The whole Kota village, rather than individual families, employed certain Kurumbas as watchmen. They were supposed to visit the village at night several times a week—oftener when there were ripening crops in the fields—to ward off animal and human predators. They were paid from the village purse and were usually those called village Kurumba who did some cultivation, in contrast with the forest Kurumba who had no settled habitation at all—or who were supposed not to have permanent homes.

Kurumbas are given short notice in the early English descriptions. Harkness came upon some Kurumba huts in August of 1829. Of the inhabitants he writes, 'Swarthy and unwholesome in countenance, small of stature, the head but thinly covered with sickly-looking hair, the only covering it has,—little or no eyelash, small eyes, always blood-shot and apparently much inflamed, pot-bellied, and with water running from their mouths, they have in most respects more the semblance of savage, than of civilized man' (Harkness 1832: 128-9). But the physician Shortt in his account of 1869 added a softer note.

> 'Whilst the appearance of this tribe is uncouth and forbidding in their own forest glens, they are open to wonderful improvement by regular work, exercise, and food: of this ample evidence is to be seen at the Government Cinchona Plantation at Nediwattum, where a gang of Kurumbas, comprising some twenty individuals, are employed as labourers, receiving their wages in grain for the most part.... They have not the pot-belly, do not gape, nor is the dribbling saliva or bloodshot eyes common to their brethren of the jungles, to be found among them (Shortt 1869: 279).

Nowadays Kurumbas are less alien in human appearance than they once were and are more like the regularly fed Kurumbas of Shortt's description. Nonetheless they still are treated with discretion tinged by some trepidation. Should a Kota grandmother spy a Kurumba or two waiting at the village bounds, she is likely to whip around instantly and shoo her grandchildren into the cover of the house. Only a very few Kotas now have regular Kurumba associates. They are mainly the few who are healers and wish to secure therapeutic learning and plants from their associates. Kurumba watchmen have not patrolled the village fields for many years. Poles for Kota funerals are still provided by the Kurumbas together with a few other ritual necessities; some Kurumbas turn up when they are needed in a few passages of Kota ceremony. (This as of the 1950s.)

But they are no longer thought to have a monopoly in the magical manipulation of the supernatural. Astrologers, amulet purveyors, exorcizers from the plains are available in town and a Kota in need is apt to try one of them first.

This change in view has come quite rapidly. In the late 1930s most elders could relate some personal encounter with Kurumba sorcery, although even then there was an occasional insistence that 'I no longer fear them'. In that decade occurred an incident which became richly embroidered in later telling. The story tells how the Kotas of Kolmel and those of Mena·r fell out over the division of the carçasses at a Toda funeral, how the Kurumbas associated with the respective villages were drawn into the affair, how the two Kurumba factions waged magical conflict against each other as well as against the opposing

Kotas over the matter, and how the dispute was finally settled at a grand meeting presided over by a Badaga headman.

A decade later, the prevailing response by men of the younger generations was 'I have had no experience with Kurumbas myself, but what is said about them is true'. It is undoubtedly true that fewer of them have felt personal terror as targets of Kurumba magic. Their children are apt to be warned that, if they are bad, they will be given to the Gollan (a gypsy-like caste of beggars) rather than to the Kurumbas.

One man, not noted for any virile bravery, had this comment. 'When I was small I was very much afraid of Kurumbas; my parents would scare me with them. But later there was less fear of them and now we hardly fear them at all. That is because there were only a few castes up here. Now there are many others and we see that they do not fear Kurumbas.' Another man (Tuǰ) had a different emphasis. 'I myself have had no experience with Kurumbas.... Now they are tamed, they are no longer jungle people but we are still afraid of them'. Then Tuǰ mentioned a large meeting called by Kotas and attended by Kurumbas and Todas in order to protest against Badaga abrogations of the traditional relationships. 'There it was said that the Badagas are now our enemies and that the Kurumbas should no longer harm the others'. While the Kurumbas have receded as a magical threat, the Badagas have increasingly loomed as material antagonists.

The Outward View: Badagas

The Badagas are now seen as the chief adversaries. Not because of pressing economic competition or gruelling power rivalry, but mainly because of the contrast, to the Kota's inward eye, between what they were to the Kota 'in our father's times' and what they are now. Then, so Kota remembrance runs, there was a harmonious working system between the two peoples, each having a sense of responsibility for the other, each giving the other its proper due of goods, of grain, and of respect. Now, some Badagas actively denigrate the Kotas (try to keep them out of the shops patronized by Badagas, for example) and—what can cut deeper—very many Badagas just ignore them.

It used to be otherwise. To begin with, Kota lore tells that the Badagas are not of the progenitor's original line. They came to the Nilgiris in later times, as refugees, and were allowed to settle in the hills by the permission of the ancient inhabitants (Emeneau 1946c: 254-61). An amicable arrangement was established (any pulling and hauling between the peoples which went on in an earlier generation tends to be smoothed over in the retrospective view); it lasted for ages; men and women now in their later years grew up into adulthood with that arrangement as an important part of village life.

Under that system as it existed in the first two or three decades of the twentieth century, each Kota family had special associations with a number of

Badaga families and with certain Toda and Kurumba families as well. Thus the members of a Kota household would call the Badagas of perhaps thirty houses in some eight or nine villages by the special term *mutgara*. 'Associate' is the best rendition of that Kota word, although Badagas, as we shall see, give a different connotation to the term and to the relationship.

A Badaga could get Kota products and services only from his associated Kota family and in return he gave grain after the harvest, money on certain occasions, the carcasses of his cattle and buffaloes.

The contract was a flexible one. If it was a year of good harvest, then the Kota associate would argue for a full ten measures of grain (about a quarter bushel, some nine litres in all) from an average household. And if the Badaga was wealthy, with a yoke of plough oxen and grown sons to help him, his Kota associate might demand twenty measures at the threshing floor. But if the next year was one of crop failure, with only half the yield of the previous year, the Kota would not press for the full amount he had received the year before.

The contract was also flexible when the Badagas were recipients and the Kotas donors. Each Kota family had the responsibility of keeping every associated Badaga family equipped with tools and pots. There was a general notion of what was standard equipment for a household—a couple of axes, a pair of pestles, a mortar, a spade, a lamp, a ladle, a scraper blade, tongs. Each woman should have her sickle, each cultivator his hoe and spading fork. But some families were larger than others and needed more tools and pots. Some were more careless of their tools and needed replacements or repairs oftener. Such matters too, were negotiated between the respective householders, with a good deal of verbal to and fro before the specific terms were settled. During February and March, at the beginning of the agricultural season, there were usually a number of Badagas around each Kota smithy, waiting for tools, discussing terms, urging their associates to hurry along with the ironwork.

It was a real contract, however, even though it was a flexible one. A Badaga could not suddenly shift his association and patronage to another Kota. Should he try to do so openly and consistently, his associate could call the village Kotas to a meeting; there all would have to pledge to stop any dealing with that Badaga. Should he then try to begin dealing with Kotas of another village, they too would eventually—under pressure of complaints from the first village—boycott him.

There was a traditional apportionment of Badaga villages among the seven Kota villages and there could be no transactions by Badagas across these divisions. A Badaga household could be reassigned from one Kota family to another, but it had to be done by mutual consent, with some discussion of the matter and with witnesses to the gesture of de-affiliation. Thus if a father died, leaving an only son with a larger Badaga clientele than he could manage single-handedly, some of his associates would be transferred to another family which had need of more Badaga associates. Or if a Kota became once and for all convinced that he wanted no more to do with a particular Badaga associate, he might bow his head to that Badaga's hand and tell bystanders that henceforth he was no longer that man's associate. Then another Kota could take him on. A Badaga comparably irate with his Kota auxiliary would usually have to induce the Kota elders to persuade the Kota associate to bow to his hand and sever the relationship.

The contract bound the Kotas too. Before many weeks passed a pair of Badaga messengers would approach the Kota village, turbans carried in hand as a token of funeral news. Their Kota associates would then drop whatever they were doing, receive the messengers and forthwith make preparations to provide a band of musicians for the funeral. Kota associates had to provide the music for the principal Badaga ceremonies.

No ordinary circumstances could curtail the Kota obligation to provide music. During the most awesome of Kota ceremonial occasions, when there was great emphasis on the totality of social participation in the village rites, even then some Kotas were segregated from the rest of the village, were exempt from the special sanctification carried by the other villagers, and so were ready and able to play, should occasion demand, at a Badaga funeral.

The musicians so provided were given food at meal times on the verandas of Badaga houses: there they slept when the rituals carried on through the night. A Kota elder might not himself attend the funeral of an unimportant youngster from an associated family but it was incumbent on him to round up a band of musicians for the rite and to provide some of the funeral goods.

The providing of music was the most visible token of association. Badagas had long supplemented Kota-made articles with purchases in the town bazaar. But under the traditional dispensation neither Kotas nor the Badagas would themselves tolerate any substitute for Kota music on ritual occasions.

Most Kota ritual occasions, in turn, entailed certain obligations from the Badagas. When there was a death in a Kota family, each associated household would contribute something (8 annas came to be the standard) towards the cost of the first funeral. For the second funeral, all the associates together contributed a buffalo for sacrifice, perhaps several animals for the final rites of a distinguished elder. Associates also contributed towards the expenses of great ceremonial events, as for the ordination of a Kota priest. Badaga associates were obliged to come and watch part of such a ceremony, to lend their presence to the occasion, and so to reaffirm the mutuality of the two peoples. The missionary Metz noted this in the God Ceremony, 'At this festival the chief men among the Badagas must attend and be spectators of the gay scene; otherwise their absence would be regarded as a break of friendship and etiquette, and the Kotas would immediately avenge themselves by refusing to make any ploughs or earthen vessels for the Badagas' (Metz 1864: 131). Not only were they obliged to be present but during one phase of the ceremony, a line of Badagas formed and danced in their own style on the Kota dance ground.

The relationship between a Kota household and a Badaga household was a kind of contractual bond, but it was indeed more than that. It often resembled a kinship tie as well. Mutual support was not confined to the exchange of goods and services; a Kota might seek and find help in any life crisis from an associate. A Badaga in need of aid might turn to his Kota associate.

Thus one Kota, in talking of a period of personal adversity in the 1920s, when he alone had to support a family of nine, relates how he obtained food from his associates—beyond the obligation of the usual contract—and kept his family going. Later, during a plague epidemic, when he was one of the few to remain in the village, 'I went around to my associates, and got gifts of grain and money as well'. Another Kota tells how he was raised in the village of his mother's brother because his own family in Kolmel village had all perished when he was a child. When he reached manhood, he was urged by a Badaga associate of his father to return to Kolmel—the Badaga offered to pay the uncle recompense for the expenses of raising him. And when he married, this Badaga disapproved of the new wife. 'He advised me to divorce her and I was willing, so after two months of marriage I went to her village and divorced her'.

Women might appeal to associates. One young wife who could not tolerate her parents-in-law, fled to the house of a Badaga associate. There she was given refuge and they eventually arranged a happier way for her.

When a Kota wanted to dress up for a great display, he borrowed from the treasures of his associates. The English official Breeks, in describing the God Ceremony, notes that 'the Kotas dress up for the occasion...borrowing jewels of all sorts from the Badagas, who are obliged to propitiate their artisans by attending and contributing on this occasion' (Breeks 1873: 44).

Between some associates there were strong bonds of affection. News of the death of a beloved Badaga associate could bring personally-felt bereavement in a Kota family. The women would burst into wails, the men, subdued with sorrow, would hurry to prepare especially elaborate funeral provisions as indication of their affection and grief. All associates were not held in close regard however. With most, the relationship was more contractual, limited rather than pervasive; with a few there was a frequent and close interchange.

The Kotas of Kolmel village had especially close ties with the Badagas of three environing villages. These villages were founded, so Kolmel tradition attests, when the people of Kolmel took in some Badaga families and settled them on their own tillage. And the tradition in these Badaga villages agrees, even adding that there was joint farming by Badaga and Kota. Hence when an important person of these villages died all Kolmel would participate in the funeral preparations, as for a similar occasion in Kolmel itself. To a funeral in a more distant Badaga village, a Kota head of family would usually dispatch a band of young men and might not attend the funeral himself. The musicians would expect (and would get) some payment for their services, in addition to the payment in grain and carcasses received by the Kota associate. But there was no such money payment at a funeral in one of the three *okl* villages, any more than there was payment to the musicians for playing at a Kota funeral.

A Badaga from one of these villages related how his people had special concern for the Kolmel people. They would, in former years, stay up all night with the Kotas at the Kolmel main ceremony, because the Kotas were ritually obliged to do so. 'We would think', he remembered, 'that our people are outside there and perhaps in danger from animals. So we go and stay with them'. And when food supplies were very low in Kolmel, as regularly happened (and still happens) after the lavish feasting of the God Ceremony, the Badagas of these three villages tided the Kotas over until food or money came in again. Other Badaga associates were dunned and might contribute; but these *okl* Badagas knew that they had to give. This was so partly because of the traditional bonds and partly because their proximity to Kolmel made Kota importuning difficult to shake off.

The Kota's approach to his Badaga associates in these and in other circumstances combined demand with entreaty. The Kota was formally the inferior; he should bow to the Badaga's hand, never the Badaga to his. The standard salutation of Kota to Badaga was that of inferior to superior in rank. The Kota could not enter any Badaga house for he would defile it, he was served his food on the veranda, from special coconut shells kept for such use. Not only with regard to formal tokens of rank, but in wealth and in numbers, the Badagas were superior. Harkness's 1832 description of the Nilgiri peoples had this opening sentence on the Badagas. 'We now come to the most numerous, the most wealthy, and what must be considered the most civilized class of the inhabitants' (Harkness 1832: 30). In later decades Badaga wealth and numbers *vis-à-vis* the others increased still further.

While the Kotas were formally and numerically the inferiors, they by no means felt totally subservient. They knew well that their Badaga associates could scarcely get along without their co-operation; they did not hesitate to use the threat of boycott. Sometimes there would be a falling out between two associates, each being irked by the demands or the recalcitrance of the other. It was not in either's real interest to be at odds. The Kota could easily dawdle and delay about fulfilling the other's regular quota of tools and in meeting emergency needs for repairs. Or he could take his time about collecting a band for a funeral, the rite would be halted in midcourse and the guests would have to be fed additionally costly meals. The corpse would be dressed up; the whole funeral concourse would grow more restless and the chief mourner more furious—until the band finally strolled up.

The Badaga, in turn, could make the Kota wait a good long time for his grain after the harvest. And when the Kota turned up with an empty sack after his God Ceremony, he might be turned back with some elaborate excuses and only a few niggardly handfuls of grain for his pains and for his hungry children. So associates had good reasons for maintaining amicable relations with one another.

Such amity did not preclude a good deal of haranguing at every harvest season—when the amount to be given was always adjusted and readjusted according to the two differing concepts of the size of the crop and of the just needs of the respective parties. But this disputation had no raw edge of bitterness to it and left no wounds to self-respect. A prosperous Badaga elder might indulgently allow a young pushing Kota, the heir of his old associate, to come away with more grain than his father had ever received. Or a Kota elder would permit a young Badaga associate, perhaps beset with the illness of his children, to give a smaller share than was common.

A Kota had something of an advantage over his Badaga associate if the two quarrelled. The Badaga had to depend on the one Kota household for tools and services, the Kota received grain from many Badaga households. Kotas also did some cultivation of their own, no Badaga provided music or tools. Hence Kotas could more readily do without Badagas—or without some Badagas—than Badagas could do without Kotas. If a Kota could convince his fellows that there was a good reason for boycotting his associates, no other Kota would supply the necessary goods to that Badaga. Then the Badaga elders would try to smooth out the affair, perhaps by arranging a transfer of associates that the Kotas would accept, and the Kota elders would, for their part, mollify the outraged Kota.

There are Kota stories of the boycott of whole Badaga villages. One account recounts that in the 1890s, the Kolmel villagers withdrew from relations with several of their associated Badaga villages for some six years. At length the Badagas sought to heal the breach and re-established the traditional association even though there were alternatives to the symbiosis by then. Badagas could buy their tools in the bazaar and procure Kurumbas or plains people to give music. The Kotas could extend their cultivation or could work for pay as artisans in town. But these alternatives were then not really satisfactory to either side. Bazaar tools were costly, money was scarce, and blacksmith repair difficult to obtain in town. Kurumbas could play music, but it was dangerous to have them too much about; the bands of Tamil-speakers were not quite right. Moreover, it was not even easy to find lowlanders ready to remove carcasses.

The Kotas, for their part, were not enamoured of agriculture or of wage labour. For both peoples the presence of the other at ceremonies was a firm part of their way of life; the absence of one made a noticeable and uncomfortable gap in the large occasions of the other.

Thus did the symbiotic system keep operating, and so was it kept in social equilibrium. The formal inferiority of the Kotas did not gall them particularly, they considered themselves as partners in the joint enterprise of Nilgiri life, partners who contributed vital elements and who deserved suitable reward and consideration from the other Nilgiri folk.

In Kolmel village it is still remembered that once the nearby Badagas had to come to pay their taxes there; it was then one of the principal villages of the sub-district. If there were disputes in the adjoining Badaga villages which could not be settled *in situ*, the opposing arguments were heard in Kolmel where Kota and Badaga elders sat in judgment together. There are stories of Kotas reversing the judgment of assembled Badagas, as in the case told me (and also mentioned in Emeneau 1946c: 261) of a Badaga girl who became a convert to Christianity. She and her whole village were outcasts. The tale does not make clear why the Kotas were so opposed to their exclusion from Badaga life, but it does pointedly relate that the Kotas used the threat of boycott to convince the Badaga elders of their view. However elaborated such tales have become by being repeated, they do reflect Kota confidence that they did and could participate in judgments over Badagas as well as over Kotas, just as a respected Badaga elder could be called in to help settle a quarrel among the Kotas.

In some of the affairs of the closest Badaga villages, Kotas did have a voice even in the late 1930s. In that period there was a bitter struggle between two Badagas, parallel cousins, as to who would have the honour of providing a road into their village. One road ran across Kota lands, the other branched off before the Kolmel boundary. The Kolmel elders discussing the quarrel—which had come to the stage of blows—urged that both roads be opened. So they were and the quarrel on that matter ceased.

Such involvement of Kotas in Badaga affairs, common earlier, was unusual later, even with the close *okl* Badagas. The trend was the other way around—of Badagas turning away more and more from association with the Kotas. One Kota reminisced that in his youth Kotas frequently helped to settle cases involving the Badagas. Then, about 1911, one of the Badagas became wealthy as a contractor and he put a stop to that. The implication—probably not far from the mark—is that the contractor's wealth opened up for him a style of life in which Kotas were unwelcome as traditional associates and that the power engendered by the wealth enabled him to bar Kotas from participation in Badaga councils.

Other circumstances strained the traditional ties. In Kolmel village one Kota, Sulli, succeeded in 1916 in qualifying as a school teacher. Badagas were then confronted with a sore problem, one which has vexed villagers of higher caste in many parts of the land. Sulli relates how in 1916 a Badaga teacher called a meeting of leading Badagas to complain that Sulli was receiving respectful salutations from Badaga school children. 'How can a low caste Kota be a guru (preceptor) to Badaga children?' Sulli's account goes on to relate how the school inspector was bribed to dismiss him, but his direct appeal to the Collector, the chief British official of the district, confounded Badaga plans and he remained in his post, though he was transferred to a predominantly Kota school. In succeeding years Sulli was engaged in a running skirmish with Badagas who could not stomach a Kota as a teacher of their children. Once, in 1923, as Sulli relates the story, he was assigned to a one-room school of Badaga children; all his pupils were promptly withdrawn and the authorities closed the school down.

Sulli's battle with the Badagas was mainly a personal one; not all his fellows approved and some had little good to say for him. But the Badagas were generally becoming more deaf to Kota entreaty and demand, and Kotas were finding the traditional relationships difficult to maintain. So when Sulli, in 1930, rose wrathful against an insult from the Badagas, many Kotas were ready to support him and to be difficult, in their turn, in their relations with Badaga associates. The insult was a double barrelled one as Sulli publicized it in the seven villages; Sulli's full account of it may be read in Emeneau (1946c: 260-73).

There was an incident at the tea party given for the Governor of Madras by the wealthy Badaga contractor who had by then been endowed by the Government with the honorific title of Rao Saheb. The symbolism of a Badaga honoured with a title and able to feast the English Governor impressed even the more unsophisticated Kotas.

At the tea party, Badagas and Todas were introduced by the host to His Excellency. Kotas, Kurumbas, and Irulas were present but were snubbed. When some tried to approach the Governor, they were told that it was not fitting for them to come up as they were not of the aboriginal tribes of the district.

Another insult was a newspaper article in which, again according to Sulli who alone among the Kotas could spell out this English article, it was said that Kotas were the offspring of miscegenation involving Badaga and Toda men and an erring woman. Sulli's account relates how the incensed Kotas and Kurumbas joined in a reprisal boycott of Badagas and how his petition for redress was rewarded when the Governor and his wife publicly received a deputation of Kotas, shook hands with them and told them that the editor of the paper had retracted his article and had apologized.

Howsoever the event may have differed from this narration of it, various Kotas do recall that the main break of traditional association with Badagas did come at this time, about 1930. Some claim that it was because certain Badaga families stopped using the menstrual seclusion hut. Kota musicians, coming into a village, might receive food from a menstruating woman and so be defiled. For Kotas, as for many peoples of the subcontinent, the threat of pollution is powerful both as a factor in determining present action and as a retrospective explanation for past behaviour.

The events about 1930 did not mark any complete severing of the traditional ties. There were Badaga families and whole villages which insisted on the old relationship, especially on the use of Kota music. Badagas split into anti-music and pro-music factions. The conflict between Badaga reformists and traditionalists became a bitterly fought contest. At the funerals of the pro-music faction, the anti-music zealots interfered. Police had to be in attendance and in 1936 a grand riot was precipitated at a funeral. The police had to fire into the embattled crowd and two Badagas were killed.

Among Kotas also the providing of music became a matter of bitter factional dispute. The reformist group in Kolmel village found it demeaning to play for those Badagas who still wanted Kota music; the traditionalists insisted on playing. By mid-century only about a score of Kota families still played for Badagas and those families were mainly in the two villages of Kolmel and Mena·r in whose environs lived most of the pro-music Badagas.

No great rewards accrue to those families, perhaps some Rs. 30 a year and 20 measures of grain from all associates—as against Rs. 100 and 50 measures of grain in the twenties. In the past a Kota collecting for the expenses of a 'dry' funeral could expect some Rs. 200, thirty years later he would be fortunate to come away with 50 of the depreciated post-war rupees.

The imperative quality of the old relationship had faded. 'There is no rule about it now' is a common comment. Dramatic expression of this was given one day in 1949 when the Kolmel elders were meeting to discuss a local problem, the matter of tea-shops in the village. In the course of the meeting, a Badaga came up and excused himself, in a restrained but clearly agitated manner, for interrupting the proceedings.

His father-in-law, a respected Badaga of the pro-music group, had just died. The son-in-law had come to get a band of musicians from his Kolmel associates. But the Kota elders, several of them strong traditionalists, went on with their discussion, giving only passing notice to the distressed Badaga. He became more and more excited, hooked the handle of his umbrella to the back of his coat collar to free both his hands for imploratory gestures. To no avail: he was forced to go back to the funeral without additional musicians and to make use of the few Kota who were there as though they were several full bands of associates.

Yet, for all such casual insouciance, Badagas are more than just casual neighbours. At Kolmel, Badaga associates still attend Kota ceremonies as required in traditional respect. Even the reformists like to see them welcomed to the village. At so awesome a ceremony as the ordination of a priest, some men still remain apart from the rite so as to be able to provide music should there be a Badaga funeral. In Kurgoǰ where the traditional services to Badagas were dropped about 1930, Badaga elders still helped solve Kota village problems in 1950.

When a new shrine was built in Kolmel in 1946 the Badagas of the close lying villages contributed a considerable share of its cost. When the Kolmel Kotas ordained a priest, they asked these Badagas to contribute for the expenses of the rite. A Badaga village leader, staunchly anti-music, rejected the requests, arguing that the Kota had no right to demand contributions as customary due. But, as he later confided, 'I know in my own mind that we must give something.' His village decided on a lump sum for Kolmel to which he contributed.

New economic relationships have developed between Kotas and Badagas which are carried through without particular friction. Badagas lease lands from Kotas: about a third of the arable land of Kolmel village is so leased from year to year. Kotas often sell their potato crop to Badaga contractors and get advances from them. The Kota makes such transactions as with a particular individual and bears no preordained rancour towards him. On the contrary, he may regard that Badaga as a cordial ally.

But such cordiality is for the individual, not for his group. Badagas as a reference category, are all antagonists, former partners who have reneged on obligations and who try to do down the Kota people. Kotas know little about differences among Badagas and generally care less. Once I heard a member of the Lingayat subcaste of Badagas spiritedly insist to a Kota that his subcaste was separate and distinct from other Badaga subcastes and was not to be classed with them. The Kota, typically, remained unconvinced: to him, a Badaga was a Badaga.

The struggles of the traditionalist Badagas to maintain the old customs do not particularly endear them to the Kotas. Kotas tend to see all Badagas as being involved in the efforts of some to bar Kotas from tea shops in town.

When a Kota reformist like Sulli was once ousted from a tea- shop, all Kotas resented the affront. The traditionalists among the Kota find no need for narrow consistency in this matter. They are content to fulfil the formal gestures of subservience when engaged, say, in the traditional role of Kota musicians at a Badaga rite. In town, they themselves may not want to take tea and food in a tea-shop, but they believe that any Kota should have the right to do so. The town situation they see as a different one, wherein subservience and exclusion take on different meanings than they traditionally do in the Badaga village. This reasoning is not without a certain logic. In the Badaga village, as we have noted, Kotas were formally the inferiors but were not without power. In town, inferiority is more total.

The right to be served in tea shops has become better established through the demands of a few reformists and especially since the country's political independence. The gain is understood as a success of Kotas over Badagas rather than as any victory of reformists, or, much less, of a political ideal. Nor does the struggle with Badagas induce fellow feeling with the disadvantaged low caste folk whose huts now lie close against each Kota village.

The Outward View: Lowland Folk

Villagers of Kolmel command a wide view from the ridge on which their village lies. Across the wide valley on one side, several distant Badaga villages can be seen. To the other side the houses of the closer Badaga hamlets can be discerned. But closer still, only some 200 yards down a slope, is the huddle of huts in which live low caste people from the plains who have attached themselves to Kolmel. Not only are these huts close, their inhabitants are constantly to be seen in the village. Yet Kota notice in this and the other villages is much more concerned with Badagas than it is with these folk, their nearest neighbours.

The Kotas class them all as Konaton, 'lowland folk'. Distinctions of caste and of provenience are only vaguely recognized. A few have opened tea-shops in vacant Kota houses and sell small articles: in Kolmel at one point there were three such shops, one run by a Hindu from Malabar, another by a Muslim from Malabar, the third by a Kannaḍa speaker of Chetti caste. They are shopkeepers, they extend credit and the villagers want to know little else about them.

Most of the families of lowland folk are labourers who live by the hire of the Kotas. They are squatters on Kota land; the word, squatter, conveys the Kota attitude towards them. Because there is no great pressure for land in Kolmel they are allowed to exist there. They have also clustered in other Kota villages where their huts do not usurp cultivated fields. Their shelters can scarcely be called shanties, crude flimsy assemblages of wood scraps, odd poles, and thatch. Their livelihood is precarious, mainly whatever they can earn by the day from Kota or Badaga farmers' needs for coolie labour. Some few are allowed to cultivate plots in Kota fields, these are generally of families who have lived near the village for years and have become attached to particular Kota families.

Lowland folk have long been known to the Kotas: trading expeditions to the plains were undertaken long before the British opened the Nilgiris to easier access from the plains. Soon after the arrival of the English, it was noted that Kotas were providing music at the ceremonies of the Hindus who had come to reside in the hills (Hough 1829: 102). By 1848 Ouchterlony observed that low country folk were settling in the hills. Many were coming in seeking employment as coolies since they could not get work in places where they lived. Many more of these, Ouchterlony (p. 23) noted, would come and stay permanently as cultivators were it not for the fact that the land rights had been vested in the indigenous tribes.

Through the century which followed, coolie labourers kept coming hopefully up to the Nilgiris from the adjoining hungry lands of Malabar and the Tamil country. Some were hired by European planters, a few managed to secure a bit of unallotted land. Others camped where they could, first from day to day, then from month to month, and eventually stayed for years and generations. Todas would not have them about; Badagas might hire them but would usually not have them stay in or near their village; Kotas have been more lenient.

They were not displeased to have them about. Kotas are not always assiduous, energetic cultivators. When fields and smithy competed for a man's time, it was quite convenient to have some lowlanders around to send into the fields. A few were hired to herd livestock, a task which the village boys had previously done. And when news of a funeral had to be transmitted to the other villages, it was easy to send a lowlander as messenger. The chore of collecting fuel for the pyre from each village house was also one which the lowlanders came to do regularly.

In return for such services, they were permitted to keep the carcasses of the cows which were sacrificed at Kota funerals (not of the sacrificed buffaloes where meat was needed for the funeral feast). They also took the cloth which ornamented the towers that were built for important funerals. At harvest, a few measures of grain were given to those lowlanders who had been of service to the village.

Thus there developed a relationship in which these lowlanders were to the Kotas as the Kotas were to the Badagas. They continue to render services at funerals and are rewarded with carcasses and funeral goods. Lowlanders are not allowed inside Kota houses lest they defile them, although—like Kotas in a Badaga village—they may sit and may eat on the veranda. They, in turn, have purity standards of their own. They will not take cooked food from a Kota woman, but food for ceremonies, cooked by men in the open, is acceptable to them.

Lowlanders may not participate in the worship of the traditional Kota deities: they may only make coin offerings to them at major ceremonies. And at the newer shrines, like those in lowland villages, they give their offerings of plantains and coconuts to the Kota officiant, then stand at some distance from the shrine, away from any Kota worshippers, while their gifts are presented before the deity. When one of these lowlanders greets a Kota, his salutation is a respectful one as from an inferior, and the Kota calls him by his personal name as a superior does.

Yet, for all this formalized interchange, Kotas vigorously deny that they have any special or regular relationship with the low country squatters. 'The Konaton have nothing to do with our rituals'. This is the important count, to the Kotas it betokens an absence of firm and continuing reciprocal obligations. Further, lowland folk do not figure in Kota myth; they are thus not ensconced as lasting fixtures in the Kota world. Relations with them are considered contractual and hence ephemeral; minor agreements to be fulfilled today with no guarantee (by Kotas at least) for tomorrow. Yet, day-to-day arrangements long renewed have a way of becoming expected patterns and disruption of them may bring consternation.

So it was in the traditionalist Kota village of Kurgoǰ. Squatter families had settled there too and were allowed to cultivate, rent-free, some village land. Each lowland family attached itself to the Kota family whose land it occupied. 'Attached' in the sense that the Kota patrons had the first call on the services of the lowlanders as hired coolies or as unpaid messengers and the lowlanders obeyed summons promptly and without demur.

Then came a change in the fortunes of these low country families. They received, through the intercession of a Badaga headman, a two year lease of some government land. The fifteen Konaton families promptly shifted their huts and their efforts to this land. They no longer submitted to the bidding of the Kota villagers. This was an inconvenience when help was needed in the fields, but it was a calamity when there was the carcass of a sacrificed cow to be hauled away. For both Kota and Konaton were by then imbued with the same pervasive belief that it was degrading to handle dead cows. Fortified by their modicum of economic independence, the lowlanders refused to do this job.

The Kotas felt that they had to exert their power as landowners. The lowland families had some fifteen head of cattle and fifty sheep which grazed on Kota fallow and wastelands. These animals could not possibly graze on the patch of leased field; when next they wandered about on Kota land, they were promptly impounded. As Sallen of Kurgoǰ relates the incident, 'Then their chief man, Sina·n, came to me and asked me if it was right. He is supposed to be my man because his father and grandfather worked for my grandfather. I said that it was right because now we are separate and you have to depend on your strength and not on ours'. Later there was a fight and talk of calling in the police. A Badaga arbitrated the affair and Sallen concludes his account with, 'The Badaga decided that Sina·n would have to pay a fine or else there would be too much trouble in the village. He did not have the money, and so I stood security for him'.

Like other Kotas, Sallen insists that there is no ritual and no regularity in the relation with these newcomers. And yet, when one of 'his' squatters is in trouble, Sallen stands surety for him as he would for an associate or a clansman. Some of the supportive quality of an old kind of relationship continues, as it does in many an Indian village, although the formal quality of such relationships has all but gone, and despite the fact that the relationship traditionally did not prevail between Kota and Konaton.

Another tendency, also widespread in village India, is apparent in this incident. Soon after low caste villagers grasp a bit of economic security, they are apt to give up the socially degrading practices in order to better themselves in the world.

The number of squatter families greatly increased in the decade of the 1940s when the needs of the Nilgiri military stations brought many coolies to these hills. In the thirties there were about a dozen such families hanging on at Kolmel village. By the end of the next decade, there were perhaps forty families, as many as in the Kota village, plus three shopkeepers. They camped in every vacant Konaton house and in the empty cowsheds. They were continuously in the village—not inside the houses or temples to be sure—but everywhere else. Some of their men who had grown up in the village spoke Kota fluently and Kota women, who had never known any but their own language, were beginning to know Tamil.

It once was that only Kotas were allowed in the village during the principal ceremonies. Now some of the Konaton squatters may be ordered to vacate a house for the duration of a great rite, but there is no attempt to exclude them from the village bounds during the whole of every ceremonial period. During these periods they, like the Kotas, stop work and with nothing to do, they are much in evidence about the village.

But Kota interest is fixed elsewhere and for the present the squatters, when noticed at all, are seen as handy labour and not as significant people whose presence may affect village life. The shopkeepers in the village are more consequential. The tea-shops attract young villagers' money away from their households and usurp their time from work in the fields. The credit extended by these shopkeepers tends quickly to become too great for the debtor to repay. The shopkeepers themselves are not considered to be members of the village even though their business is with the villagers.

As for other plains folk, they enter the orbit of a Kota's life only sporadically. In the towns, there are moneylenders, amulet sellers, or merchants who may be patronized with some regularity. The government officials of the district may be petitioned, as one would go on a pilgrimage nowadays to a great Hindu temple in the hope of securing some special boon, but government officers and similar educated personages are not, as a Kota usually views them, for the likes of him. At the other end of the social scale, the villages are periodically visited by bands of Gollan, gypsy-like people from the plains who beg and tell fortunes. They come one day and are gone the next, to turn up again in a year or two.

A few low country people are now available to the Kota for personal services. Local barbers refused to shave Kotas for a long time, lest they lose their Badaga customers. Now some barbers, especially in town, ask no questions when a Kota comes for a shave. A washerman calls at a few of the more prosperous Kota houses, where the women have learned that it is not quite right for them to wash clothes.

The three Nilgiri towns are full of people from the low country. When the heat is on the plains, vacationists abound in these cool hill towns. At all times of the year, soldiers from the Wellington camp are about and also a miscellany of traders and government officials, plantation managers and plantation coolies. They come within a Kota's sight when he is in town but they impinge little on his social view.

The view of those outside the communality of Kotas is linked to the view of those inside as the outer pair of binocular lenses is linked with the inner. We have already noticed something of the Kota view of themselves. They are not given to self-deprecation or to tame submission. In myth, there are episodes of Kotas taking on the soldiers of a Raja. A modern echo of the old tradition was heard when some Kotas discovered, to their indignation, that a Tamil schoolbook mentioned only the Todas as the original inhabitants of the Nilgiris. 'The Todas had to come to us for all the things they needed to live by', complained one man, and added, 'they were our servants'. In ritual, no outsiders intervened between the Kota and their Gods. Indeed, when the whole village was sanctified for a great ceremonial occasion, no outsider was pure enough to come into it. All aliens whether Toda, Badaga, or Brahmin had to stay out; none but Kotas were fit to come within.

In economic and social interchange, the Kotas—acting as one—did not shrink from using their prime economic weapon, the boycott, when they found that their rights and dignity had been insufferably transgressed. Now that this weapon is useless, they have begun to try different means of asserting their due—such as litigation, petitions to government and aggressive argument—when they feel wronged. And they feel wronged whenever others (mainly and especially Badagas) disparage what Kotas believe are their proper rights and dignity.

As Seen from Without: Their Neighbours' View

As the enquirer talks with Badagas and Todas about Kotas, he gains the impression from them that Kotas can lay claim to few rights and small dignity. So it appeared in the first English account of the Nilgiri peoples, by William Keys in 1812, in which they are termed as 'the lowest class of inhabitants'. In 1820, Lieutenant Evans Macpherson, he who craved the reader's indulgence, wrote of them, 'This is esteemed a low caste—the paria of the hills and none of the other castes will eat with them or even enter their houses'. The memoir of a survey of the Nilgiris made in 1821 by Captain B.S. Ward tells how they are of 'a very inferior caste' whose 'filthy propensities render them so peculiarly disgusting to their neighbours that a Badaga will not drink of the stream that flows in the vicinity of their villages, polluted, as it is supposed to be, with the flesh of dead animals and their raw hides, these being generally dressed on the side of a stream'. Yet Captain Ward also had some kind words for the Kotas. He called them 'the most industrious race in the Neilgherries' and thought that they were good husbandmen (Grigg 1880: xlviii, lviii, lxxvi-lxxvii).

Later European observers agreed that the Kotas were commendably industrious and unconscionably low. One of the best of the nineteenth-century reports of the Nilgiri peoples is by the German missionary, J.F. Metz, who laboured for many years among them and came to know them very well. His section on Kotas notes that they are the only one of the hill tribes who practise the industrial arts 'and they are therefore essential almost to the very existence of the other classes'. Yet this gives them no pre-eminence in the eyes of the missionary or of the other classes. 'They are, however, a squalid race, living chiefly on carrion and are on this account a bye-word among the other castes, who, while they feel that they cannot do without them, nevertheless abhor them for their filthy habits' (Metz 1864: 127-8).

Since grain, not meat, was really their staple food, and since the meat often was from the fresh kill of a sacrificed animal rather than carrion, Metz's indictment that they live chiefly on carrion was not literally true. What was eminently true was that the others thought of the Kotas as carrion-eaters. Not simply eaters of meat: most Badaga subgroups would eat goat meat on occasion. But Kotas would eat the flesh of cows and buffaloes and would not care how a beast had perished. This above all their dubious practices (as the playing of music) marked them as an unclean and polluting lot.

Metz tells a story about a Badaga of the Chittre subcaste who quarrelled 'many years ago' with Kotas of Tičga·r village over some land. The Chittre Badagas are of the Lingayat sect and each wears an amulet around his neck containing the sacrosanct lingam, symbol of Siva. Then '...during the altercation, one of the Kotas touched the Chittre's lingam. The latter felt so polluted by the circumstance that he killed himself on the spot with a hatchet'. Notwithstanding this terrible self-punishment, Metz sadly relates, all the lineal descendants of this Badaga have since then been excommunicated by the other Chittre (Metz 1864: 53).

Another aspect of traditional Badaga attitude toward Kotas is given in two of a list of Badaga proverbs which Metz recorded. 'If I make the Kotas my friends, my rice will be eaten'. And in similar vein, 'If you lend to the Kotas, you will become poor; if to the Todas, you are an idiot; if to the Kurumbas, they will kill you' (ibid.: 93, 97). Here the emphasis is on the persistence and enterprise of Kota demands.

Kota enterprise was further shown to Metz when they proposed to keep buffaloes as they had not theretofore done. The headmen of the Todas and Badagas opposed this because they could not permit the unclean Kotas to have anything to do with so sacred an occupation as that of milking—even the milking of animals owned by Kotas. The Kotas appealed to Metz who said that he would employ them as milkmen—but only if they materially changed their habits. This was too hard a bargain and they abandoned the idea (though in later years, after Metz's time, they succeeded in keeping milch cows and buffaloes). Metz comments 'The circumstance shows, however, that a spirit of independence and enterprise prevails among them, and that they are prepared to put their hands to anything that may offer a prospect of bettering their condition, notwithstanding that by so doing they would be subverting the customs and social usages by which their forefathers had from time immemorial been guided' (ibid.: 128-9).

But in sorrow the doughty missionary records the obverse aspect of Kota independence.

> The *Kotas* are somewhat more intellectual than either the Todas or Kurumbas, but their living on carrion [again!] renders them so savage and revolting in their habits, that it is difficult to approach them. When I endeavour to address them they drown my voice with their dreadful music, or compel me to retire by abusing me in the most obscene and offensive language or barking at me in the style of their own half-wild dogs. Thus I have often had to leave their villages, with a heavy heart at their apparently hopeless condition (ibid.: 134-5).

It was not only the missionary who faced the rude independence of the Kotas. Even the Todas, admired as they were by the other Nilgiri peoples and even feared as sorcerers, at that time, by the Badagas, were not always given a courteous reception when they came to collect grain and tools. Metz reported that Todas, appearing in a Kota village, were often taunted with these words:

'What! beggar, have you come again!' (ibid.: 129).

Thus the Kotas appeared to the German missionary, and to various English observers, as a people of considerable spirit and independence, of praiseworthy enterprise and industry, albeit of foul habit and careless ways. Such consideration and hint of praise is notably absent from the testimony of Todas and Badagas about Kotas.

Toda opinion is succinctly given in the classic anthropological description of Toda culture by W.H.R. Rivers. 'While a Toda regards a Badaga as his equal, or perhaps even as his superior, he looks down on the Kota as inferior, as hardly to be classed as a man with himself' (Rivers 1906: 636).

To Todas, Kotas are defiling. In greeting a Badaga elder a Toda may bow his head respectfully and the Badaga will touch his bowed head as salutation and blessing. No such touch is permissible between Kota and Toda: if it should accidentally occur, the Toda must purify himself ritually. The sole exception to this touch taboo was when a Kota came to the funeral of a Toda associate. He would then touch his bowed head to the upraised foot of the principal Toda mourner.

One Toda respondent who came to give linguistic information objected to sitting in the same chair that had been occupied earlier by a Kota respondent. Whereupon the linguist took that chair and all was well.

Another Toda said that when a Toda approaches a Kota village, he first sits some distance off from it so that his spirit (*teu*) may remain there while he goes closer to the defiling place. And he does not go within the village proper, much less into a house. He visits the smithy only—save at Gudalur village where one of the Toda deities once entered, so that mortal Todas may also come into a Kota street, only there.

The Toda tale of how the Kotas came to the Nilgiris bears no resemblance to the Kota origin story of the differentiation of three brothers. As a Toda, Kaṇfïs̱ody, related it, his people first met Kotas while returning from an expedition for salt. The Kotas shot arrows and the Todas fled. But soon the Kotas turned up at a Toda settlement and were given food. Eventually the Todas, pleased with the tools and utensils which these Kotas made for them, gave land and buffaloes to them and allowed them to settle in the hills in return for yearly payments. The two peoples agreed that the Kotas would regularly give grain, handicrafts, and music; the Todas, in return, would give ghee and buffalo carcasses.

The story reflects the current attitude of Todas. Kotas as a class are polluting and rather despicable, but withal are useful and sharers of an old agreement. Hence Todas still provide ghee for the Kota God Ceremony (Peter 1955:90). Despicable though the class of Kota may be, individual Kotas may be quite humanly treated. So when a Kota woman, the wife of his father's old associate, came up to Kaṇfïs̱ody at a ceremony and smilingly badgered him for a little money, he did not ignore her or drive her off peremptorily. Trousered and Christianized Toda though he was, Kaṇfïs̱ody gently humoured her and gave her some annas.

Latterly the Todas have been under pressure from influential Badagas to give up all traditional relations with Kotas. The payments of grain which Badagas customarily gave to Todas have since 1939 been stopped by many Badagas. When Prince Peter carried on ethnological studies in the Nilgiris in 1949, only five of the ten Toda clans were still receiving their traditional due from Badagas.

Under the leadership of one H.B. Ari Gowder, who had risen to become a municipal commissioner of Ootacamund town, Badaga reformists had objected because Kotas attended and played at Toda funerals. Their complaint was that Kotas were so low and defiling that their participation in Toda ceremonies lowered the Toda which, in turn, lowered the Badaga who maintained association with the Toda. As we have seen, this Badaga group has been largely successful in barring Kotas from Badaga rituals. But they have not succeeded in keeping them out of Toda ceremonies. Some Todas still want to have Kota participation. Prince Peter quotes the Toda question, 'Who will take the sacrificial buffaloes away if they don't come?' (Peter 1955: 90).

The Todas, like the Badaga traditionalists, need Kota services, shun the polluting aura which those services induce, yet may deal with particular Kota men and women as with people rather than as with members of a defiling class. But to the Badaga reformists, Kotas constitute something else again. They are a threat to the social status of the whole Badaga community. So much so that these Badagas will not continue traditional relations with Todas, if Todas continue to have a place for Kotas in their rituals, even though it is the place of menials.

Why do so many Badagas see a menace in the few Kotas? There are a hundred Badagas for every Kota. No economic competition accounts for Badaga sensitivity. The Badagas, by and large, are energetic and careful farmers: the Kotas are generally indifferent cultivators, their main interest and absorption is not in land and crops and they readily lease or sell farm plots to expansive Badagas. Kotas are not rivals to Badagas in traditional, ceremonial spheres nor in the newer arenas of political campaigns and elections.

Nor did the Badagas have such apprehensions under the traditional social interchange. One Badaga term for a Kota associate is 'veranda son'; that is, a kind of son who cannot come into the house but must sit on the veranda. The reference denoted a subordinate with whom there was something of a kinship relation but it was a relationship that could be no more intimate than was possible in the public view. There was no question in a Badaga's mind but that his associated Kota was an inferior and a subordinate. While the Kota considered the Badaga as his associate, the Badaga looked on the Kota as his dependent and on himself as the patron.

The Kota sitting on the veranda was a potential source of pollution for the Badaga household, but there was little anxiety manifest on this score. It was such an everyday occurrence, so well channelled, understood, and managed that it presented no dread of danger. The Kota visitor had to be fed when he came to provide music, but he was not given milk, too sacred a substance to be consumed by so impure a person. The Badaga woman came out to pour water over his musical instruments by way of purifying and blessing them for the ceremonial tasks, but neither she nor any Badaga would touch a Kota; if they did so inadvertently, a purifying bath had to be taken quickly. The Kota could wander about the village, visit many a veranda, but never go into the temple area or into the house itself.

So long as the proper social and ritual distance was thus maintained, Badagas did not worry about Kota ways. Indeed, both groups seemed little interested in each other's culture. We have noted before that Kotas generally know and care little about such aspects of Badaga culture as the differences among the Badaga subcastes—a matter of great moment to the members of these subcastes. And Badagas show little interest in particulars of Kota culture. One Badaga whose land adjoins Kota lands and who has dealt with Kotas almost daily since childhood, has only vague notions of the purposes of Kota ceremonies and no knowledge of Kota fraternal arrangements in marriage. So it was, and is, with other Badagas and reciprocally with Kotas. A Badaga and a Kota may know each other well as persons and be able to judge accurately each others fluctuations of mood and motivation in striking a bargain about a ploughshare, but they are mutually ignorant about much in each other's way of life.

Badagas became concerned about certain Kota customs only when the Kota seemed to be taking on some new usage which might betoken a higher social station. We have noted how Badagas (and Todas) objected when Kotas first proposed to keep milch animals. In the same period of the 1850s 'A school for Kotas was established by the missionaries, but it had to be closed through jealousy of the Badagas'. So wrote Grigg in his *Manual* and added 'The promise was good, as the Kotas are an intelligent race' (Grigg 1880: 426). Badagas were then aroused to protest against any Kota innovation which, to them, symbolized unduly high aspiration, as when, about the turn of the century, Kotas began using tile instead of thatch for roofing and when a Kota bought a horse. If the Kotas were to rise, the Badagas by so much would sink. So believed the Badagas, from those of the lowliest subcaste to those of the highest.

In former days, they felt threatened by anything regarding the Kota which might disturb the balance of the traditional relations. In later times, many Badagas felt threatened by anyone, Kota or Badaga, who wanted to maintain the traditional relations. Then innovating, aspiring Kotas were the danger, now the traditionalist, conservative ones are.

One reason for this apparent reversal is simple and straightforward. Most Badagas do not need Kota services any longer and the old social obligations have become a nuisance to maintain. Tools and pots may be bought in the bazaar. Money is available, so are the goods. A purchase is made and there ends the transaction. With Kota tools and pots, the utensil is apt to be only one part of an involved relationship which entails payments in grain at harvest, in money at ceremonies, in animals for rituals, and in bothersome details and haranguing throughout the year and through all the years. Latterly money has become more abundant and grain relatively more scarce, so Badagas are all the more reluctant to keep up payments in grain rather than in cash. They do not object to buying utensils from a Kota as they would from a town merchant, with a cash-and-carry termination to the transaction.

So is it also with Kota music. A Badaga who wants to do a wedding or a funeral in the proper traditional form must scurry about making arrangements with Kotas of varying degrees of reliability but with uniformly high and continuing demands. It is much easier to hire a band of Kurumbas, Irulas, or Tamils, or, for a bit more, a modern ensemble of Tamil Christians complete with trumpets and other bass instruments. A Badaga village leader, comparing musicians, said 'The reason why some people will take another band rather than Kota is because you fix a rate with the others and they play and go away. But if you get Kotas then they want to follow all the old customs. They want gifts and they want to give tools and get grain and all that sort of thing'.

He went on to talk of his own attitude. 'I myself am more anti-Kota than I am anti-music, but there are some who are completely anti-music'. Those who are against all funeral music are also the most zealous for heightened group status.

A Badaga village-uplift volunteer, an enthusiast of Gandhian reform, told with manifest approval how a Badaga assembly had agreed, about 1930, that Kota music should be prohibited at their funerals. 'The reason is that funerals should be times of mourning without the traditional dancing. There should be no rejoicing with music at funerals'.

This remark gives a clue to another reason for Badaga antipathy to Kotas, an emotional reason which builds up into revulsion. 'Rejoicing with music at funerals', many Badagas have become convinced, is not suitable for people of their social rank. Badagas know that neither high caste Hindus nor Europeans countenance such behaviour at funerals. And in recent decades, as they have been increasingly exposed to formal education (mainly in Tamil) and have increasingly participated in the general South Indian social sphere, they have become anxious to take a place in that sphere which would not be a lowly one. They believe that folk who rejoice at funerals, who practise animal sacrifice, who entertain carrion eaters on their verandas, who follow other depressing practices of the traditional Badaga culture, are not likely to be given high rank and so they are determined to slough off the demeaning customs.

Sloughing the customs means disavowing the Kotas. Hence when Kotas demand all their traditional dues, they are a menace to these Badagas. Not because the few Kotas offer any clear and present danger, but because they remind them of what the Badagas used to be and what they fervently no longer want to be. They are a reminder of what should be forgotten; they raise the spectre of a former and lower morality.

We get a glimpse of how this aversion was fostered in the account of the missionary, Metz. He relates how the Badagas of two localities were being boycotted by the Kotas and '. . . feeling the inconvenience acutely, have desired me to get them some ploughs from the low country. In one case I promised to help them, because I was in a measure the cause of the quarrel by dissuading the Badagas from sacrificing a cow to the Kota idol' (Metz 1864: 131). For several passages, Metz regrets that he could persuade few Badagas to become converts, but, in this instance, he could readily dissuade Badagas from a traditional service for the Kotas.

At the time of this incident, after lowland folk had regularly been coming up to the plateau for some thirty years, Badagas may well have become more sensitive to the general Hindu aversion to cow killing than they had formerly been. The roads which the British had built into the Nilgiris undoubtedly intensified Badaga contacts with plains folk, especially in Mysore.

Even in 1820, before these roads were made, James Hough reported that Badagas maintain a 'constant intercourse' with the people of their original homeland in Mysore State, 'even to an occasional intermarriage with their families' (Hough 1829: 88). Forty years later, Shortt observed that 'many of them, at the present time, have connections and friends in the plains' (Shortt 1869: 287). Thus it looks as if in this particular the European influence exerted through the missionary reinforced the Hindu influence from the plains.

From such beginnings, and under this dual influence, developed the current Badaga view of the Kota. Badagas have consistently held that Kotas are a people who must be kept in their place, in their subordinate, lowly place. Earlier that place was one which involved interdependence and inter-responsibility between families of the two groups; now most Badagas want to push the lowly ones clear out of any special relationship in the Badagas' social orbit.

Such watchful jealousy of status also occurs, and has long existed, within Badaga society. The higher subcastes of Badagas, as the clan which is vegetarian and wears the sacred thread, are alert to keep the lower, non-vegetarian subcastes in their proper place. Even among the Christian converts there is similar status sensitivity. A Bishop of the Church of South India told me of the problem he had in dissuading Badaga congregations from cutting off all connections with those other Indian Christians in the Nilgiris who are mainly Tamils of an upper-class, urbanized kind. This, because the others follow the custom, deemed improper by the rural Badaga Christians, of wearing shoes in church.

Badaga sensitivity about Kotas is part of a psychological stance that also, as we shall later observe, is assumed by Kotas. Badaga sensitivity is focused on Kotas because most Badagas see them as a particular block to a desired social orientation.

There are some countervailing pressures against cutting off Kotas completely. A minority of Badagas, principally the priests and ritual leaders for the other subcastes, want to preserve the old Badaga ways of life and conceive of themselves as the keepers of the traditional culture. They make up the pro-music faction and try to keep alive the former relationship with Kotas. And at the other pole of Badaga opinion, those Badagas who have been most highly influenced by Gandhian ideas say that Kotas, like other Harijans, should be uplifted. But the verbal avowals of these few are not put into practice. 'I would not care if I entered a Kota house', said one such Badaga young man in characteristic manner, 'but if I did it would trouble my parents and others would make trouble for me.'

Thus when young Badagas of this persuasion equate Kotas with the Harijans, the beloved of Gandhi, their professions now come to little. But when their fellows equate Kotas with the untouchable pariahs from the plains, as most now do, it leads to their renouncing the former Kota relationship. Some Badagas renounce and yet carry on a vestige of the old reciprocal affiliation; others feel a revulsion against any contact whatsoever with the Kotas.

Even Todas are beginning to acquire such attitudes. Now that Todas—after a century and more of persuasion—are sending their children to school, they object to having Kota children in the same schools. Badagas made such objection decades ago; most of them now see the Kotas as, at least, a nuisance; in their more fervent moments and among the more fervid Badagas the Kotas loom as a menace.

As Seen from Without: the Ethnologist's Perspective

So much to explain what a man means when he says 'We are Kotas of the Nilgiris', and what his Badaga and Toda neighbours understand the phrase to mean. Thus do the Kotas present themselves to their world. No matter how one differs from another in personality, acculturation, or wealth, all have this identification in common. All whom I know, some in great gusts of argument, others in less articulate insistence, assert that they are one of the ancient tribes of the place, and as such they have certain inalienable rights. Such is the inside view.

From the outside, their Badaga and Toda neighbours perceive them differently. How Kurumbas see the Kotas we cannot say. As for others who now live in the Nilgiris, such as plantation coolies or soldiers or retired government officials, they generally know little more about Kotas than that there are such people in the hills. But Badagas and Todas have a decided view of them as useful, though lowly and despised adjuncts to society or as low, despicable creatures to be avoided.

There is yet another outside view, that of the ethnologist. It takes into acount both self-image and neighbours' appraisal and attempts to place both in wider reference. The Kotas may be seen in various lights and at different levels—as participants in one variety of South Indian society, as sharers in certain all-India culture traits, as citizens of a Republic, as manifesting general human patterns of conduct. At each level certain data will be more apposite than others and so it is well to make clear from what distance and with what definition the ethnologist is looking at his people.

Since we have been discussing the Kotas as a functioning group we may, in this first overview, consider the Kotas as a community and their relations with their neighbours as part of a social system, particularly of that kind of social order called a caste system.

The Kotas are clearly a community. There is the awareness of 'we Kotas', the distinctive criteria of identity, the interaction, the mutuality of culture, the self-perpetuation, the clear location in time and space—all the characteristics which are usually listed in defining community. Each of the seven villages is a community in its own right and is also a part, a subcommunity, of the larger Kota community.

But what of the important relations among the indigenous peoples? How to characterize the four together, in terms of a generic human process? They can hardly be grouped as a single community, not in their ancient state and latterly even less. Yet they did, and do, participate together in a kind of social system.

What, then, is meant by 'system'? The term is frequently used and is an important one for the anthropologist's analyses. Sir Raymond Firth says 'The notion of a system is basic to our study of society' (1951:27). For our study of Kota society, we may paraphrase—and perhaps oversimplify—the definition of system suggested by Parsons and Shils (1951:107-8), and that of equilibrium given by Homans (1950: 301 ff.).

A system consists of a set of parts which are interdependent and are interdependent in a specific order of behaviour toward each other. That order involves a tendency towards equilibrium. This tendency is the process by which a change in the regular behaviour of one part towards the others is followed by reactions from the other parts which tend to reduce the effect and consequences of that change. This reduction of deviant effect may result in a return to the former state of interdependence and order ('steady state') or may result in a different state of relations but one in which interdependence and order are still mutually maintained ('dynamic' or 'moving equilibrium').

By these criteria, a true social system was maintained by the four Nilgiri peoples. Each was dependent on the others and every community had a particular kind and order of dependence on the others. None could carry on its way of life without the aid of the others.

Equilibrium was maintained by the devices of social control noted above. As between Kota and Badaga families, tools or grain could be withheld if one party thought that relations were out of proper balance. As between Kota and Badaga societies, group boycott and other pressures could be exerted by one or the other. Nor were the forces making for equilibrium entirely punitive: the system was a part of the accustomed and established order of life for those of each group, they had been—as was noted above—raised in it, they liked it and wanted to have it go on in its steady state.

This local system presumably developed after Badagas came to the Nilgiris. It was in operation when the first European, Father Yacome Finicio, reported on the Nilgiri inhabitants in 1603 (Rivers 1906: 723). It then operated in relative isolation—though not complete isolation—from the influence of extraneous peoples. We have seen how this isolation was broken about 1820 when Englishmen and plains people began coming in some numbers. For the next hundred years the traditional societal equilibrium was maintained, despite increasingly disruptive factors, until it was drastically modified about 1930.

In some basic ways the Nilgiri system was like the classic caste system which prevailed over most of India during the nineteenth century and before. The main system is here called 'classic' in the sense of typical, standard, and traditional. The Sanskrit scriptural classics are involved in it, but do not provide a complete blueprint for it. It is 'classic' also in that certain themes remain recognizably constant through many regional variations. Yet though much of the Nilgiri arrangement resembled the usual model of the main tradition, it was also different from it in certain aspects.

It was like the classic caste system in that there were a set of separate social groups which were interdependent within a locality. They were interdependent in multiple ways—economic, ritual, societal. Each group was, and is, hereditary and endogamous: a person is a member of the group into which he is born and marries only another person of this same group. Each group has a traditional occupational specialization which need not be the sole occupation; it follows customs which, in the eyes of its neighbours, are unique to it. In the Nilgiris, such cultural visibility is stark; the four groups of the enclave differ markedly in language, ritual, and, until recently, in dress and appearance.

Each group is made up of families which are typically arranged into lineages or clans, then exogamous units operate within the endogamous group. Those of one group can, indeed must, co-operate in certain ways with people of other castes; in all other ways they must avoid contact and co-operation with them. The co-operative activities are mainly between families, in the form of interchange which is often called the *jajmani* system (Wiser 1936). And as among the Kotas and Badagas, the exchange of goods and services has some of the characteristics of a contract but it also may entail a wider and warmer relationship in which there is mutual support.

Families of different groups reciprocate aid and even share friendship—within certain limits: beyond these limits there should be avoidance, not only as between the families, but as between one's own group and all others on the social horizon. Interdependence and co-operation are essential but circumscribed.

The limits of permissible interchange vary according to the social status of the actors. This is true throughout caste society but the limits may also be drawn a bit differently, in one respect or another, in the several culture areas of the land.

The activity involved, for the Nilgiri and other Indic peoples, is an important factor in determining the scope and stringency of avoidance. Certain relations outside one's group are absolutely prohibited; other kinds of relations are permitted within a somewhat larger circle and so on through degrees of avoidance and widening circles of permissibility. Thus, marriage out of the group is an absolute prohibition and one that has been maintained with remarkable consistency despite all the stresses on the system. Eating in the company of those of another group is also restricted, but a man may often eat with a person of a close-ranking group although he may not marry his daughter. That is, a man's commensal circle may be larger than his endogamous group. A still wider circle of co-operation is permitted in other activities. Badagas will certainly not intermarry or interdine with Kotas but they will on occasion work side by side with Kotas in the fields and some still attend ceremonies in the Kota villages.

Not only the nature of the activity but also the social status of the participants regulates any joint action, both in the classic system and in the Nilgiri case. Each group ranks itself and is ranked by its neighbours in an order of superiority and inferiority among the interdependent groups of a locality. There is rarely close agreement among villagers on all details of ranking and precedent but there is usually general agreement among them—whatever their own place may be—as to whether a particular caste is of the higher, or middle, or lower rank. The village consensus may mark off two, or four, such classes rather than three but there is a wide agreement on these groupings of castes.

A superior group tends to restrict and minimize the permitted relations of its members with those of a lower group. The lower group tends to be more permissive about relations with those of a higher group, but always maintains certain avoidance standards towards all other groups, low or high.

Thus Badagas would not take cooked food from Kotas while Kotas take food from Badagas when they come to play in a Badaga village. Yet no Badaga could enter a Kota kitchen or temple. So it is, for example, in the Tanjore village observed by Gough: there orthodox Brahmins sternly dominate the lower castes, yet a Brahmin is not allowed by the lowest caste to enter their street (Gough 1955:85).

Each of the lowliest groups typically considers that it is higher than some other group of the locality. The Kotas know that they are higher in status than the Kurumbas and the squatters; Kurumbas, in all likelihood, know themselves to be above Kotas. The squatters, as we have seen, observe certain avoidance *vis-à-vis* Kotas. In that sense there is no bottom to the hierarchy.

But, as we have noted, there need not be close agreement concerning ranking at any level of the order. In fact, there is frequently a difference of opinion, especially as between two groups of generally similar station concerning their relative rank positions. Many an argument, sometimes a riot, has raged about the symbols of rank. While Kotas did not dispute Badaga superiority, they also did not meekly accept Badaga rulings. Thus they thought themselves quite entitled to keep milch animals, to use tile roofs, to send their children to school—all innovations which Badagas said were not permissible for so low a group.

One index to the gradations in a village is the order in which food is distributed on ritual occasions. Even this could be disputed and the order of ritual precedence could change in course of time (cf. Dubois 1906: 23; Ibbetson 1916: 174-5).

In all, the rank order is not and perhaps never has been rigorously calibrated for all life situations, nor does any one table of precedence command unanimous consent in a locality. There was room for flexibility and disagreement in the system so that it could be adapted to changing circumstances and so that it provided for the satisfaction of ambitious groups—all of which made for the stability of the system, as a system, over time.

There are also certain common understandings among participants in the system which are essential to its operation. Thus all in the Nilgiris and in Indian caste society know that certain experiences and activities make a man unfit for free relations with his fellows. People who customarily have such experiences and perform such activities are unfit for close social relations with the other groups. Those who have to do with death and corpses and carcasses are to be avoided. And 'having to do with' a carcass includes eating the meat. Vegetarians must shun close relations with eaters of, say, goat meat. The latter must keep at a social distance from those who eat the still more debasing meat of the carcass of a cow.

Music comes within the aura of death. Not all music and musicians: not devotees extolling deity in hymn, nor the sophisticated professionals whose intricate and bittersweet poesy daily recounts hopeless love over the All-India Radio. But those musicians, like Kotas, who are summoned and must come to play at funerals—by this funeral token they are lowly and socially disabled.

The theme that contact with death disables and demeans the living, is not restricted to the ordering of relations among groups. It is given dominance in other parts of the culture, in Kota as well as the classic caste. For example, Kota widows and widowers must undergo long ritual purification before they can be readmitted to normal social life. Again, when Kotas celebrate their most sacrosanct ceremony, only vegetarian food is eaten at the feast. The fundamental values which govern relations among groups also obtain for conduct within the group.

There are other, related themes concerning social interchange which prevail in the Nilgiris as they do in Benares. Thus contact with bodily emissions and with birth is defiling, hence the abasement of those groups of the classic system who are scavengers of human refuse, such as washermen and midwives. These criteria do not much concern us here, since they are not major elements in the relations of Kotas with their neighbours.

Another shared understanding is that when a person incurs social stigma, his whole local group is contagiously affected. Usually it is his own group which is most sensitive to this. And should one person take on a flagrant token of higher rank, all of his fellows are thought to be claiming higher status: usually greatest resistance to his behaviour comes from the higher strata. For example, when one Kota put in a tile roof, Badagas saw, in this act, all Kotas as a group putting on airs.

Women especially, must give no cause for offence and group debasement. No Kota woman, all of whom I asked were agreed, has ever taken a meal in the home of a lowland squatter. But if one should (the response to my hypothetical question was immediate), and thereby bring shame to the whole community, every Kota, in redress would cut off all social relations with her. She would not even be given a Kota funeral when she died. By every means, tangible and symbolic, the other Kotas would demonstrate to themselves and to their neighbours that the perpetratess of the shameful deed was not a Kota and so her calumny did not taint the group.

Whether this severity would really be imposed on such an errant woman, whether in the event ideal justice would be tempered by expedient mercy, is another matter; the Kota attitude in this is like that of classic caste societies—one tainted member sullies the whole. The stain must either be ritually expunged or the defiled member excised. However low a group may be, certain practices are thought demeaning beyond toleration by those belonging to it.

Why then did Kotas and other low groups carry on some work which they knew was degrading? For the Kotas the answer must be compounded of several reasons. One is simply that they liked to play music, they enjoyed a good meat meal, they liked working in the smithy. Many a Kota in the nineteenth century could have made his living by agriculture alone on his own land and so be freed of dependence on Badagas for food: very few did. To a Kota, to that manner born, the dull monotony of work in the fields was not nearly as attractive as the solid pleasure of taming the hot iron on the anvil or the lilt in spirit he could evoke with the melodic skirl of his clarinet.

As for eating carrion meat—our fathers ate meat as the progenitor commanded. Kith and kin now do so and so do I. Not to do so is to set myself apart from my people and what then should I be? We shall note later how one man from the conservative village of Kurgoǰ has aroused his fellow villagers' wrath by stubbornly being a vegetarian.

Nor did the Kotas believe, as did the Badagas, that their dealings with carcasses and their other low practices necessarily corrupted all their culture and person. As we have noted, they tended to be quite sure of their general worth. Similar ego-strength may have been characteristic of other low status groups of the main civilization. Perhaps those Kotas who now argue to carry on the full roster of traditional relations do so because under that régime they were important in the lives of their neighbours.

Another reason for continuing degrading customs is certainly as prevalent in the classic system as in the Nilgiri version. Those of the lower depths could not readily give up such practices because those of the higher ranks would not permit them to do so. Those who command wealth in a locality generally command power. They may use that power to force poorer groups to perform the menial functions necessary for the communal life. Power also may reside in numbers. When a group is both wealthier and more numerous than one subservient to it, the inferior peoples are kept firmly in their inferior places. D. N. Majumdar cogently writes, 'Where the depressed castes are numerically small, the disabilities are rigid' (1945: 132).

Thus the Badagas, far more numerous than the Kotas, occasionally waylaid and beat up Kotas whom they thought were getting above themselves. Usually force was not necessary. Badaga influence, for a time at least, curtailed education and some other Kota moves toward higher status. Yet for all Badaga wealth and dominance, Kotas were not cowed, nor were they completely defeated in their activities. In some localities of village India, the low castes are and were as spirited as the Kotas. But in other places the lowest castes are so starveling that their aspiration seeks little more than food for now and for the close future.

When a lower group gets an upper hand, turnabout is often the play. Kotas, as we have noted, can play the part of the wealthier and more numerous group. Villagers of Kurgoǰ forced their squatters to remove cow carcasses when the squatters, on attaining a little economic independence, refused to do so. How a villager views such exercise of power depends on whose ox is being beaten.

Still another feature of classic caste obtains in the Nilgiris. The boundaries of a group are differently perceived from the inside than they are from the outside. This too we have noted above: to a Kota, all Badagas are of one class and rank. A Kota will know that there are village differences among them. He has some idea that there are five or six main endogamous groups who speak the Badaga language and who share the main elements of Badaga culture. But to him, they are just Badagas and the divisions among them are of little importance. But to a Badaga, the internal divisions are of very great importance.

Those of the Wodeya division, for example, who are vegetarians and of the Lingayat sect, consider themselves a separate and segregate group from the Toreyas, Badagas who may eat goat meat, who traditionally performed menial tasks, and are not Lingayats. True, a Wodeya will agree that the Toreyas, low as they are, are above the Kotas. But if he is of traditional bent and not much swayed by modern trends, he considers himself to be a Wodeya and not a Badaga. (In a book about Badagas the blanket term Badaga, could scarcely be used as freely as is possible and necessary in a work on the Kotas.)

So it is with regard to caste generally. When a man refers to this own social ties, then the endogamous division, the *jati*, is of paramount significance. But when he refers to the social relations of others, then a generic class of *jati* is generally the important reference category. This generic term is either what is usually called a caste name, as Badaga, Chamar, Jat, or is a varna classification, as mentioned in scripture—Brahmin, Kshatriya, Vaishya, or Sudra. G. S.

Ghurye succinctly formulated this feature of caste. 'Stated generally, though it is the caste that is recognized by the society at large it is the sub-caste that is regarded by the particular caste and the individual' (1950: 20). Thus Kotas are quite undiscriminating when they use the single term, lowlander, for squatters of several castes, hailing from different linguistic areas.

Both the Nilgiri and classic caste systems, finally, use similar processes for maintaining the system. The means of social control are similar; the values for which control is exercized are, in considerable part, common to both; the personal motivations to uphold the values seem broadly comparable.

Control of errant members is enforced within the group by various pressures; the supreme penalty—for a maximal transgression such as taking a spouse from a lower group—being outcasting. Similarly, control of errant groups is by boycott and avoidance of the whole transgressor group. Note well that it is not, as in other civilizations, by hammering the transgressors until the fault is expunged.

In this system certain choices are regularly made from among potential alternatives; that is, certain values are consistently maintained. These are mainly choices which have to do with carrying on limited interdependence and upholding residual avoidance among groups that are ranked in hierarchic order. Social distances are calculated by the criteria we have mentioned, some dealing with death and diet, others with relative status position, still others with relative wealth and numbers.

These values are also asserted within the group, in the village, clan, and family. Among Kotas, such values appear to rest upon certain personal motivations which I have discussed elsewhere under the term 'aggressive defence'.

While this social system affects much of ordinary life and large occasion, it is still not all of life. Thus a Kota elder may not only be respected by his own people but may also be acknowledged, even by Badagas who know him, as a wise and competent man whose advice they may, on occasion, seek. The hierarchy counts for much, but still individual merit can be judged by criteria which transcend caste; individual influence may run counter to the caste order. In many a conservative village, a man of high caste may receive from his inferiors the respectful salutation due to his rank, but be judged a fool by them and so teated in everyday commerce.

Thoroughgoing as the systems are, they are also flexible, they are amenable to changes of personnel. Within the classic system new groups were often formed, either by breaking off from an old one (this process is evident among Kotas now) or by the incorporation of peoples not formerly in the system. A particular group can rise in the system, as is exemplified by those Kotas and Badagas who are zealous to abandon debasing practices now that they are adapting to the main system.

What has remained constant has been the maintenance of a hierarchic system and the broad values entailed in it. The Badagas who now feel menaced by tradition-maintaining Kotas are not completely different in this from their fathers who were threatened by innovating Kotas. The two generations stand agreed that Kotas are lowly and should not contaminate the social status of Badagas; only the means of social prophylaxis have changed as the social horizon has broadened.

The Nilgiri peoples understood the kind of social system which the lowlanders brought to the hills and the lowlanders saw nothing especially outlandish in the relations among the indigenes. Kotas and low country squatters fell easily into a ranked relationship. This is an important aspect of the caste system of India: the component groups differ from region to region, the details of grouping and of social distance vary in time and in place, but the outlines of the system are constant, so that a visitor from one part of the land can readily orient himself to the particular social establishment in another part.

There are differences as well as similarities between the Nilgiri system and the more prevalent classic tradition. As we examine these differences, we find the essential elements of the main system more clearly highlighted among the Nilgiri peoples—as their system operated between around 1830 and 1930—there was no corpus of scripture to sanctify the higher groups. One high division of Badagas did wear the sacred thread, but to the Todas and Kotas they were only Badagas, not sacrosanct scholars and priests. While Badagas worship much the same gods as are worshipped in Mysorean villages, the deities of the Kotas and of the Todas were not those of any lowland temple. The traditional Kota gods are for Kotas only, the Kota temples are their only abodes. The key scriptural doctrine of Karma or reincarnation was only hazily known to the Kotas and was not an important belief. But this appears to be a doctrine not strong among low castes in the plains, nor among Badagas either.

Nor was the Nilgiri system supported by a state and guarded by a panoply of rulers and lawgivers. In pre-British times, the Nilgiri peoples occasionally paid taxes, irregular tribute may be a better term, to one feudal lord or another from the environing gentry, but the hand of the government was not otherwise felt in the hills.

The hand of the British government did not directly tamper with the relations among the hill folk, although in time its indirect and cumulative effect was great. The influence of cities and of centres of pilgrimage, important in the classic system, was not a factor in the isolated hills.

The Kotas were not abysmally impoverished as are many of the lowest in the classic hierarchy. In earlier centuries there was apparently no dearth of arable land near their villages and under the British many Kotas owned more land than they farmed. Thus sheer inability to get a living otherwise was not the reason for the Kota adherence to their traditional occupations.

The Nilgiri system did not entail as many and as complicated rules of food acceptance and of other details as did the classic system. Moreover, a Nilgiri village did not include people of several groups, as plains villages usually do. Kotas live only in their own villages (they are a single *jati* in the classic sense), Todas live only in their own settlements and Kurumbas in theirs. Badaga villages may have more than an endogamous section, but all are Badagas in the village.

Thus the Nilgiri arrangement lacked several elements which have been deemed essential to the classic system. It did not have a priesthood acknowledged to be superior by all ranks; it did not have a shared pantheon or a common belief in reincarnation; it lacked the regulation of the state and the great poverty of the lower ranks. Yet, for all these absences, it operated much like the classic system in certain basic ways.

The Nilgiri system may have been established on the model of the classic system by the Badagas, as some writers have thought. The Badagas did come from an adjacent area where they must have participated in some form of the main tradition. The significant fact is that the Nilgiri order operated for centuries without benefit of these components. Under modern conditions some of these same components are losing strength as underpinnings of an hierarchic order: it would appear from the Nilgiri example that some modification of the classic system can be maintained without them.

The Nilgiri peoples have usually been called tribes. In some respects they did resemble tribes. Those who are properly called tribesmen in India are peoples who have been relatively isolated from the main tradition, who have generally lived in remote, inaccessible places where the writ of scripture and the edict of government did not reach, or reached in reduced form.

The Kotas were also like a tribe—as the term is often used in anthropological studies—in that they maintained a distinctive society and culture, they had their own language (and no writing), and shared a feeling of unity as against other peoples. But they did not have, as tribal societies usually have, a common, contiguous territory. They were neither as culturally or socially self-sufficient as tribes tend to be. Tribesmen sometimes classify their own people as men and dub all others as non-men: this dichotomy could scarcely be made by Kotas.

The distinction between tribe and caste is not an easy one in India; census commissioners who had to classify peoples as one or another have testified to the difficulty of the task and the frequent irrelevance of the distinction. Peoples of the same name, provenience and language may include sections which are fully part of an agricultural village, caste order; other sections will live by hunting and gathering, will exist in small bands scattered through a matted terrain.

The sociologist Max Weber, in his perceptive work on Indian social organization, listed several differences between caste and tribe. A tribe, he notes, has a fixed territory, includes all socially significant ranks, has the full range of useful occupations, is a firm political entity. A caste is characterized by none of these. Conversely, a caste must be endogamous, must have rules of diet and commensality while a tribe need not be strict about such matters (Weber 1946 [originally 1916]: 398-9).

Using more recent sources, Surajit Sinha has further distinguished castes from tribes (Sinha 1957). Kotas are, and were, decidedly more akin to what Sinha calls the Hindu peasantry of the plains than they are like the tribesmen of the central hill belt. Kota traits from the list of tribal characteristics are mainly that they had relatively little to do with markets before British times, that their social horizon was limited to the Nilgiri area, that they did not have the temples and deities of Hinduism. Yet even in these matters many low caste villagers of the plains were not entirely different, a century ago, from the Kotas. Both Kotas and low castes followed customs not approved of in the Brahmanical culture.

In all, the nineteenth century Kotas were a tribe in that they were not only non-literate but were not much influenced directly by literate peoples and cultures (as low caste villagers were). Their social system was not geared into the main system and was less elaborate. But withal, they shared a great many traits of general Indic culture, more with groups in the lower strata of the main system than with those of the upper, Sanskritized levels. In the list of underprivileged communities eligible for help from the Welfare Department of Tamil Nadu State, Kotas, Kuṟumbas and Todas appear under 'Scheduled Tribes'. But note that both tribes and low castes come within the jurisdiction and uplifting programme of the Welfare Department. Kota life has as yet been little influenced by the activities of welfare officers. But it has been and is being vastly affected by forces which play on low and high caste alike, in the Nilgiris as on the plains.

If the Kotas have not been a caste before, they are becoming more and more like the lower castes of the plains below their homeland. The distress which some Kotas feel about their lowly rank and demeaning customs is similarly felt in a good many low caste communities. The bitterness which Kotas have towards Badagas and the menace which Badagas so often see in Kotas can be found in many another milieu in India (and far elsewhere) where peoples, caught up in the press of world forces, can only know to blame their neighbours for their discomfort.

Caste is not disappearing from India, but its patterns and scope are shifting. Whatever caste is becoming, the Kotas are becoming that. Whatever major tides rise in the affairs of the nation, they will carry the tiny coterie of Kotas as well. Whatever fate affects the great centres of civilization and government, it will come to touch these villages also.

The Kotas are now part of the modern world and part of world history. They and the millions like them help shape that history. There are great multitudes of men who are like Kotas in being close to land and village, in being newly invited to take part in national decisions, in trying to fit their old order and their former beliefs to the life they now must live, in which the old does not fit as neatly and happily as once it did.

As ethnologists we try to know the Kotas from near, as persons and personalities, and from afar, as a group working out a particular phrasing of the common human destiny, as people caught up in world history and as repre senting a force acting on world history. By such observation we may get to see fresh facts about mankind and to test theories about human conduct.

•

NOTES AND REFERENCES

1. This paper appeared in *Introduction to the Civilization of India* published in 1956 by The College, The University of Chicago. It is reproduced here as originally published with a few minor corrections and deletions. The trends described as of 1956 have continued to hold true, as I found during a brief visit with Kotas in 1979.

Toda Society Between Tradition and Modernity

ANTHONY R. WALKER

7

We came upon two...huts at the foot of a mountain. They were like a large barrel half buried in the ground....The hoops of the barrel were of thick reeds...with both ends fixed in the ground....The front was made of stakes set on end, like organ pipes....The door was...[so small that we] could scarcely enter, and inside we had to kneel....Beside these houses was a pen for the buffaloes, and close by another little house where they make the butter....[The Todas] are clothed in a large sheet with no other covering but a small loincloth....They wear long beards, and rather long hair....The women wear nothing but a long sheet like the men....Their hair hangs loose, but their faces are uncovered.

—J. Finicio (1603)

Introduction[1]

The words cited above (Rivers 1906: 726-7)[2] were penned, in Portuguese, by an Italian Jesuit priest nearly four centuries ago. The photographs which accompany this chapter (Pls. 18-19) were taken by me in the 1970s. The comparison is instructive: in some respects Toda culture is remarkably resilient.

Jacome Finicio journeyed to the Nilgiri toplands in 1603, following rumours that there existed in these mountains an ancient community of backsliding Christians. In the cleric's view the journey was a failure because he did not find among the Nilgiri peoples any of his own religion. 'Thanks be to God,' he wrote, 'I am returned from Todamalâ [the Toda mountains], though with great labour and little satisfaction, for I did not find there what we hoped and were led to expect' (Rivers 1906: 721). To subsequent students of Toda society and culture, however, his expedition was extremely valuable, resulting as it did in our first authoritative description of this now well-known Nilgiri people. Nothing more is heard of them until the outpouring of reports from British civil servants and tourists in the nineteenth century, from 1812 onwards.

Perhaps largely because of their distinctive hairstyles and toga-like shawls, the Todas have captured the imagination of generations of foreign visitors to the Nilgiris. 'Very prepossessing', 'remarkably striking', 'tall, well-proportioned and athletic', of 'bold independent carriage', 'a noble race of men', 'remarkably muscular...possessing herculean strength' were some of the amazed comments recorded by early British chroniclers (cf. Hough 1829: 63, 65; Harkness 1832: 6; Ouchterlony 1848: 51). Indeed, the striking appearance of the Todas convinced many observers of their racial distinction from other Indians, setting minds awork on wildly speculative theories of Toda origins. 'I cannot but think that they may be found to be the remains of an ancient Roman colony', wrote the Reverend James Hough of Madras (1829: 63). They have 'so decidedly Jewish an appearance, that no beholder can fail to be impressed with the idea that they must, in some way, however remote, be connected with one of the lost and wandering tribes of ancient Israelites', opined the surveyor Captain John Ouchterlony (1848: 51). Such speculations have continued into modern times, with the anthropologist Prince Peter of Greece and Denmark espousing a possible Sumerian connection (cf. Peter 1951) and a former professor of bacteriology and hygiene, J.T. Cornelius (1963), equating the Todas with the Miro, a people who occupied the Danube basin in neolithic times. Many other notions, equally improbable, have been put forward to explain their origin.

Today, with many a young (and not so young) Toda dressed no differently from the Tamilian majority, the truth of the matter is more evident to the casual observer than in bygone years. Far from being racially distinct, the Todas share a wide range of physical characteristics with neighbouring South Indian peoples. But the question of their origin continues to fascinate, as it will probably do for years to come. Given the long isolation of the Nilgiris from centres of South Indian learning, and the former lack of writing among all of the Nilgiri peoples, it is unlikely that a definitive answer will ever be found. This does not bother the Todas, however, for those of the older generation, at least, are certain that their ancestors were created on these mountains by their great goddess Tö·kiṣy.[3]

Perhaps the most promising clue to the Toda past is found in linguistic studies. Murray Emeneau, the foremost authority on the Toda language, has demonstrated it to be an independent language of the Dravidian family, separated from a common Tamil-Malayalam-Toda background before the beginnings of Tamil records, that is before the beginning of the Christian era (Emeneau 1958: 47-50). Thus we can be fairly sure that the Todas are South Indian in origin, though whether they came from the west or the east of the Nilgiris is open to debate. Probably by the twelfth century there were Todas living in or near the Nilgiri mountains, for an extant stone inscription of A.D. 1117 relates how one Puṇisa, 'Minister of War and Peace' under the Hoysala king Visnuvardhana (died 1141), 'frightened the Toda...and entering into the Nîla mountain offered up its peak to the Lakshmî, of victory....(Rice 1898; inscription no. 83 of Chamrajnagar taluk). After this momentary appearance on the stage of history, Todas fade again into obscurity, to reappear briefly in Finicio's manuscript five centuries later.

A reading of Finicio's account today reveals the remarkable continuity in Toda culture. But changes have occurred as well, particularly since the arrival in the Nilgiris of British officials, settlers and planters from the second decade of the nineteenth century onwards. Half a century ago Todas sang (Emeneau 1971b: 516),

They are saying: 'Change has increased.'
The horn that was raised high is becoming low.
The horn that was low is rising high.

A whole new world had encroached upon them, and it was characteristic that they expressed their sense of the upheaval in the idiom of the buffalo herdsman, their chief occupation through the centuries.

It is true that *some* Toda men and women, *some* Toda houses and *some* Toda institutions of the 1980s look remarkably like those described by Father Finicio in 1603. But one cannot ignore another sight: the Toda man dressed in trousers, shirt, jacket and raincoat going off in a taxi to supervise the planting of his potato crop, while his wife stays in town to choose a new sari and blouse-piece for herself, a school uniform for her son and a fancy frock for her daughter. The Toda farm manager is increasingly representative of this formerly pastoral people.

In this chapter my major endeavour will be to portray Toda society as it is today, poised—still a little uneasily—between old and new, conservation and change. I hope thereby to rectify the rather static picture of the Todas drawn in many modern anthropological writings.[4] Another purpose is to suggest that Toda society, not only now but through previous centuries, is best viewed within the essentially Hindu world of South India. Too many writers, both in the past and today, have over-emphasized the isolated, unique and so-called 'tribal' nature of Toda social institutions.

Settlements and Economic Base: Some Tradition, Much Change

Always a small community (in 1603 Finicio estimated their population at no more than a thousand [Rivers 1906: 729]), Toda today number about 1000 traditionalists. There are also a couple of hundred Toda Christians, mostly inter-married with co-religionists of other South Indian backgrounds; this community, which grew out of efforts by Anglican missionaries from 1890 onwards (cf. Ling 1910, 1934), is scarcely considered as Toda by the traditionalists. Several factors may account for the small Toda population. In the past, female infanticide kept their numbers low; although it was officially banned in 1819 (cf. Hough 1829: 70), the practice continued long after that. Then, with the influx of strangers to the hills in the nineteenth century Todas fell prey to diseases from which their physical isolation probably had preserved them hitherto. Influenza and relapsing fever epidemics claimed many lives, and venereal infections seriously impeded conception (cf. Pandit 1927: 20). In the early 1950s the Toda community, not counting Christians, reached what was doubtless an all-time low of 488 persons (Peter 1963: 243). Fears were then expressed that the Todas might soon be extinct, but a great improvement in medical facilities since 1952, coupled with a major campaign against venereal disease, brought the hoped-for result: a slow but steady increase in the birth-rate. There is now no fear for the physical survival of the Todas; the question is whether they will maintain their ethnic distinctiveness.

As in Finicio's time, Todas today live in small settlements known as *moḏ* in their language but more popularly called 'munds', an anglicization of the Badaga term. These hamlets, widely scattered across the Nilgiri grasslands, are concentrated on the northwestern side of the high mass—particularly in that area of gently rolling pastures which the British called the 'Wenlock Downs'. For purposes of land registration there are 122 Toda *moḏ* recognized by the government, but many of these are either abandoned or occupied only for short periods each year. (Todas may continue to use land registered in their name around uninhabited settlements.) In 1976 there were 56 hamlets under permanent occupation, with an average of seventeen persons per settlement; four other *moḏ* were occupied during the dry-season migration of buffaloes to greener pastures.

A traditional Toda settlement, such as Finicio described, comprises one to five barrel-vaulted huts, a buffalo pen and calf shed (sometimes more than one of each), and occasionally a separate calf pen. Most hamlets have at least one dairy building, which looks just like the traditional dwellings; a few hamlets have up to three dairies. Surrounded by ample grazing land for the buffaloes (much of it today may be cultivated instead), the settlement must also have running water and a shola (Nilgiri copse) nearby, the latter to supply firewood and, in the past, building materials. But for more than a century Toda hamlets have been undergoing change, and many traditional huts have been replaced by front-gabled wooden houses in the old Badaga style, side-gabled brick or stone buildings with tiled roof and concrete floor, and other innovations. Most recently, to encourage modern house-building among the Todas, the state government of Tamil Nadu has constructed, in two showcase hamlets, 'model' houses in pseudo-traditional style, half-barrel-shaped but made of concrete blocks on a poured concrete floor. Only the dairy huts remain virtually unchanged; although those that have been rebuilt recently have walls of stones instead of wooden planks, the shape of the buildings and their thatched roof look just as they did centuries ago. Almost all regularly-occupied settlements now have electricity, meaning that there is one outside pole light and the possibility of interior lighting when people feel they can afford it.

Toda economic life for centuries has revolved around their handsome herds of long-horned, short-legged and rather ferocious buffaloes. Being vegetarians, Todas keep these animals for their milk products, not for meat. Finicio in 1603 observed, 'They have no crops of any kind, and no occupation but the breeding of buffaloes, on whose milk and butter they live' (Rivers 1906:727). Roughly speaking there are two categories of Toda buffaloes: 'temple' or 'sacred' animals and the ordinary domestic beasts which are the mainstay of the Toda pastoral economy. The former are buffaloes whose pedigree in the female line sets them apart as possessing special ritual importance. They may be tended only by a 'dairyman-priest', who has undergone special purificatory rites of ordination, and their milk is processed only in the dairy huts. The dairies are the 'temples' of the Todas; each one is located in a special place away from the dwellings, in order to preserve the building and its contents from pollution by women or domestic implements. The milk of temple buffaloes is churned, according to a ritually prescribed routine, to make butter; buttermilk from most dairies may be distributed for domestic consumption, and clarified butter, ghee, may even be sold to outsiders. Domestic buffaloes, on the other hand, which today comprise two-thirds of the Toda herds, may be milked by any layman (women may not touch a buffalo) and the milk is sold raw or else is processed, without ceremony, in the dwelling huts.

In the past, the easy rhythms of the pastoral life characterized the daily routine of the Todas. Morning and evening the buffaloes had to be milked, but for most of the day they ranged freely and often untended over their grazing grounds, being remarkably hardy animals. Once the milk was churned and the butter clarified to make ghee, the Toda man was at leisure. His requirements other than buttermilk and ghee were supplied by neighbours of other castes in a complex trading-cum-ritual relationship. Toda women, prohibited from contact with the buffaloes or their milk, cooked the meals and cleaned the dwellings; the pestle, sieve and broom have long been the ritual symbols of womanhood. These tasks finished, they sat under a tree and embroidered new shawls (*pu·txuḷy*) or dressed each other's hair in long ringlets. Some, but not all, of this routine has changed. Nowadays the grazing buffaloes must be watched, if only by a small boy, lest they stray into cultivated fields or afforestation projects; and grain, pots and implements do not come from neighbours in exchange for dairy products but must be purchased, requiring cash. Todas now keep fewer buffaloes and generally sell the milk raw rather than process it. Toda women embroider tablecloths, placemats and bedcovers for sale to visitors and Toda men are exploring occupations other than herding, although they have long resisted the more arduous and grubby tasks of farming for a living. Vestiges of the more carefree pastoral life are still apparent: much visiting between hamlets, large attendance at the almost unending round of ceremonies, and frequent trips to the Nilgiri markets not only to buy and sell but often just to meet friends and acquaintances.

Even today buffaloes are still very important to the Todas, both for economic and (as we shall see later) ritual reasons. Todas derive income from their herds by selling raw milk, ghee, dung and, occasionally, a buffalo or calf. Milk is sold either to the government-sponsored Nilgiri Co-operative Milk

Society or to the privately-run Neela Malai Milk Society, or directly to tea and coffee shops. Ghee is sold mostly to Chettiar ghee merchants in the Nilgiri markets or to any private party who puts in an order, but only when there is a large quantity of milk does the owner trouble to process it. Dung, formerly put to no use, now is sold as manure to Nilgiri cultivators; for the few Todas who own very large herds (80 head or more) it is a potent money-maker, up to Rs. 12,000 to 16,000 a year. Male buffalo calves and occasionally a full-grown animal are sold to butchers, mostly Muslim; Badagas sometimes buy Toda female buffaloes to augment their herds.

But in order to make an adequate income solely from buffaloes, an average Toda family of husband, wife and three to four children needs a herd of some twelve to fifteen head (four to five in milk, four to five dry and four to five calves). In 1975 more than half of all the Toda families had too few buffaloes to subsist on herding alone; indeed a full 25 per cent of all families owned no buffaloes at all. To be wealthy while relying entirely on herding, a family needs to own upwards of fifty buffaloes. Only about 6 per cent of all Toda families had such large herds in 1975.

Several factors may account for the decline in pastoralism, but probably the most significant is shrinking pasture-lands. Afforestation of grassy slopes and flooding of valleys for hydro-electric projects by the government have consumed huge tracts, and the Toda practice of leasing land to cultivators has also diminished their former pasturage. To many Todas the large areas of grassland required for range-feeding buffaloes are no longer available, and neither the animals nor their masters are used to stall-feeding. The unpopularity of stall-feeding is also a factor in the repeated failure of attempts by interested outsiders to introduce better-milking plains buffaloes to Toda herdsmen. Toda buffaloes are finely adapted to these hills, able to withstand extremes of weather without shelter and to flourish on the coarse Nilgiri grasses, but other breeds require more care and have not fared well under the Toda *laissez-faire* style of herding.

Given the dim prospects for pastoralism, those in political authority attempted many times in the past to make agriculturalists of the Toda herdsmen. Time and again such attempts came to nothing. Traditional Toda society was so totally oriented — economically, socially and ritually too—towards the buffalo that it is hardly surprising that the people would resist such a thoroughgoing change. Their usual response to official inducements in the form of seed, tools and fertilizers was to sell these to others more enthusiastic than themselves about farming. Alternatively they hired somebody to do the work for them or, more commonly still, simply leased their land to a cultivating family of another caste. Cash subsidies were gratefully received and promptly used for other purposes. This impasse might have continued indefinitely had not the pastoral base of the Toda economy, for the reasons cited above, begun to show signs of serious trouble from the late 1960s onwards.

The turning point was 1975 when, under the fifth Five Year Plan, the central government allocated Rs 2,025,000 (approximately US $227,000) for the social and economic development of the Toda community as a part of its Hill Area Development Programme. With this financial backing a Toda Welfare Scheme was organized under the auspices of the Indo-German Nilgiris Development Project (Raman 1978). The object of the scheme was to educate the Todas in the practice of scientific agriculture so that the community would never again be economically dependent on buffalo herding and leasing of land. The idea was by no means new, but for the first time sufficient funds and expertise had been brought to bear, and the effect of these was enhanced by the absence of any really viable alternative. More and more Todas were at least managing if not actually cultivating their own potato plots, and in all probability it will never again be appropriate to talk or write, except in the past tense, of the 'purely pastoral Todas'.

The Organization of Toda Society: Tradition Preserved

If there has been much change in the material culture of the Todas, and a good deal in the economic sphere as well, in matters of social organization there has been little challenge to orthodoxy except that represented by the breakaway or outcaste (depending on one's perspective) Christian Toda community. In much of the literature on the Todas their social institutions have been regarded as quite specific to the community and thus portrayed as essentially 'tribal' in character. But I shall argue here that Toda social organization is better seen as a variant of the more general Hindu model of social relations, and Toda traditions indicate that this variant is an ancient one, long predating the modern influences which have so greatly affected other spheres of Toda life.

The person who, to an outsider, is simply a Toda, is among his own people a member of several different social groups and categories. First of all he belongs to one of the two endogamous subcastes into which the community is divided: Tö·r̠θas̠ (probably with the original meaning of 'the important ones', cf. Emeneau 1966: 26-8) and Töwfiḷy (probably 'servants of the gods', cf. Emeneau 1974b: 6). The Toda language has no general term for these two divisions and in the ethnographic record they are usually called 'moieties'. I prefer to call them subcastes because there is so much in this bifurcation of Toda society that resembles the internal organization of caste groups all over India (which are also divided into endogamous subunits, but not necessarily only two). Besides the restriction on formal intermarriage, the relations between the Toda subcastes are characterized by other, typically Hindu, features of social organization: ritual specialization, hierarchy determined on the basis of relative purity, a prescribed degree of separation and a prescribed degree of co-operation.

The most striking contrast between the Toda subcastes is their differing relationship to the sacred dairies. Tö·r̠θas̠ people alone own the highest grades of dairy temple and their associated buffalo herds; but only Töwfiḷy men may fulfill the highest priestly tasks connected with them. The owners of the most sacred dairies are generally considered ritually the higher of the two subcastes.

Each subcaste is again divided into a number of named exogamous patrilineal clans; such division is a common feature of Hindu social organization. Presently Tö·ṟθas̠ has ten such clans, Töwfiḷy only five. Each patriclan owns a number of hamlets, and usually a clan takes its name (cf. Emeneau 1974b: 48-82) from its chief hamlet; the clan's name may change if the chief hamlet is abandoned in favour of another. Patriclans also own at least two funeral places (one for males and the other for females) and possibly an isolated dairy which is of such sanctity that it may not be near an ordinary residential area. Dwelling huts are owned by individual families but the hamlet site, its associated dairies and more sacred grades of buffalo are the joint property of the clansmen. So too was the surrounding grazing ground (insofar as it was 'owned' by anyone) until the land registration exercises of British times, when it was listed under the names of individual clansmen. Today, when the expansion of agriculture has given land titles an importance never imagined by the herdsmen of the past, almost all Toda lands are individually registered. But nobody outside the patrilineal clan can own land traditionally associated with that particular clan.

Divisions of the patriclan are four: *kwï·r*, *po·lm*, hamlet and household. The *kwï·r* (literally 'horn'; the name indicates the binary nature of the segmentation) divisions become operative on ritual occasions when it is necessary to expiate offences or counteract misfortune. At such times the *kwï·r* of the offender or sufferer has certain duties to perform, principally the presentation of an expiatory offering, while the opposite *kwï·r* has complementary functions, notably to receive the offering (cf. Rivers 1906: 294-312). The second type of segmentation, the *po·lm* (meaning 'portion') is not a binary division; some clans have only two *po·lm* but others have three, four or five. These segments become operative when clan expenses arise, each *po·lm* having to share equally in the financial outlay, regardless of its numbers.

The third level of patriclan segmentation is the hamlet. As mentioned above, each patriclan owns a number of hamlet sites, some occupied, others long abandoned. Clansmen have the right to live in any hamlet belonging to their clan, while non-clansmen may live there only by invitation, which is rare. Thus we may term the hamlet a residential unit of the clan. The hamlet is also an economic unit of sorts, the care of the buffaloes being for the most part hamlet-oriented. Only the highest grades of buffalo are the common property of the owning patriclan; less sacred beasts are owned by individual families. But all buffaloes in a hamlet are penned and pastured together and, although each family milks its own domestic buffaloes, all the sacred animals are milked, if at all, by the dairyman-priest of the hamlet.

It may be noted here that Toda society functions without formal headmen at any level, except perhaps the household, as we shall see. Yet the community possesses well-defined procedures for ensuring that its members observe caste norms, as well as for settling disputes between individuals or factions and for deciding upon united caste action. Whenever it becomes necessary for the community as a whole to take action, the adult males of the caste constitute themselves into a caste council which Todas call a *no·ym*. In the discussions of the *no·ym* the unofficial but clearly recognized leaders of the community have a decisive role to play. These men, always in late middle to old age, sometimes wealthy and (increasingly) with some education, listen quietly at first while others shout, and then slowly begin to take control of the assembly, directing it to an eventual consensus embracing compromise and, in dispute, reconciliation. Similar assemblies, also called *no·ym*, may be held within the subcaste, within the clan and within the hamlet, depending upon the nature of the affair to be discussed.

At the lowest level of patriclan segmentation we come to the individual household, those people occupying a single dwelling. Today a Toda household generally comprises a nuclear family of husband, wife and unmarried children. It may also include a widowed parent of the husband, and sometimes the young family of a married son who has not yet built his own house. In the past, the Toda household usually comprised a polyandrous family: two or more brothers married to the same woman, together with their offspring, the brothers taking turns to accept paternity. Although there were still one or two polyandrous households in the early 1960s, today this form of marriage and residential arrangement has disappeared. Monogamy is now both the ideal and the norm, although a few remnants of polyandrous practices may still be found, particularly among the older generation. Thus it is quite in order for a man to enjoy sexual access to his brother's wife and when a man dies, a brother may well take over his role of husband. As in the past, there are a few wealthy Toda men who maintain polygynous households but these are rare. Paternity of children today, as in the past, is determined ritually rather than biologically; the man who presents a symbolic bow and arrow to a woman in her seventh month of pregnancy thereby becomes the father of her child.

Except in the case of a widow with small children, the household head is always a man: the husband and father of the nuclear family, or the grown son of a widow. He is the owner of the household property: domestic equipment, family heirlooms, buffaloes and, in recent years, a defined portion of the hamlet lands, all of which will be divided among his sons when he dies. Buffaloes may be apportioned when he retires from active herding. Daughters inherit nothing except a small dowry, usually jewellery, but this belongs to a girl alone and her husband has no claim on it. Within the home the household head's voice is supreme, arbitrating disputes and delegating duties. In the decision and authority structure of Toda society he is the elemental component, responsible to his patriclan and ultimately to the community at large for the good behaviour of all who dwell under his roof. Women and minors are never summoned before a *no·ym*, and so if anyone breaches norms of conduct it is the head of that household who must appear.

In addition to the affiliation with a patriclan and thereby with the various segments of a patriclan described here, a Toda man or woman is also a member of an exogamous matrilineal clan. Tö·r̲θas̲ subcaste has five such matriclans, Töwfiḷy six. But these are descent categories rather than groups, since they have no corporate unity; there is no occasion when matriclansmen meet for joint action. The matriclans are not without their importance, however, for they limit sexual and marital partners and determine ritual obligations in certain cases where matrilineal links between persons override patrilineal affiliations. It has been said in the past (cf. Emeneau 1941: 168), and I was told too, that Todas view sexual relationships within the matriclan with even greater abhorrence than those within the patriclan. Nonetheless, since the 1950s there have been two breaches of matriclan exogamy. Both unions met with fierce resistance within the community but the marriages were eventually allowed, on payment of a fine; undoubtedly they were permitted only because of the considerable influence of the families concerned. Among members of a patriclan no breach of the rules of incest and marriage has ever been allowed. Christian converts, however, have not observed the rules of exogamy because of the shortage of potential spouses in their community.

If the major divisions of Toda society—the two subcastes and the patrilineal and matrilineal clans—reflect, as I suggest, a typically Hindu background, then the Toda kinship system further emphasizes this people's affiliations with the rest of South India. In reckoning who are their relatives Todas follow a classificatory system common to the majority of Dravidian-speaking peoples (cf. Karve 1953: 196-202). It is not possible here to go into all the technicalities of South Indian kinship, but perhaps its most important characteristic, and one that is quite alien to the average English-speaker, is the distinction made between parents' siblings of the same sex and those of the opposite sex. Those of the same sex—that is, mother's sisters and father's brothers—are regarded as parents. Parents' siblings of the opposite sex—mother's brothers and father's sisters—are in a completely different category and are called by special terms which we may translate as 'uncle' and 'aunt'. Given this distinction, it follows that the children of parents' same-sex siblings (the offspring of father's brothers and mother's sisters) are not first cousins, as in English usage, but siblings, for they are the children of classificatory 'parents'. With these relatives, marriage or sexual contact is absolutely prohibited. The children of parents' opposite-sex siblings (the offsprings of father's sisters and mother's brothers) are a very different sort of relative. So far from being siblings, hence sexually forbidden, they are the preferred marriage mates. And their parents, the 'uncles' and 'aunts', are thus not only kin but also the potential fathers-in-law and mothers-in-law, affines *par excellence*.

Relations with Neighbours: Tradition Abandoned

The relations of Toda with their neighbours in the Nilgiris, through centuries when they were largely isolated by topography from other South Indians, form an important component of any inquiry into the nature of Toda society. These hill neighbours included an artisan caste of potters, blacksmiths and leather-workers called the Kotas; an immigrant group of castes with the common name of Badaga, who became the dominant food-producers, hence political overlords, of the Nilgiris; and two jungle-dwelling communities, Kuṟumbas and Irulas, both of whom were upward extensions of peoples whose main population was (and is) widely scattered over the South Indian lowlands. But these Nilgiri peoples, despite much writing to the contrary, were hardly a collection of 'hill tribes' who just happened to interact with one another. Rather, they maintained a system of interdependence in a variety of ways—economic, ritual and social—very much within the tradition of multi-caste rural communities throughout Hindu India. At least, this was the situation that obtained after the advent of the Hindu Badagas, as Hockings mentions in the following chapter. We have no information from earlier times.

From the Toda viewpoint the *sine qua non* of traditional Nilgiri inter-relationships was economic dependence. With no occupation but the herding of buffaloes, the Todas had to rely on other peoples for several necessities of life. As elsewhere in India the lines of supply ran between families rather than between castes as a whole. From hereditary Badaga friends a Toda family obtained subsistence grain and a variety of other items: cloth, salt, crude sugar, etc., which the Badagas themselves procured from the plains. From Kota partners Toda families received pots, axes, knives, iron jewellery and, occasionally, some grain too. (Unlike Todas whose energies were focused entirely on their herds, Kotas kept buffaloes and farmed in addition to operating as chief artisans of the hills.) From Kuṟumbas, and rarely Irulas also, Todas got forest products: bamboo, honey and the like. The Toda contribution to this network was mostly dairy products (cf. Hockings 1980a: 111-22).

But relations were far from being simply economic. They involved the provision of ritual services (for example, Kotas were the principal musicians at Toda funerals), reciprocal attendance at each other's major ceremonies, and the presentation of gifts on appropriate occasions such as marriages and deaths. And all these interactions involved notions of hierarchy based on ritual criteria. Todas saw themselves in ritual terms at the top of the local Nilgiri hierarchy, although they were prepared to accede to the political dominance of the numerically superior Badagas. Others—except perhaps the Badagas—mostly accepted the premier ritual status of the Todas. The Todas and the Badagas agreed that the Kotas were at the bottom of the hierarchy, but the Kotas probably regarded the forest-dwelling peoples as the lowest (cf. Mandelbaum 1956: 324).

In the early nineteenth century the relative isolation of the Toda homeland was shattered with the coming of the British and their entourage: servants, labourers, minor government officials and others in search of a living. As early as 1826 it was reported that the immigrants 'now form a large village [at Ootacamund], where a bazaar is established' (Hough 1829: 44). It was not long before there were bazaars at several places in the toplands, bringing with them the first intimations of a cash economy. Meanwhile, British commercial interests had discovered how suitable the Nilgiri soils were to a whole range of cool-weather crops and trees, and large-scale plantation agriculture (beginning with tea and coffee) was gaining a foothold.

The establishment of market centres, the visible economic advantages of cash crops, and the opportunity to earn wages disrupted the economic interdependence of the Nilgiri peoples. The Badagas were the first to take advantage of the new economic milieu, turning quickly from subsistence to cash-cropping and the professions; the Todas were the last. Today the old system of interrelations between the Todas and their traditional Nilgiri neighbours is mostly history. Todas still have their hereditary Badaga and Kota friends (very few Toda families were ever linked with Kuṟumba and Irula ones), and invitations are still passed back and forth between communities to attend each other's important calendrical and life-cycle ceremonies. But economic interdependence has disappeared. The modern Toda is much more familiar with the workings of the market economy of the Nilgiris than he is with the former system of inter-community relationships.

The Cult of the Sacred Dairies: Tradition Truncated

The Todas have traditionally populated their supernatural world with a number of deities generically termed *töwθït* 'gods of the mountains', because most of them are thought to have their residence on particular Nilgiri peaks. Of these anthropomorphic mountain deities the most important by far is a goddess called Tö·kiṣy: creator of the Toda and their herds and ordainer of their more important social and ritual institutions. In addition to the *töwθït* there are less anthropomorphic *töwno·r* or 'gods of the sacred places', representing the divine essences of the dairy-complexes (buildings, associated paraphernalia, herds and pastures). Not all Toda dairies fall into the *töwno·r* or 'divine' category; those that do include the most sacred dairies isolated from the domestic settlements and those located within the *ïtwïdmoḏ* or 'principal hamlets' of each patriclan. These diffused forces are also sometimes conceived in anthropomorphic terms, so that it may be said that a *töwno·r* has become angry with someone or that the *töwno·r* attend the council of the gods (cf. Emeneau 1938: 114; 1971b: xli).

Although all Toda traditionalists seem to regard the goddess Tö·kiṣy with reverence, in terms of ritual neither she nor any other 'gods of the mountains' are the object of their most devoted concern and care. Rather, it is the sacred dairies that are all-important.

Emeneau's succinct summary of the dairy cult, published fifty years ago, is still well worth quoting. He wrote (1938: 111-12):

> The religion of the Todas is a highly ritualized buffalo-cult. Every important operation connected with the buffaloes is conducted according to rule, milking and converting the milk successively into butter and ghee, giving salt to the buffaloes, taking them on migration to fresh pastures, burning over the pastures, giving a buffalo a name when it has calved for the first time, introducing new utensils into the dairy and preparing new coagulant for the milk, rebuilding and rethatching the dairy, consecrating dairymen, and even drinking buttermilk from the dairy.... Infractions of the rules involve pollution, and most of the precautions surrounding the cult seem designed to prevent pollution of the milk by contact with profane persons or utensils. The milk, as the primary product, is most liable to pollution and the successive operations finally result in ghee, which possesses so little sanctity that it can be sold to outsiders.

Many pages could easily be written on each of the elements mentioned in Emeneau's synopsis, which bears careful re-reading. Rivers (1906: 38-181, 231-48) devoted seven chapters of *The Todas* to the sacred dairy cult, and the reader wishing for a full treatment can do no better than take up that work.

Among the Todas, as Emeneau hints, the rather straightforward tasks of the dairyman have been surrounded by a system of ritual incredible in its complexity and detail. This is so because the buffaloes, dairies and dairymen are by no means of uniform ritual status. Both the dairies and the buffaloes are graded into a complex hierarchy according to relative sanctity, and different rules pertain to each grade. The higher dairies and buffaloes are in the hierarchy, the more elaborate is the ritual surrounding the daily tasks of the associated dairyman, and the greater must be the purity in which the dairy, its appurtenances and the dairyman himself are maintained. This in turn requires ever more stringent precautions against defilement and, correspondingly, ever more rigid rules of conduct for the dairyman's daily life. It also requires increasingly elaborate rites of ordination in order to raise a man of ordinary ritual status to the level of ritual purity necessary for him to take charge of the sacred dairy work.

I have already noted a rough distinction between Toda 'non-sacred' and 'sacred', or domestic and temple, buffaloes; the Toda terms are *pïtyïr* and *postïr* respectively (*ïr* meaning female buffalo). For a number of reasons, however, this dichotomy grossly oversimplifies the actual situation. First, all Toda buffalo cows (but not the bulls) are to some degree sacred (witness the ban on women having anything to do with their care). We should say then that the domestic *pïtyïr* are the lowest grade in the hierarchy, rather than a different kind of buffalo. Secondly, although Toda do sometimes refer in general to the higher grades of buffalo (*i.e.* all those above *pïtyïr*) as *postïr*, the latter term properly belongs only to those animals associated with dairies belonging to members of Töwfiḷy subcastes. Most Tö·ṟθaṣ patriclans own several grades of buffalo above the domestic level, and each grade has its own name (*e.g. peṣosïr. nošpepïr, moṟtïr, wiṣoḷyïr, kogfoḷyïr*). Thirdly, at all grades of Toda dairy, barring only the very highest, lower (but never higher) grade animals—including even the domestic grade—may be milked by the officiating dairyman and their milk, mixed with that of the higher-grade animals, churned in the dairy building. It seems clear enough then that sanctity pertains, in the first instance, to the dairies rather than to the buffaloes. A dairy confers its sanctity on the grade of buffaloes particularly associated with it; hence these animals may not be tended at lower-grade dairies. But because all female buffaloes are essentially sacred and pure, a higher-grade dairy is not defiled when a lower-grade buffalo is tended at it.

To give some idea of the nature of Toda dairy rituals I shall take as an example just one grade of dairy and outline the ordination rites for the priest who is to operate it. For the sake of simplicity this example is one of the lower grades, namely the *poḷy* belonging to clans of Töwfiḷy subcaste. The dairy building has two rooms; the dairyman sleeps and eats in the outer room, and the dairy equipment—two separate categories of it—is kept in the inner room where the all-important churning is performed. No one except a properly-ordained dairyman may enter this inner room. The man who operates this grade of dairy is called *poḷyxarp oḷ* 'the man (*oḷ*) who milks at (*xarp*) the *poḷy*'. An ordinary layman acquires the ritual purity required for this post through an ordination ceremony that begins early in the morning at the dairy where he is to serve. Here the candidate washes his hands and changes to a dairyman's black loincloth. He then goes to the sacred dairy stream where he collects seven leaves of a certain plant and a handful of young shoots. After pulping the shoots on a stone, he dips them into the stream and squeezes the juice three times onto one of the leaves, from which he drinks after raising it to his forehead. He tosses the leaf back over his head and repeats the process with each of the other leaves. Then he dips the pulped shoots again in water, rubs them over himself three times, and puts them into his hair. This done, he returns to the dairy where he must ritually sweep the threshold and bow to a certain dairy pot. Then he can enter the dairy, where he salutes the two categories of dairy equipment in turn. Finally he picks up the sacred milking vessel and goes out for his first duty, milking the *post ïr* given into his care.

At the top of the Toda dairy hierarchy were the *ti·* complexes (the word derived ultimately from the Sanskrit *śrī* 'holiness, sacredness' [Emeneau 1953: 106-7; 1971b: xliii, n. 21]). Now all abandoned, the *ti·* were once isolated settlements inhabited only by the dairyman-priest and his assistant. At this highest grade of dairy the Toda dairy cult reached its greatest complexity, resulting in an almost forbidding elaboration of ritual detail (cf. Rivers 1906: 83-122, 130-43, 153-65). But it is important to note that these *ti·* constituted more than just another, higher, grade of Toda dairy. In fact they were microcosms, maintained at the highest level of ritual purity, of the wider Toda dairy cult in both its sacred and domestic aspects. Each of the *ti·* complexes comprised a number of settlements, often far apart, and each settlement—like any ordinary hamlet—included the dairy, a dwelling hut for the dairyman and his assistant, calf-sheds, buffalo pens and two water sources: a sacred one for the dairyman to use and a non-sacred one for his assistant. The associated buffalo herds, which included highly sacred animals and also those of a grade similar to the domestic beasts, were driven from one *ti·* settlement to another of the same complex at prescribed times and over prescribed routes.

Today the whole *ti·* institution is practically extinct. A generation has passed since the last *ti·* complex ceased to function, and the *ti·* buffaloes of all owning patriclans except one have disappeared. The remaining *ti·* herd, having gone untended for a quarter of a century, is completely wild. There is nobody still living who has served as dairyman at a *ti·*, and there seems little likelihood of the Todas ever reactivating the *ti·* rituals. This collapse of the most sacred and the most complex of the ritual institutions of the Toda dairy cult is probably the most startling event in the ritual life of this community in well over four centuries.

All other grades of Toda dairy remain operative, although most of them probably are manned for shorter periods now than in times past. Priesthood is not, and never has been, a lifetime occupation.

The Passage Through Life: Tradition Modified

From womb to funeral pyre, the ritual high-points of a Toda's passage through life are oriented towards the sacred dairy cult. The primary concern is always to maintain the purity of the dairy institutions. Thus a life-event which is least likely to endanger the dairies' purity, such as name-giving, ear-piercing or marriage, receives in this society the minimum of ritualization. Events which do endanger purity, such as birth and death, are highly ritualized. Again, the multiplicity of ritual detail associated with these events is beyond the scope of this chapter (cf. Rivers 1906: 313-404), and what follows is only a skeletal description.

Like other communities throughout Hindu India, Toda regard childbirth as a particularly defiling event. Not only the mother and child but also people and things in contact with them suffer ritual pollution. Until about 35 years ago a Toda woman was required, both before and after her delivery, to be isolated from the hamlet for some time in a special 'pollution hut'. Clearly this removal was meant to protect the purity of the dairy or dairies, for the greater the sanctity of a dairy, the further away the hut was situated. At about the fifth month of pregnancy a woman would go to the pollution hut and remain there for a lunar month before returning home. After her delivery, almost as soon as she could physically manage the move, the woman would again go to the pollution hut, with her child, and stay there for up to one month, sometimes longer in the case of a first child. Various rites marked her removal and return on both occasions (cf. Rivers 1906: 313- 31).

Thirty-five years or so ago, the Toda caste council abolished the institution of the pollution hut because of its discomfort to both mother and child, especially during the wet, cold monsoon months. But the rituals which formerly preceded and followed the period of segregation are still performed, often amalgamated into a single ceremony which includes a symbolic departure and stay at the old site of the seclusion hut.

After birth, for at least one month and up to three, a Toda infant must be completely covered whenever it is taken out-of-doors. A ritual 'face opening', when its face is uncovered for the first time outside, takes place sometime before the end of the third month (cf. Rivers 1906: 331-3). A boy baby's first sight must be of the dairy; a girl's face is uncovered at the entrance to the hamlet. Later in the morning the child's name is announced—ceremonially for a boy, quite informally for a girl.

All Todas are expected to have their ears pierced. Traditionally this event, like name-giving, was ritualized for boys but not for girls, and the piercing ceremony usually was performed for several youths of the same patriclan at the same time. Each boy had one ear pierced by his maternal uncle and the other by a member of the subcaste opposite to his own. Today the rite is performed mostly for individuals and the venue has shifted from the Toda hamlets to Hindu temples, and the Mariamman temple in the Ootacamund bazaar has become particularly popular. Within the temple premises the maternal uncle and opposite-subcaste man still do the piercing, after which the youth pays obeisance to the presiding deity and receives blessings from the officiating priest. Ear-piercing, which may be done at any age, marks the attainment of ritual adulthood; only after this rite is a youth entitled to serve as dairyman in any but the very lowest grade of dairy. Interestingly, although girls' ears used to be pierced quite informally, today they too are frequently taken to a Hindu temple, where the job is done—as never in the past—by their maternal uncle and an opposite-subcaste man.

Marriage, in Toda society the formal recognition of a relationship between a male and a female which could lead to their living together, is usually arranged in infancy. A cross-cousin, on either the maternal or paternal side, is the preferred mate. Todas are adamant that this formal link between infants who will not cohabit until they reach maturity, if at all, is marriage and not simply betrothal. Child marriages are essential to meet ritual requirements, they stress; being married is the ideal ritual status and dying unwed, even as an infant, is a calamity. The ritual which seals the union is brief: the boy's father presents the girl's father with a white loincloth for the girl, and the infant boy, held in his father's arms, is made to bow ritually to all members of the girl's patriclan. The babies are now married, and the girl's father may not give her to another husband without paying compensation in buffaloes to the first husband's father.

Physical maturity in boys has never been marked by ritual. In girls it used to be preceded by ritual defloration, performed by a man of the subcaste opposite to the girl, but this custom has long been abandoned. After maturity the husband and wife may live together; a feast in the girl's hamlet is part of the boy's ritual 'taking away of the wife'.

In establishing a regular marriage alliance the endogamy of the subcaste is strictly enforced. There exists, however, a parallel institution of sexual relations between a man and a woman who are in opposite subcastes. After receiving permission from the woman's regular husband, the man gives her an embroidered shawl to establish this formal liaison. Any offspring of such a union belong, always, to the legal husband of the woman, necessarily a man in her own subcaste.

In a society which used to practise fraternal polyandry and which still allows these formal inter-subcaste sexual alliances, the number of men who could have fathered a woman's child is sometimes considerable. But biology does not determine paternity among the Todas. The father of the child-to-be is established in the bow-and-arrow rite which, with several others, takes place during the woman's sixth or seventh month of pregnancy (cf. Rivers 1906: 31-3).

Among all Toda life-crises it is death which generates the greatest profusion of ritual; doubtless this is because death threatens the purity of the sacred dairies more than any other of life's principal events. Rivers (1906: 337-404) produced 67 pages of funerary detail, but I must be brief. The deceased is cremated at the end of a first funeral ceremony during which buffaloes are sacrificed so that they may accompany the spirit to the afterworld. A second funeral, sometimes months after the first, used to be held, in which a relic of the deceased (a lock of hair and, some say, a fragment of the skull) was cremated. In this second ceremony more buffaloes were sacrificed and the rites were concluded by the breaking of a pot at the cremation site. Today second funerals are extremely rare (the last was in 1966, the next-to-last in 1963) and the pot-breaking rite is performed at the first funeral after the body has been consigned to the flames.

During a funeral Todas manifest in ritual drama the full complexity of their social organization and of the sacred dairy cult. Every major social division—subcaste, patriclan and matriclan—and every major kinship and affinal role comes into play at a funeral. The reciprocal obligations of the subcastes on this occasion make it impossible for a funeral to take place unless members of both are present; for instance, men of the subcaste opposite to the deceased must catch the sacrificial buffaloes. Specific duties must be performed by patriclansmen of the dead person, while matriclansmen are obliged to observe certain important restrictions consequent upon their incurring death-pollution. Affines have particular duties to perform at a funeral, including the provision of a buffalo for sacrifice. The institutions of primary marriage within the subcaste and of official sexual liaisons between subcastes are emphasized by the important roles of regular spouse and recognized consort of the deceased.

Important differences between male and female funerals stem from the fact that only males are associated with the dairies; for instance, only the lowest 'domestic' grade of buffalo may be sacrificed for a woman, while at least one higher-grade animal is killed for a man. And the use of higher-grade animals at male funerals requires a corresponding elaboration of the ritual, including the presence of a dairyman-priest of appropriate rank to perform the sacrifice.

For the Todas, death is an initiation into a new life and status as an ancestor in the afterworld. As in many other societies, a change in social status necessitates separation from one's old status, a transitional period and, finally, incorporation into the new status (cf. Hertz 1907). The Todas' first funeral marks the separation of the deceased from the community of the living. The interval between the first and second funerals clearly was seen by Todas as a period when the dead person was neither of this world nor of the next, and special ritual care had to be taken by the surviving kinsmen during this time (cf. Rivers 1906: 367-71). The breaking of a pot, which used to end the second funeral and now concludes the single ceremony, is said to coincide with the arrival of the deceased at the threshold of the afterworld.

Into the Future: The Todas Without Buffaloes?

'Just like the Eskimos', declared a Nilgiri tourist guide a few years back as he sought to show me the strange and wonderful sight of the Toda hamlet above the Ootacamund botanical gardens— that much-visited hamlet which, so far from being out-of-the-way, has been within the municipal limits ever since Ootacamund was established. To tourists it may seem romantic to view modern Todas still as an isolated and exotic tribal people, as did the early British visitors; but times have changed and so—eventually—have the Todas.

Those whose knowledge of the community goes beyond a tourist brochure and a casual photographic expedition will appreciate how far behind the times (at least a century) is the following account, which appeared recently in a learned journal:

> The Todas practice no agriculture. They devote nearly all their attention to their buffaloes, supplying the other tribes dairy products in return for various goods and services. Badagas...provide...grain and other farm products. The Kotas...make and supply most of their pottery and ironware. They also provide music and objects of ceremonial importance to the Todas.

These academics (Agasthialingam and Sakthivel 1977: 119-20) either have not seen or have ignored the numbers of Todas in the Nilgiri bazaars, purchasing their family provisions with the proceeds of their potato sales.

Modern Todas, some of whom own no buffaloes at all, do retain much of their traditional pastorally-oriented social and cultural heritage, as we have seen. At the same time they are part of a complex modern society. Though dominated by the Hindu culture of South India, as mediated through the Tamil language, this modern Nilgiri society is socially, culturally and linguistically heterogeneous. It is further characterized by a cash economy and by political affiliation with the state of Tamil Nadu and the wider federal Republic of India. The Toda individual today has friends, acquaintances and business associates, not only among the indigenous hill peoples but also among the immigrant plainsfolk, and he is frequently involved with a number of government agencies: administrative, judicial, developmental and others.

The sacred Toda dairies of every ritual grade except the highest, the *ti·*, are still operated at least sporadically. And the major ceremonies associated with important life events, birth and death in particular, continue to impress upon the Toda the importance of their ancient dairy institutions. The goddess Tö·kisy is still honoured as the principal Toda deity. These elements constitute a ritual idiom that is specifically Toda, even though many of the ideas which support it—purity, hierarchy, ritual specialization, etc.—may be derived from a broader South Indian Hindu background. Co-existing with the Toda tradition there has long been recognition of the power and importance of various Hindu deities and their associated cults. This process of adopting the religious beliefs and rituals of wider Hinduism, already well advanced when Rivers investigated the Todas eighty years ago (cf. Rivers 1906: 457-8), continues apace. Lithographic icons of all-India deities such as Śiva, his consort Párvatī, his sons Ganeśa and Subramanyam, of Visnu and his consort Laksmī and his several incarnations, particularly Rāma and Krsna, are now common in Toda homes. I have even seen one Toda house with a complete 'God's room': a kind of domestic chapel for worship, in the normal Hindu style (lighting lamps, offering flowers, fruit and incense), of the important gods of all-India Hinduism. One Toda temple bears stone carvings of the five Pāndava brothers of the Sanskritic Mahābhārata epic who, because of their polyandrous union with one woman, are now claimed by some Todas as their ancestors; and pilgrimage to Hindu temples, practised in Rivers's time (1906: 457-8, 705), is becoming more and more widespread in the Toda community. In recent years several young Todas have become devotees of the Kerala-based deity Ayyappan and they perform annual pilgrimages with the multitudes to the god's shrine in Kerala's Sabari hills.

The modern Toda seems fully to accept the efficacy of two parallel ritual systems: his own and that of popular South Indian Hinduism, itself a variant of all-India Hinduism. And there is nothing which is strange, at least in India, about this situation. Hinduism seldom converts in the manner of proselytizing faiths like Christianity and Islam, which generally demand that the individual break with past beliefs and practices. It rather absorbs groups, often over long periods of time, first tolerating then reinterpreting past ideological and liturgical traditions. There is no need, then, to expect any sudden abandonment of the ritual aspects of Toda pastoralism.

And yet, as fewer and fewer Todas herd buffaloes, it is certainly possible to conceive of a time when the few remaining animals are kept solely for ritual rather than economic purposes. Beyond that, there may be a slow withering away of the dairy cult, a process which has already begun, in fact, as evidenced by the termination of the complex *ti·* dairy operations. If the dairy cult wanes there will surely be a further curtailment of other traditional rituals (as has already happened to pregnancy, birth and funeral rites) and a slow redefinition of these, almost certainly in the idiom of a more popular South Indian Hinduism. This process too has already started, although only with regard to such minor rites as those associated with ear-piercing.

If these several processes follow through to their logical conclusion, the Todas of tomorrow may well be unrecognizable as heirs to the pastoral society described by Finicio (1603), Rivers (1906), Emeneau (1930s), and still more or less intact during my field studies in the 1960s and 70s. But Todas, whose physical extinction has so fortunately been arrested, are unlikely to disappear without a trace. India absorbs and changes her constituent castes; she seldom eliminates these endogamous units which constitute the basic bricks of Hindu society.

NOTES

1. This chapter is a partial précis of my book, *The Toda of South India: A New Look*. My original fieldwork among the Todas was for one year in 1962-3; I have been back to the Nilgiris on brief visits in 1965-66, 1969-70, 1974, 1976, 1978, 1981, 1984 and 1987-8. I am indebted to the Worshipful Company of Goldsmiths of the City of London and to my mother, the late Dr Joan Hazelden Walker, for financial assistance during my initial fieldwork. Among the Todas my principal debt is to T. Muthicane (Mutxe·n) of Ka·s̱ patriclan, although a great many others of this community have assisted me over the years. I am, as always, deeply indebted to my wife, Pauline Hetland Walker, for her editorial assistance.
2. The translation in W.H.R. Rivers's book is by Miss A. de Alberti. One original manuscript, which I have consulted, is in the British Museum (B.M. Add. MS. 9853, p. 479, MS. 40 vol.). For further bibliographic data on the Finicio manuscripts, see Hockings (1978).
3. Toda words in this chapter are transcribed according to the system devised by Murray B. Emeneau; *cf.* Emeneau (1958). For the sake of stylistic consistency throughout this volume, the editor has added the English plural marker in referring to the Todas; the author's preference is to omit this.
4. This is especially true of works published in India (cf. Bhowmik 1971: 181-92; Fuchs 1973: 285-8; Agasthialingam and Sakthivel 1977; Bahadur 1978: 27 *et passim*), which simply repeat the findings of that great Cambridge University anthropologist, William H.R. Rivers (1864-1922; cf. Slobodin 1978), or still earlier ethnographers like Shortt (1868), Breeks (1873) and Thurston and Rangachari (1909).

The Badagas

PAUL HOCKINGS

8

Often called by the English Burghers.... The most numerous, wealthy, and civilized of natives of Neilgherries. Speak an old Canarese dialect. Originally from North Coimbatore and South Mysore.... They are almost all Sheiveites.... They are an active race, of moderate stature, with the usual Dravidian features, a prepossessing expression, and light complexion. They are accustomed to labour from their earliest youth....

C. Macleane (ed., 1893: 64)

The Setting

Scattered over the eastern half of the Nilgiri Massif are nearly 350 hamlets of the Badaga community; nine more in the Nilgiri Wynaad and a dozen further outliers in the Ha:sanu:ru region off to the northeast give a grand total of about 370 hamlets. They are organized into communes, groupings which take the name of the head hamlet and are always made up of a few contiguous hamlets. The smallest commune however has just one big hamlet (Ta:mbatti), while the largest, (Me:lu:ru, close by it), has thirty-three. A village typically consists of several lines of houses ranged one above the other on the protected easterly slopes of a hill. Every hamlet is surrounded by its fields and plantations, many of them now terraced to prevent loss of soil. A scatter of cattle sheds, an isolated temple or two, and stands of eucalyptus complete the rural scene in most areas of Badaga habitation today.

Important as this particular setting has been for their economic pursuits, first as refugee buffalo herders and millet cultivators, then more recently as potato, cabbage and tea producers, it was never the total environment in any sense for the Badagas. It has always been complemented by a social environment lying beyond their villages, but still close by (cf. the endpapers); an environment consisting of seven Kota villages interspersed between Badaga ones, and, only slightly more distant, over fifty Toda hamlets. It was from these two Dravidian tribal groups that the Badagas initially acquired land to settle on, back in the sixteenth or seventeenth century; and it has been these groups too who have supplied almost all of the subsistence and ritual requirements which the Badagas were unable or unwilling to provide themselves with; for until the British established such towns as Ootacamund there were no markets in the district. As has often been noted, the Todas regularly supplied dairy produce and jungle-made artifacts to their Badaga associates in return for grain; while the Kotas proved even more crucial, since they alone acted as blacksmiths, carpenters, gold- and silversmiths, potters, leather-workers, musicians and builders to the Badagas, also in return for regular gifts of grain and cloth.

History

The social and economic history of the Badagas falls into three distinct periods (Hockings 1980a). Although space does not permit any presentation of their interesting though unevenly documented history, it is important for an understanding of what they have become today to know what those three phases were.

The first period was also the longest. It ran from about AD 1565, the year when the Vijayanagar Empire fell apart, to 1819, the year of John Sullivan's first exploration of the Nilgiris. The former date marks the beginning of refugee movements from villages lying just to the south of Mysore City which soon placed a scattering of settlers, mainly from the Okkaliga castes, in Nilgiri forest clearings. These people thus came to be known as Badagas, i.e. 'Northerners', and the term was actually used by Father J. Finicio to identify them in 1603. Their occupancy expanded slowly, from perhaps two or three hundred settlers in Finicio's day to about 2,200 in 1812, the year of the first British attempt at a census in this district.

John Sullivan, as we shall see below (pp. 337-8), was not the first British explorer of the Hills, but he was the first European settler there. The train of events which he initiated brought towns, markets, highways, cash and governmental administration almost overnight to a people who had previously lived without the benefit of any such facilities.

Although the British administrators and settlers were concentrated in the Nilgiri towns and on plantations, the network of roads, police-stations, schools and sub-post offices which they developed could not but have affected Badaga life in some measure. It was during this second period, of British imperial expansion, that all previous ethnographic accounts of Badaga society were written. Even the 'memory culture' recorded by myself from contemporary Badaga informants hardly goes back beyond the lives of their grandparents. One can thus say that virtually everything we know today of Badaga cultural traditions relates to a period when the British government, its facilities and interests, were already somewhere in the background. Most importantly, the cash economy and the local bazaars were factors in Badaga daily life because of their development under British control.

Nonetheless it is possible to talk about Badaga culture as a separate reality, a style of living which characterized only those South Indians who called themselves Badagas and were readily identified as such. This culture never became noticeably English in the way that the culture of Parsis, many Maharajahs, and the Anglo-Indians did. The British presence was always an outside presence for Badagas, and there is no evidence of their internalizing British values during the century-and-a-quarter that they were directly exposed to them. Instead we find that during that period they continued to be millet-farmers and cattle-herders, and they continued to rely on the Todas, Kotas, and Kuṟumbas for those goods and services which the Badaga community was not in a position to furnish for itself.

Of the three phases in the history of the community, the latest began in about 1930, when most Badagas broke off this traditional exchanging relationship with the Kotas, and soon after let those they had maintained with the Todas and Kuṟumbas fall into abeyance too. The break with the Kotas, and the local meetings which inspired it, were charged with considerable animosity at the time; but whatever may have been the immediate cause for the discontinuance of what had been a complex, traditional interchange of goods and services, the underlying cause is clear in the decennial censuses.

TABLE 8.1
The Proportion of Badagas to Two Other Local Communities

Proportion of Badagas to:	Year of Census										
	1871	1881	1891	1901	1911	1921	1931	1941	1951	1961	1971
Kotas	18:1	22:1	25:1	27:1	33:1	33:1	38:1	59:1	56:1	89:1	82:1
Todas	28:1	35:1	40:1	42:1	51:1	63:1	71:1	89:1	98:1	112:1	129:1

(Source: *Census of India*; Hockings 1980a: 249)

These show that, over the past century, the proportion of Badagas to Todas, Kotas and Kuṟumbas has been growing prodigiously (see Table 8.1). What this means is that, with the passage of years, it became increasingly difficult and finally impossible for any one Kota family (and any Toda or Kuṟumba individual) to continue to satisfy the various customary needs of ever increasing numbers of Badaga associates: something had to give way. The presence of local bazaars with trading connections in the South Indian plains offered an alternative, and ultimately a cheaper, quicker source for those goods which Kotas and Todas had previously supplied. Not that these tribesmen had been paid in cash: the clarified butter of a Toda friend, the pottery or iron goods of a Kota associate, the nets and baskets made by a Kuṟumba village watchman *(u:r Kurumba),* were all produced upon request, and reimbursed through a general distribution of Badaga grain at every harvest. Yet the market value of the bucketful of grain was steadily increasing over the years, while the market cost of comparable ghee, nets, pottery and so on was low enough that sooner or later most Badagas realized they could satisfy their needs more cheaply and rapidly by buying from a 'resident stranger' (to use Simmel's term) in the market-place.

Today therefore we find most Badaga farmers have given up millet and are cultivating potatoes, cabbages or tea for a national market. The Kotas likewise have adopted these crops and abandoned most of their traditional crafts. Many hundreds of Badagas have left off cultivation altogether, and now work in such urbanized professions as law, medicine, administration, teaching and agricultural technology.

Environmental and Economic Constraints

The goals of daily action are not often stated, and have to be inferred from what people are seen to do and heard to say in conversation. It is difficult to be sure that the pursuit of happiness, so widely valued in our modern world, is perceived as a worthwhile goal by the Badagas. Yet they do have four goals which, taken together, amount to much the same thing: the striving for spiritual release, the desire for good cooked food, the lust for land, and the need for progeny.

Such goals are pursued against environmental restrictions that we may review at this point. Availability of land is clearly one factor. From the sixteenth to the nineteenth centuries land was seemingly there for the clearing, and as late as the 1860s it was still possible for the Badaga farmer in the Kunda region to clear a tract of jungle land and cultivate millets on it without hindrance. But since that time slash-and-burn cultivation has become impossible, and the Nilgiri population (both Badagas and others) has increased so rapidly that today it is only the rising fertility of the soil, achieved with chemical fertilizers, that can support the heavier rural population density. The nature of the soil itself, and its tendency to be eroded or flooded, are associated environmental considerations of great importance.

Timber resources present a changing pattern in the Nilgiris. A century ago the Badagas' main timber requirements were for building houses, cattle-sheds, beds and ploughs. A certain amount of rough-cut timber was also used in fencing and funerals. There were then many extensive government-protected forests on the hills. Today these forests have diminished, although others have sprung up within the past twenty or so years. An ambitious, almost devastating afforestation programme has seized vast tracts of Toda grazing ground (without any compensation to these herders) and has covered them with acacia (16,000 ha) and eucalyptus (8,000 ha). These are very quick-growing trees, the timber of which is useless for building or furniture. Its sole value has been as firewood and pulp, with the leaves fuelling a 'cottage industry' of eucalyptus oil distillation. Another specialized pattern of tree-planting has involved the growing of shade-trees in tea and coffee plantations, necessary for the well-being of these commercial crops but otherwise useless.

The availability of credit for the purchase of seeds, cuttings and fertilizers has made an impressive difference to Badaga cultivation in recent years (Hockings 1980a). Not only has credit eased farmers through difficult parts of the year before crops were harvested and sold: availability of credit, primarily through co-operative societies, has even helped farmers to switch from potatoes to tea or cabbages when disease was ruining the first-mentioned crop in the 1960s.

The unit of farm-labour has traditionally been the Badaga family: men did the heavier work, including ploughing, while women did most seeding, weeding and harvesting. This basic work-force could be augmented by several means: birth of more sons and daughters; taking a second or third wife; contracting with a young man to work for a few years, in return for which he would marry the land-owner's daughter and (in the event that she had no brother) inherit the land; hiring the labour of landless Badagas. These alternatives are not so feasible today as they were in the past: recent economic change has made it less complicated and probably cheaper to hire non-Badaga day-labourers whenever needed. Those who have tea or coffee plantations may well have to hire skilled labour for many tasks that their families cannot perform adequately themselves. This labour normally requires a prompt cash payment, a demand that can put a severe restriction on the amount of labour that can be hired.

Yet another economic constraint relates to the market for a crop surplus. The tea is processed locally and then auctioned in Coonoor or perhaps shipped to Cochin for marketing if it is in large enough amounts. In general it is not of a quality to compare with the best Darjeeling and Ceylon tea in the world market. Badaga tea prices are determined in a market that is usually depressed, and totally beyond their control. By eliminating the middleman some farmers who grow potatoes and cabbages as cash crops do exert rather more control over the selling price, and considerable profits have been made with a newly introduced variety of cabbage.

A further constraint on maximum production in agriculture is the age and stamina of the farmer. By the time a man has reached seventy he is often lacking in the necessary strength to plough, hoe, or do much of the other heavy work associated with agriculture: these duties have to be shared by younger people. Badaga society however, gives the elderly useful roles as baby guardians, child educators and village councillors that are both valuable and more in keeping with the stamina of old men and women.

Disease too can take its toll on the labour force. Consequently many villages have a herbal therapist who works part-time as a curer. In addition those epidemic diseases which are believed to be caused by goddesses are kept at bay by annual ceremonies performed for the protection of the entire community by a Badaga priest. Not all diseases however are caused by supernatural agents. A combination of priests, shamans, therapists, midwives, and veterinarians put up the best efforts Badaga culture can make to keep the work force healthy and strong (Hockings 1980b). During most calendrical ceremonies a prayer is uttered requesting that the deity bring fertility to crops and cattle, and health to the children.

Here then are some of the main factors that act as constraints on the maximization of the economy. The Nilgiri climate should also be considered a constraint, particularly as the main monsoon (around July) can at its mildest make fieldwork impossible and at its worst create mud-slides.

Family and Household

As with any society, the family is the basic social unit within which each individual lives out his life. It is certainly not the only social unit, but is the one on which other larger and more extensive units are founded. The Badaga family, like those we would find elsewhere in Hindu India, is patrilineal, patrilocal and sometimes joint. At marriage, therefore, a woman moves into her husband's household, normally in another village, to spend her married life in the company of his parents, his brothers, and their wives and children. Of course it can happen that the Badaga house becomes too small for all these people; in that case the elder brothers move out with their wives and children to other houses in the same village. So they are still neighbours and, for many purposes, still a joint family.

This is the backdrop against which Badagas follow a tradition of preferential cross-cousin marriage, most commonly with a mother's brother's daughter or a father's sister's daughter. Since each village was founded by one particular family several centuries ago, and their descendants have continued their patrilocal occupation of it, we find that in modern times each village belongs primarily to one clan only, the clan descended patrilineally from the founders. Clans are exogamous, and hence villages too are exogamous. The patrilocality was advantageous to farmers in the highly variable Nilgiri landscape (cf. von Lengerke and Blasco, above), because their economy was traditionally centred on buffalo pastoralism and the cultivation of several varieties of amaranth and millet, in earlier times by a system of swidden agriculture.

To keep a herd of buffaloes required adequate grazing and certain other conditions. One of them was the availability of several active men, typically brothers, who would keep the herd together and fend off tigers when possible. It was the kind of work that women, burdened as they often were with motherhood, could not be expected to do. Both swidden and ordinary plough agriculture also called for the combined efforts of several men (as well as a number of women), and so here too Badagas found good reason for cultivating their land as a joint family. Many still do so today. It is noteworthy that buffalo herding, swidden cultivation and ox-ploughing were economic conditions that favoured patrilocality, the joint family, and brother/sister exchange at marriage (Hockings 1982).

Marriage Arrangements

The Badaga kinship system, as would be expected from their geographical location and language, is of a type that is commonly known as Dravidian. The preferred bride for a Badaga is stated to be his father's sister's daughter, and if she is not available to him the most acceptable alternative is his mother's brother's daughter. From a girl's point of view, this simply means she is most likely to be married to her mother's brother's son, or failing that her father's sister's son. Thus Badagas follow the widespread practice of choosing a cross-cousin; but it is felt that a father's sister's daughter is a better choice than a mother's brother's daughter, and we must explore why this is so.

The mother's brother is called guru, which means that he is the religious preceptor of a boy or girl. He has a crucial role in all of the childhood ceremonies, and is certainly accorded more importance than a child's father in them. Should he become a boy's father-in-law by giving his daughter away in marriage, the relatively high status of this guru becomes compromised. This follows because Badagas are explicit that bride-givers are of lower status relative to bride-receivers. Of course, a boy could take the daughter of another man than the uncle who is his personal guru, and this does happen sometimes. Occasionally too a girl marries her mother's brother. Such uncle-niece marriage is commoner in other Dravidian societies, but is very rare among Badagas other than the Toreyas, the lowest-status phratry.

The relationship between bride-givers and bride-receivers has to be seen against the background of payments that are made at the time of a wedding. In 1930 a bridewealth (*honnu*) of Rs 100-150 was paid to the bride's parents

GROOM'S FAMILY		**BRIDE'S FAMILY**
Rs 200 for ornaments	→	
		Bride, wearing Rs 200 or more in gold
Expectation of providing another bride at some later date	→	
	←	Foodstuffs and tobacco
	←	Clothes for the bride from her mother's brother
	←	An ox, buffalo, and later a cow
Food and tobacco for all the wedding guests	→	
Cloak given at bride's father's funeral	→	

before the wedding could take place. As I have shown elsewhere (Hockings 1980a: 217-18), this sum had increased gradually since the beginning of the nineteenth century, with the increase in the price of gold. At no time could it be construed as purchase of the bride, however: the expense was made for gold body-ornaments with which the bride was to be decked out when she first came into the groom's home. Should the couple later divorce, these ornaments were retained by the husband's family, while the bride's wedding dress was returned to her. In the custom of bridewealth there is implicit the suggestion of inferiority on the part of the bride's family, insofar as seemingly they could not be relied upon to bedeck the bride suitably for her new status.

When a marriage is arranged—and all marriages that do not begin with elopement are arranged—a series of exchanges links the two families. The exchanges can be expressed diagrammatically as shown on page 212.

The expectation of another bride in return is an important feature of Badaga kinship, and constitutes evidence that brides are indeed seen as items to be reciprocally exchanged between hamlets that customarily intermarry. The village of Hulla:da, for example, had (in 1963) the following set of marriage relations with other hamlets (see Table 8.2).

TABLE 8.2

Marriage Reciprocity

Name of Affinal Hamlet:	Number of wives originally from:	Number of living Hulla:ḍa girls married to:
Tummanaṭṭi	29	13
Mainele	21	12
Kundasappe	11	13
Tu:ne:ri	7	9
15 other known hamlets	20	13
unknown hamlets	10	?
	(*N*=98)	(*N*=60+?)

At the time I collected these data I was told that the villagers felt at ease with those hamlets with which, in living memory, they had provided roughly as many brides as they had taken; but that they felt uncomfortable towards those hamlets (Tumanaṭṭi and Mainele) where the reciprocity was marked by imbalance in the number of brides exchanged (Hockings 1982).

Larger Social Units

A village consists of one or several minor lineages, each taking its name from its founding ancestor; and since those founding ancestors would be descended from an earlier known ancestor, the minor lineages would cumulatively belong to one major lineage bearing his name. In many instances two or more major lineages similarly constitute one maximal lineage, itself named after one founding ancestor who lived back around the seventeenth century. In general a lineage is a group of patrilineally related persons who can trace their descent from one founding ancestor, a Badaga man. This unit is exogamous.

There are also larger exogamous units, the clans. A clan is made up of several maximal lineages, but even when its founder is known by name his actual kin relationship with present-day Badagas is only a presumption for them. Some clans do bear a founder's name; others, e.g. Ma:ri clan, have another designation. The name Ma:ri, though common among women, refers in this instance not to an ancestress but to the smallpox goddess Ma:riamma, whose image the clan founders carried up to the Hills at the time of their flight from Mysore.

Several clans—minimally two and maximally fifteen—constitute a phratry (some earlier writers called this a sub-caste). It is basically an endogamous unit, unlike the clans, and it occupies a rank-position in Badaga society. The smaller phratries with only two clans each operate through a simple exchange of marriage partners between those clans; but the larger phratries have much more complex, though still regular, patterns for marriage exchange. Aside from this regulation of marriage, neither the clans nor the phratries appear to have any other functions of a more political or juridical character in this society.

Cross-cutting this kin-based system of lineages, clans and phratries is a hierarchy of authority based as much on locality as on descent. Each village (*haṭṭi*) has its own traditional headman (*gauḍa*), who works with a small council made up of male elders of each lineage and/or dominant family living there. All of the village headmen involved also constitute a commune (*u:r*) council under the leadership of a common headman (*u:r gauḍa* or *maneka:ra*). These headmen in turn constitute a regional (*na:ḍu*) council which now rarely meets, under the leadership of its *pa:rpati*. Finally, the headmen of all four *na:ḍus* occasionally meet formally in Ootacamund when called together by the paramount chief of the Badagas (*na:ku beṭṭa gauḍa*, literally 'four mountains headman'). He, like all the lower-level headmen, occupies a hereditary status; and he is always the headman of Tu:ne:ri village, since it was in all probability the first Badaga settlement on the Nilgiri Hills.

When a crime has been committed or a dispute breaks out, it is—or was, until recently—adjudicated by the relevant village headman and council. Punishment of the guilty would follow, but appeals against a conviction could be heard by the next higher level of headman and council. Theoretically this process of appeal against a judgement could be continued right on up to the paramount chief, though this rarely happened. As the magistrates' courts have become more and more popular with the Badagas, use of their own headmen and councils in adjudication has been proportionately limited.

The Conceptualizing of Time

The advent of cheap alarm-clocks has, instead of totally eradicating traditional conceptions, made it easier nowadays for some conservative Badagas to observe the puranic periods of auspiciousness throughout each day. It is an idea imported from the literate civilization of the plains that the day (approximately 0600 to 1800 hours) is divided into eight periods or *mukūrttas*) of equal length but variable auspiciousness, which were supposedly established by the ancient Vedic sage Agástya over two thousand years ago after collecting records of when good and bad things happened. One should avoid an inauspicious period when *starting* any new kind of work, *pu:ja* or ceremony, and especially a journey or a wedding. Most Badagas, however, are unaware of these calendrical details.

The days of the week have long been classified by Badagas as follows:

Monday—first day of the week and a holiday; God's day; the most auspicious day; vegetarian diet universally; bad direction, east.

Tuesday—inauspicious for ceremonies; bad direction, north.

Wednesday—third most auspicious day; bad direction, north.

Thursday—inauspicious for ceremonies; bad direction, south.

Friday—second most auspicious day; goddesses's day; bad direction, west.

Saturday—also a weekly holiday, but not widely observed as such for the past century; inauspicious for ceremonies; bad direction, east.

Sunday—somewhat inauspicious for ceremonies; bad direction, west. Furthermore, during the lunar month the period after the full moon is never the time for ceremonies other than funerals: all should be performed on appropriately auspicious days between the new moon and the full moon. In any ritual, turning to the east means that people want wealth; to the west, long life; to the north, rest. No one would turn ritually to the south, as it is the abode of demons. I summarize with a diagram (Fig. 8.1).

Clocks, almanacs and printed calendars were of course a recent introduction to the Badaga community. Nonetheless they had over the centuries developed adequate measures of the passage of time, basing their observations on the nature surrounding them. The day and night are thus divided up by the passage of certain celestial phenomena and by bird calls, according to local tradition.

Very late in the night there comes a period marked by the appearance of a star called *ettu kaḍasi mi:n* ('ox driving star'), because it is the period when oxen should be given hay before being driven out for ploughing (Rivers 1906: 595). Then the 'six stars' (Ursa major) appear. Shortly after this the first call of the pied bushchat bird is heard (or in non-vegetarian villages, it is the first cockcrow). This time varies with the season, being around 0400 hours much of the year, but coming somewhat earlier in the sowing season (the European winter) and as late as 0430 hours in the rainy season (the European summer). Next the 'silver star' (or Venus) becomes visible, followed a few minutes later

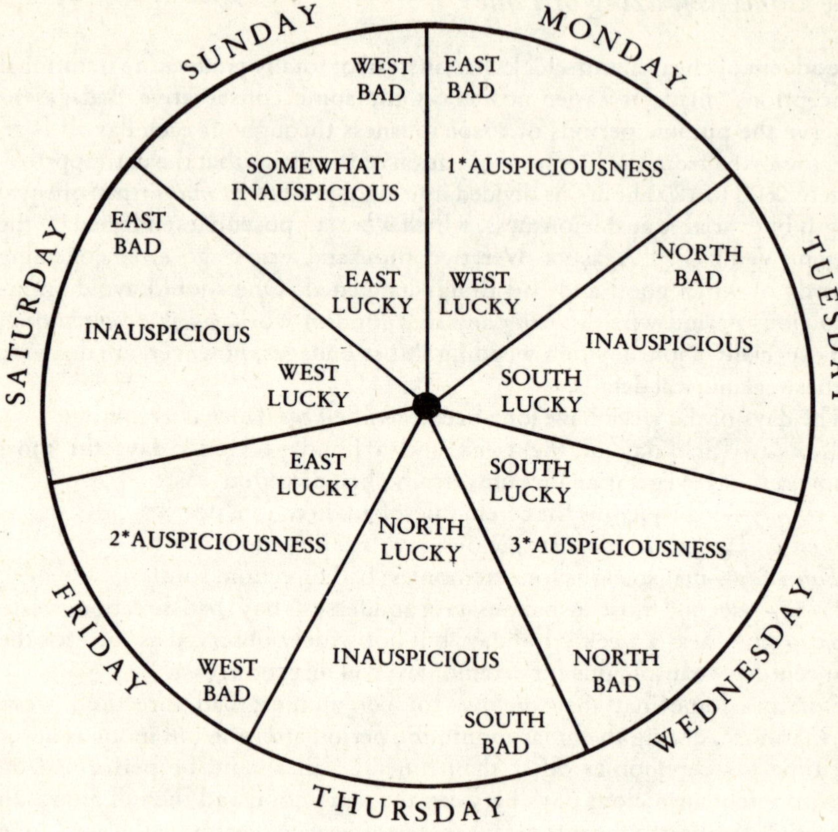

FIG. 8.1 The Badaga Week.

by a lesser star, called *padikka:ra* ('rivalling crystal in lustre'). Then a slight light emerges: it is called *mu:de birigira* ('darkness disappears'). Next comes *mabbu ti:dira*, 'the division between night and day.' Then 'it becomes dawn' (*ba:lgua:ra*), and finally 'the sun rises' (*huṭṭira*). This is when the new day begins. All of these phases between the song of the first bird and the sunrise consume about two hours, and are divided by these various indicators into six periods that are presumed to be of equal length, lasting some 20 minutes each.

For the most part the passage of daytime is marked by cultural rather than celestial phenomena, high noon and sunset being the exceptions. The sun was traditionally believed to circle round the earth.) Thus the significant times after sunrise are (1) the first suckling of infants; (2) high noon, the point when the sun is highest in the sky; (3) as men notice the sun is past its height, it is time for them to stop ploughing and untie the oxen; (4) the second suckling; (5) return to the village of those cows which have calves; (6) sunset, the time when milch buffaloes return to their calves in the village; (7) dusk, the last moment when things are still visible, which is also the time when the remaining cows and buffaloes are brought back from pasture. Then for six months of the year the evening star (Venus) can be seen just after sunset. A group of four stars called 'cot legs' (probably Gemini) rises at around 2100 hours. In addition to these natural phenomena, the evening is also marked by cultural events, namely *hiṭṭu ja:ma*, the evening mealtime, *hakka ja:ma*, when the last couples go to bed, and *nadu ja:ma*, the 'middle of the night'. It is against this agreed framework of everyday observations that people organize their activities.

The Badaga calendar is quite different from the Tamil and Kanarese ones, even though they all divide the year into twelve zodiacal or solar months, and share some of the names of months. The Badaga month called *A:di*, for example, falls two months prior to the Tamilians' *A:di* (Cancer). There is no evidence that the Badagas have inserted an intercalary month every ten years during recent times, as is done elsewhere in India; but something else has been happening to change their calendar. In 1918 Richards recorded (1920:24) the sequence of Badaga months, and identified *A:ḍi*, for example, as May-June. Some 23 years later the Badaga folklorist M.K. Belli Gowder equated *A:di* with June (Belli Gowder 1923-41:109). Some twenty-three years later yet I recorded *A:di* as equivalent to June-July. What has been happening? I conclude that the Badagas have been recognizing a 366-day year, with the result that every 23 years their calendar becomes further advanced against the Gregorian one by about 17 days, because a year is actually much closer to 365-1/4 days. Hence in the 46 years between 1918 and 1964 we have seen their calendar advance by 35 days, or just over one month.

This observation leads to a quite illuminating result. As we have just seen, Badaga *A:ḍi* in 1918 was 2 months ahead of Tamil *A:ḍi.* If, as was suggested, the Badagas have consistently been reckoning 366 days to each year, this has been 18 hours 11 minutes (or 0.7576 day) in excess of the correct solar figure. This means that their *A:ḍi* would last have been synchronous with the Dravidian *A:ḍi* approximately 402 years prior to 1918, *i.e.* just a few decades prior to when they first left Mysore for the Nilgiri Hills.

The same kind of argument can be based on the fact that in 1918 Badaga *A:ḍi, A:vani,* and *Perita:ti* were two months ahead of the eponymous Tamil months. A problem arises, however, with *Tai* and *A:ni*, also occurring in both languages, because the Badaga *Tai* and *A:ni* are *three* months ahead of their Tamil eponym. Perhaps at some time in the last century an attempt *was* made to add an intercalary month, namely *Hemma:ti*, with these confusing results. *Hemma:ti* derives from a Sanskrit word ('cold weather') which does not seem to occur in other Dravidian calendars. It has been positioned immediately after the Badaga *Tai*.

Although the year appears climatically to be made up of three seasons, hot,

TABLE 8.3
The Badaga Year

Ka:r (March 16–June 15)	Muṅka:r (March 16 – April 15)	sowing season	Ka:r bo:ga
		cultivation	
Ko:ḍa (June 16–Sept. 15)	Muṅko:ḍe (June 16 – July 15)	'main monsoon;' no work possible	
	Moda Ko:ḍe (July 16 – August 15)	'subsiding monsoon;' cultivation continues; harvesting starts.	
	Mora Ko:de (August 16 – September 15)	'tree monsoon;' wind in trees.	
Ka:rtige (Sept. 16–Dec. 15)		cultivation	Kade bo:ga
	Kiru (part of November)	rain; cultivation difficult.	
Be:sage (Dec. 16–March 15)		dry, cultivation possible with some irrigation; harvesting second crop.	

wet and cold, the Badagas in fact recognize four seasons, each lasting for three months, and within these further distinctions are made according to the agricultural calendar; as in Table 8.3.

(Although the terms in Table 8.3 are also found in the Kannada language, they apply to different periods in Mysore than they do in the Nilgiris.)

Daily Life

Women are the first people to get up in the morning, as early as 0300-0400 hours in poor households where they have to grind or pound grain. All women will be up before sunrise, but before getting out of bed they will pray to the village god. A woman will not touch any cooking vessel until she has washed her hands, feet, face and teeth, for fear of polluting it. This bathing is usually done with cold water in a small back room. Next the woman will change out of the clothes she has slept in, and will sweep out the house, veranda and work-yard. She then takes a handful of cowdung and, mixing it with water, sprinkles it over the floors and the yard. This should be done before sunrise, but if the woman is late she will sprinkle the cowdung within her own shadow, not in direct sunlight: to let sun fall on the sprinkling would be a sin.

In earlier times the next task for a housewife would be to visit the Great House (that of the founder of the village) and bring fire from the hearth of that family. Today matches are used instead, and then the cooking of breakfast begins. Water left over from the previous day is heated for the family's bathing. Generally the sons are still sleeping with their father, but the daughters, who sleep with their mother, get up ten or fifteen minutes after she does. Mature daughters bathe in cold water as their mother do; small girls use warm water. The daughters are then expected to help with the housework. Fresh water is first to be brought for the cooking.

A man rises somewhat later than his wife, uttering the name of God or some other prayer before he gets out of bed. He then bathes, puts on clean clothes, and goes out to inspect his fields. He wants to check on how the plants are growing and whether any damage has been done by pigs, rats or cattle. When the sun rises he faces it and prays to Siva; and he does so again when he brings his cows and buffaloes out of the shed and takes them to the house for milking. (In earlier times they would be milked at the shed itself, or on the village green.) Nowadays they are driven to the green after milking and there, at about 0730-0800 hours, a cowherd takes charge of all the village cattle together. Herds are not as large now as they once were.

Generally the entire family takes a heavy meal between 0800 and 0900 hours, the sons and young daughters eating with their father but the mature daughters eating with their mother afterwards.

Nowadays nearly all children go to school, which usually starts at 0930 hours. Children aged between two and five will spend the day playing with each other around the house. If there are even younger children, their mothers will stay away from the fields all day so as to care for them, or else will entrust the children to some old man or woman in the family.

Generally the adults go to work in their fields, but if they have none, or there is no work left to do, they may go to work for wages in some neighbour's field. The day's work begins when people arrive in the field, touch the earth and then the forehead. Basically men do the heavier work, including forking (or in former times ploughing) and the removal of heavy stones. Women do the weeding, pick up smaller stones, and build stone walls around some fields.

At around 1300-1400 hours these fieldworkers stop for a light meal, often just coffee and bread. If near home they may return there to eat, or have a snack brought to them. Women are the first to leave the fields later in the afternoon, since even those who do not go home once or twice to suckle an infant have to go there to begin the preparation of the supper which is eaten around 1600-1700 hours.

After a while the men return too, after dismissing coolies and picking up their tools. On the way home a man will try to wash his tools, for it is said that if he fails to do so he will be ill the next day. On entering the house a man worships Siva by addressing the household lamp. Once home, his first duty is to tend the cattle while his wife continues with the cooking. Meanwhile old members (if any) of the family take a thorough bath, and then bathe and feed the children.

Lingayats offer a *pu:ja* each evening before supper; others may engage in some prayer at home before breakfast. The supper is eaten at around 1930-2030 hours, the housewife taking hers after everyone else has finished. Afterwards the women clean up all the dishes, perhaps pound some grain, and then everyone goes to bed immediately. They sleep in total darkness, and in earlier times the entire family would occupy a single large bed. Now several rooms may be used for sleeping, although young children still sleep with their parents (Hockings 1980b:94).

This then is the traditional pattern of daily life for the Badagas. It does of course vary somewhat in individual cases, particularly when a man is a school-teacher, bus-driver, village councillor or policeman, or when a woman has no children. (Jagor 1876: 201-2; Rhiem 1900: 507-8; Ketti Medical Mission 1940: 4).

World View

There is little evidence that the dominant orientation of the early Badagas was other-worldy. They did not claim that they had been 'saved' from the Muslim invaders by any particular god; or if they did, the claim was not broached strongly enough to have found any place in their oral folk-literature. It was quite obvious to them that they had saved themselves by the powers of their own legs, and by good fortune. This down-to-earth attitude was reflected to some extent in the kind of stories they told—and still tell— about their flight. None of the clans has a myth of origin that involves gods and demons and miraculous events the way many other Hindu castes do elsewhere. Instead they have pedestrian stories about leaving a certain village in the plains, hurrying up the escarpment, and creating a new settlement on the plateau. It is true that one village (Tu:ne:ri), the first to be settled, enshrined an image of the goddess Ma:ri that the refugees had carried up from the plains with them; but even here we find no tale of their having been saved by Ma:ri.

In such a this-worldly climate of thought, a few refugees of priestly background found themselves in a somewhat uncomfortable and ambiguous situation. On the one hand, these men had come somewhat later, and thus could not share in whatever little glory might have been felt about being a first settler. On the other hand, their sacerdotal powers had not been enough to save them or other refugees from a total disruption of their former society and lives once the God of Islam had come to the region. The old Badaga gods seemed close to impotent in this anomic and unprecedented situation.

There were two sacerdotal groups, the Wodeyas and the Ha:ruvas, both extremely small. The former were Lingayat gurus, whose asceticism seems to have been acceptable to other Badagas and whose journey into the mountains was not out of character with it. The other group claimed to be Brahmin priests, a much more debatable proposition. They appeared not to have been

Brahmins when they lived in the plains, they wore no sacred thread there, they could read no sacred (or other) texts, and there were dark stories about early misalliances in the marriage field. The general Badaga attitude towards sacerdotal upstarts like this was expressed in the proverb: 'If a crow drops something on you, it is a bad sign! If a Ha:ruva shouts at you, it is a bad sign!' But even the other-worldliness of the Wodeya was called into question in one proverb: 'If the population is increasing, it is good for the Wodeya'—the meaning being that he gets more free meals and more income from performing his *pu:jas*. Yet in the long run the newly settled Badagas had a need for a regular priesthood, and so both of these groups were allowed to function in that capacity.

Under these circumstances it is not surprising that the Badagas found it very difficult to understand the need for a God of Salvation, of the kind that missionaries tried to introduce them to during the later nineteenth and early twentieth centuries. Gods were known to be vengeful and protective, but never Saviours.

Badaga religion (with the obvious exception of the small new Christian community) is not a messianic faith; indeed it would be difficult to argue that it is an ethical faith. The dominant religious practice, in what has loosely been called 'village Hinduism', differs but little from the popular Dravidian worship described for other parts of South India by Elmore (1915) and Whitehead (1921). The most important acts of worship are the *pu:jas* performed by a priest on behalf of a village or a family on the occasion of an annual ceremony or funeral (in the case of the village community), or any other life-cycle ritual (in the case of the family).

This is not to suggest that the Badagas are an unethical people. There are indeed normative values that find expression in the prayers said at most ceremonies, where an emphasis on the fertility of cattle, crops and mothers is often evident. Yet the ethical values of these people are most commonly expressed in a secular context, as for example during the deliberations of a village council, or the delicate negotiations which lead to a marriage, or just in the give-and-take of semi-public conversation, punctuated as it is with proverbs. Badaga ethics appear to have arisen from the historic experience of the community itself, and are shared with no other community. Their proverbs are thus unique, and there is certainly no evidence of their having been given a set of ethical prescriptions by a god, as in the Christian instance. We must not on the other hand rule out the possibility of itinerant *sádhus* and resident philosophers, particularly those of the Lingayat persuasion, having had an important role in the formulation of Badaga ethics. Yet beyond knowing that a few such people were to be found in the Nilgiri Hills from time to time, we know nothing whatsoever of their impact, and their names are now forgotten.

The generalization that Badagas practise popular Hinduism can be modified in certain details. First they are one of the few South Indian communities whose temples (until a century ago) were sometimes circular. This unusual architecture finds no charter in the extensive prescriptive writings about the form of the Indian temple. Secondly, and perhaps more importantly, one must note that many of the *minor* Badaga deities are unknown elsewhere in the Hindu world. They originated as folk heroes in the early decades of Badaga settlement on the Nilgiris. Thirdly, there is no clear authority on religious matters within the Badaga community. Elsewhere Hindu villagers look up to a local Brahmin priest for religious guidance. The Badagas, it is generally agreed, have no Brahmins (or ex-Brahmins) among their various clans and phratries.

This last point has however been a matter of some contention since 1897. The Ha:ruva clan and some Gauḍas (the largest phratry) who had close relations with them have been advancing the claim that Ha:ruvas *are* really Brahmins. It is arguable that they do function as temple priests, are wholly vegetarian, and wear a sort of sacred thread. Against this can be juxtaposed the facts that they are ignorant of Sanskrit, perform no *śraddhâ* ceremony for their dead, have no *upanâyana* ceremony for their sons, and wear an unconventional type of thread. Most Badagas would add that Ha:ruvas are temple priests only because the rest of the community appointed them to act as such. One of the most persistent factional disputes during the present century has been founded on this particular difference of interpretation.

The vast majority of Badagas accept Siva as the dominant deity, and hence nearly every village has a major temple devoted to this god and his family. A minority of Badagas (the Kanakka, Adikiri and Kongaru clans, and the Wodeya phratry) are adherents of the Lingayat sect, which was founded in Karnataka in the twelfth century AD. This sect also acknowledges Siva as the Supreme Deity, and is supposedly henotheistic. In point of fact though one finds Badaga Lingayats worshipping a range of other godlings also, just as other Hindus do. Ma:riamma, the goddess responsible for smallpox, has been worshipped by most Badagas for the past century, and even the Vaishnavite god Rańgasva:mi has a following. Tamilian influence on habits of worship is nowadays quite evident.

Social Integration

The domains of Badaga action are, as we have seen, several: they include worship, eating, work, sexual relations, and conversation. Individuals naturally vary in the amount of emphasis they place on each domain. But these are the social performances that everybody will engage in during at least his adult years; and these are the performances which, if not adequately managed, can produce stress.

The tensions of daily life are regulated by various means. Formal legal pronouncements of guilt or innocence can be made by a village council, while informal public interaction is an effective and more widely used means of social control. Abuse and innuendo have their parts to play in this; gossip and disparagement are always to be heard. Mental and spiritual release from tensions is sought at the time of village festivals and even more frequently during Monday, the weekly holiday, when many villagers manage to visit a town and perhaps see a film.

The sources of social integration, the values which operate to hold Badaga society together in its distinctive conformation, are embedded in the cultural traditions (*ma:mu:le*) of the community and the clan, and are openly expressed in prayers, in proverbs, and notably in the long recitation of possible sins that accompanies every funeral (among non-Lingayat Badagas; see Appendix).

Something of the complex manner in which their mythology operates as an integrating force in their society, and expresses the deepest concerns of their life, can perhaps be understood from the following analysis of a brief but crucial myth in all its variation, the myth of their ancestress Hette, whom all Badagas venerate.

The Myth of Hette

There are at least eleven Badaga villages which worship an ancestress called Hette, and hold an annual festival in her honour (Harkness 1832: 106n). In each of these exogamous villages she is regarded as the mythical founder, with her husband Hiriodea, of at least some of the village lineages. Both of them have become minor deities, associated particularly with the fertility of women and crops. As we shall see, there are numerous local variations in the tale of Hette, and these all lead to three broad conclusions: she was the ideal Badaga woman; she exemplifies contradictions in the life of all Badaga women which cannot be resolved except through a ritual; her behaviour represented a popular flight from enforced Vaishnavite orthodoxy in the seventeenth century.

It has recently been noted by Claude Lévi-Strauss and others that

> The essential themes in a myth, impossible to identify from a simple reading of one version, emerge upon consideration of a number of other versions of that myth in which, despite various changes and reversals, certain elements persist. What is important is what is repeated, reworked to fit different circumstances, transformed even to the point of apparent meaninglessness, but always retained (O'Flaherty 1981: 18).

In the present instance we find that Hette was the ideal woman for a number of reasons that are stated in parts of her legend, namely: she was so given to the ideal of chastity that she did not want to go to the bed of an elderly husband and so drowned herself instead (Whitehead 1921:126; Karl 1945:8); she selflessly did not want to become pregnant until a senior but childless co-wife had already conceived; her husband was once out guarding a heap of grain when someone started a brush-fire—Hette went to look for him, thought he had perished in it, and so immolated herself in the flames; her husband was at home with a Chetti woman whom he loved when the house caught fire, and his ever-faithful wife returned to throw herself into the flames (Noble 1976: 115-16); because her husband died she committed *sati* by strangulation, drowning, or self-immolation (Harkness 1832: 106-7n; Metz 1864: 64; Breeks 1873: 44; Grigg 1880: 226; Belli Gowder 1923-41: 21; Karl 1945: 8); her son Batrabala kept a girl he was not married to, and also ate meat, which so upset his mother Hette, a vegetarian Ha:ruva herself, that she committed suicide (Belli Gowder 1923-41: 54); old and devoid of any relations, she killed herself so as not to be a burden on the village (Benbow 1930: 29-30); this was done by jumping down a well where she 'saw' her husband and dead relatives (Rhiem 1900: 501; Tignous 1911:119); and after her death she possessed another woman, during which time her ghost explained that she had really been an incarnation of Parvati, the wife of Siva, and hence a goddess who was universally the model of Hindu womanly virtue; Hette undertook to protect the Badagas, and asked for a temple to be built for her. An annual ceremony has subsequently been celebrated in her honour (Rhiem 1900: 501; Thurston and Rangachari 1909, I: 90; Whitehead 1921: 126).

In addition to all these versions of her story, two others have been reported recently by a Badaga writer (Dharmalingam 1971). In one case, a man had three sons by his first wife, and after her death he married again, the second wife giving him three daughters. Once the man had died his eldest son began pestering the surviving wife for money; and his servant was once so obstreperous in his demands that the woman threw herself down a well. She is said to have cursed the eldest son and blessed her other children; and later somebody dreamt she was the incarnation of Maha Śakti, who is equated by Badagas with Hette. Another such incarnation arose under different circumstances. A man had promised his elder daughter to a man working for him, but she was enticed into another marriage and the jilted man was heartbroken. Her younger sister Ma:si (i.e. Maha Śakti) consoled the man and promised to marry him, but he became ill with grief and soon died. At his funeral Ma:si had a vision that he was calling her, and so immolated herself on the pyre. Such then are some of the variant legends, and we shall return to them shortly.

Perhaps the commonest version of the tale is that Hette was a very young girl whom her father promised in marriage to his own sister's husband. That sister was an old and barren woman. The idea that Hette should be offered as a young co-wife accords well with the Badaga idea of marriage: her father's lineage still owed a fertile wife to her father's sister's husband. On the day when Hette reached physical maturity she knew that her aging husband would soon sleep with her, and the idea did not appeal to her. She went out with some other girls to gather firewood, and, once in the woods, left them and built a pyre. She then asked God to grant a child to the older co-wife, her father's sister, and to accept Hette's self-immolation if He judged her to have been pure. Oblivious of all this, the other girls meanwhile returned home. When the villagers saw the column of smoke they went to investigate, and found Hette's cremation. Belatedly they performed the funeral ceremonies; and before the final ritual which releases the soul from the village, some eleven days after death, it was found that the old co-wife was pregnant.

In all these legends about Hette and all the villages where she is venerated, Hette is seen as the epitome of womanly perfection; yet underlying that image are some contradictions that no Badaga seeks to explain, contradictions which beset the life of every Badaga woman to this day, but which are resolved in due course. Let us look more closely at the virginity-and-marriage theme, and the contradictions will be apparent. On the one hand Hette committed suicide while a virgin, rather than upstage a senior co-wife and have possibly fruitful relations with the old father's sister's husband; alternatively she gave birth only to three daughters, no sons. But on the other hand she *was* the ancestress of Badaga patrilineages. Furthermore, though she was a virgin she was nonetheless held to be the model of a wife devoted to her husband. Yet another contradiction lies in one story that she committed suicide after all her relations were dead, lest she become a burden on the villagers. This variant, and a message from her ghost that she would protect that village, are somewhat at odds with her desire to join her dead husband through *sati* and go with his soul to paradise.

In summary, we have found four contradictions inherent in the story of Hette:

Virginity	↔	Lineage Ancestress
↕		↕
Devotion to her husband	↔	All her relations dead: devotion to the community

It is a situation that seemingly has no resolution: this is the predicament of a woman in Badaga society.

Yet one solution to the conundrum does present itself as an obvious possibility. It is that the four elements diagrammed above represent major stresses during four stages in the life and death of any women, viz.: young virgin; devoted wife; old lady without close living relatives; deceased ancestress of a lineage. Each status is at once typical and ideal. The transition from one to another is marked by a radical shift in personal standards, and also by an appropriate ritual which mitigates the anomalous situation, viz.: the first two stages are marked by wedding; the second and third stages by the husband's funeral; the third and fourth stages by *sati*. All these transitions have been summarized and truncated in the brief myth of Hette, with whom all women can identify.

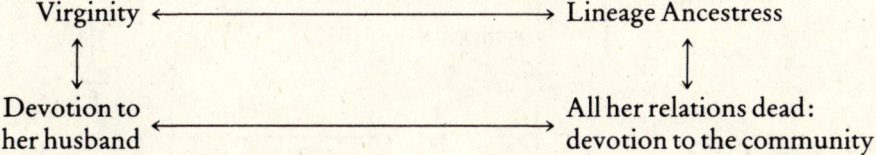

In the sequence of statuses just referred to, the contradiction between being a virgin and being a lineage ancestress remains unexplained. Yet here too a cyclical view of time yields some insights.

In Badaga and similar patrilineal, virilocal societies of India, a young girl is far removed, both mentally and biologically, from a lineage ancestress. Devotion to her own lineage ancestress, i.e. her father's, will be abandoned once the girl marries and joins another clan and lineage. In that lineage, i.e. her husband's, the girl will forge a new, subservient relationship with its ancestress, for her duty will henceforth be to produce a new generation of descendants for that ancestress. In a sense this ancestress is really a mythical mother-in-law, a personage with whom any young Badaga bride is likely to have strained, subservient relations. That the legend presents Hette as both virgin and ancestress is not then an inexplicable conundrum, but rather a reference to the common reversal that with time the fearful bride becomes a mother-in-law herself, and then with death moves towards the status of a lineage ancestress. The word *Hette* after all means 'old woman, grandmother'.

A more sinister understanding of Hette's marital situation emerges when we look at two segments of the myth in combination. Reflection on her chagrin at her son Batrabala's behaviour in keeping a mistress and eating meat (Belli Gowder 1923-41: 54), and on another version of the legend which makes her stepmother to three sons and mother to three daughters (Dharmalingam 1971) leads us to the following kinship diagram (Fig. 8.2).

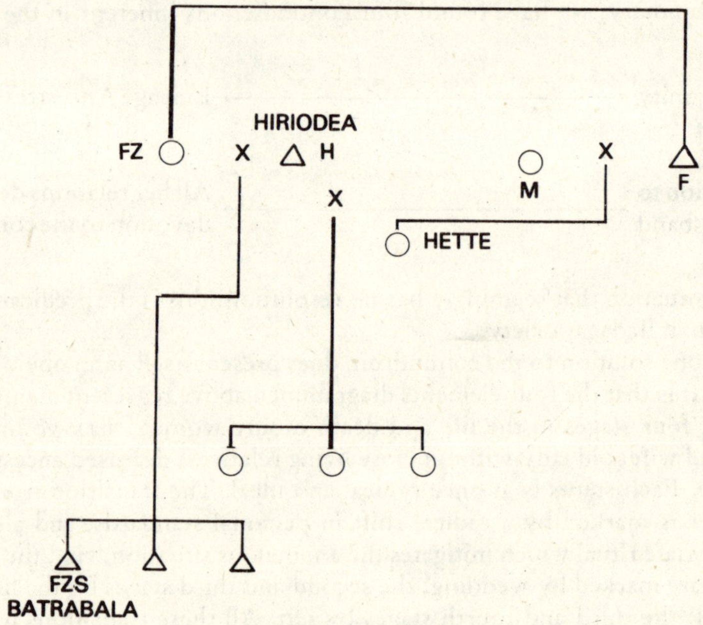

FIG. 8.2 The Family of Hette.

It will now be noticed that Batrabala is not only Hette's step-son: the two are potential spouses, for Batrabala could marry his widowed mother's brother's daughter and this union would be permissible cross-cousin marriage among the Badagas. Thus when he came pestering her for money was he asking her *in loco parentis* for the bridewealth so that he could marry her, his cross-cousin, or was he possibly seeking a dowry, which is not a Badaga practice? At any rate we can see that for Hette therefore this man represented a dilemma: he was at the same time her own husband's son and an ideal cross-cousin mate; she was Jocasta to his Oedipus. No wonder she cursed him!

A grander dimension to the whole Hette myth appears if we refer back to its template in the Bhavishyottara Purāna, for the Hette story is actually a parochialized version of one recorded in the Hindu scripture. There it is related that Parvati had, since a child, vowed to marry none but Siva. When she came of age and was told that her father Himachala had been advised to marry her to another god, Vishṇu, she became very indignant. Taking one companion she ran off into the forest, and there on a river bank she made three *liṅgas* (the phallic emblem of Siva) which she worshipped all night, while singing the praises of Siva. The god was gratified by this and appeared to her, promising to grant any request she made. So she asked to become his wife, and he agreed, then disappeared. After their vigil Parvati and her maid lay down to sleep, and were later found by her father, who was so relieved that he consented to her marriage wish.

Now there was in medieval Badaga history a sort of parallel to this situation; for when the earliest Shaivite Badagas were fleeing into the Nilgiri Hills from Mysore at the turn of the seventeenth century (when 'Sri-Vaishnavism reached its high water-mark'), the Wodeya lineage on the throne of Mysore was converted (1617) to Vaishnavism, and made the worship of Vishnu thenceforth a state cult. There followed some widespread disturbances between the Vaishnavites and the Shaivites (Hayavadana Rao 1930: 2368). Given that Badagas did recognize Hette as an incarnation of Parvati, her myth becomes a political statement about a people, the Badaga refugees, symbolized by a pure virgin, Hette, fleeing to the forest to avoid the fatherly command of the Mysore Woḍeya rulers to institute the cult of Vishnu. They wished to remain faithful to Siva, their *Iśṭa dēvade* and caste deity. The imperious royal gurus, Tātāchārya *et al.*, in the early seventeenth century actually had the power 'to punish people who swerved from the right path' of Sri Vaishnavism (Hayavadana Rao 1930: 2368). In marital terms this change could have led to Shaivite daughters being married off to Vaishnavites. Like Parvati (the Mountaineer), these girls sought instead to be constant in their devotion to Siva, joining their families as refugees to the Nilgiri Hills in order to achieve this aim.

The variant legend recorded by Noble (1976: 115-16), in which Hette's husband is described as having a Chetti mistress, is isomorphic with this political situation. Since the Chettis of Coimbatore District came from Mysore and settled there, and were Vaishnavites, this illicit relationship between the Badaga man and the Chetti woman exemplifies turning away from the Saivite faith of the Badagas. In the face of it, Hette's reaction was to commit *sati*, an expression of her devotion and purity. It is not surprising therefore that of over a thousand Badaga temples and shrines that exist now, only one or two are devoted to a Vaishnavite deity, Rangasva:mi (Jagor 1876: 202). Nor is it surprising to find some of the Hette legends contrasting her purity and selflessness with the meat-eating, mistress-keeping, and monetary greed of her relatives. Indeed, an underlying theme in those legends is that Hette is an outstanding individual because she over-values certain ideals that those around her under-value: she over-values virginity, whereas her elder sister under-values it to the point of marrying (eloping?) a man other than the one her father had designated for her as a husband. Hette over-values marital fidelity, in direct contrast to the behaviour of her own husband. She over-values vegetarianism, in contrast with the meat-eating stepson, Batrabala. She over-values the equivalence of sisters in marriage, in contrast with Batrabala again, who seems to view Hette and his own mother in quite different lights. Finally, Hette alive under-values the obligation of the community to support her in widowhood, which is in direct contrast with the way she, when dead, over-values her own ability to protect and support the community in response to their building her a temple and worshipping her as a goddess.

As a goddess, her association with the fertility of women and crops is appropriate for one who had promised to protect future generations of Badagas. Yet one might justifiably question why in a patrilineal society a minor goddess should be given such importance, especially when her husband Hiriodea plays what is undeniably a secondary role in the entire story. The answer to this question lies in the interconnected facts that Hette was an ancestress of Badagas only, while her husband kept a Chetti—and hence a non-Badaga—mistress. Only a married Badaga woman can confer legitimacy on the children she and her husband raise: children of the husband by a Chetti mistress would not really be Badagas. It is in this light that we can now understand the capital importance of pre-marital virginity, and of wifely chastity, purity, and devotion, in a patrilineal society where the husbands have been known to stray. Hette's legend accounts for the legitimacy of people who consider themselves to be Badagas.

Appendix

Absolution of Sins at a Funeral

Although there is some individual variation in this recitation (seemingly an effect of faulty memory), the complete prayer is easily reconstructed, and can loosely be translated in the following words, all of which would be used in an ideal absolution of sins:

This is the death of———:
In his memory the calf Bassava is set free,
Bassava the holy ox, born of the brindled cow Barrige.

From this world to the other one,
He goes in a chariot.
Let the man's body return to the earth;
Let the breath given by Siva go back to Siva.
He [the dead] has indeed sinned thirteen hundred times.
All the sins committed by his ancestors;
All the sins committed by his forefathers;
All the sins committed by his parents;
All the sins committed by himself—
May they fall at the feet of Bassava!
If———had carnal enjoyment in this world, it is a sin. [*Response*: A sin]
If———did evil toward his father or his grandfather, it is a sin. "
If he wronged all the past generations, it is a sin. "
If he was sinful towards his parents, it is a sin. "
If he did evil towards his father-in-law or mother-in-law, it is a sin. "
If he had carnal relations with his daughter-in-law, or his own children, it is a sin. "
If he brought about enmity between brothers, it is a sin. "
If he broke a bond of friendship, it is a sin. "
If he has killed a lizard, it is a sin. "
If he has killed a great lizard, it is a sin. "
If he has killed an ant-eating lizard, it is a sin. "
If he has killed a frog, it is a sin. "
If he moved a boundary-stone over, it is a sin. "
If he removed the field fences and let animals in, it is a sin. "
If he removed thorny branches around a field to let animals in, it is a sin. "
If he broke the growing plant, it is a sin. "
If he wasted dried firewood, it is a sin. "
If he cut the field pea stealthily, it is a sin. "
If he cut the raspberry outside his boundary, it is a sin. "
If he dragged away the sharp branches of holly, it is a sin. "
If he plucked young plants and threw them in the sunlight, it is a sin. "
If he swept with a broom, it is a sin. "
If he discarded seeds of grain, it is a sin. "
If he used a cow to plough the land, it is a sin. "
If he milked a cow liberated as a calf at a funeral, it is a sin. "
If he coveted a cow or buffalo yielding milk abundantly, it is a sin. "
If he coveted the good crops of others, it is a sin. "
If he was jealous of another village, it is a sin. "
If he spoke evil of another region (*na:ḍu*), it is a sin. "
If he welcomed strangers instead of friends, it is a sin. "
If he refused food to the hungry, it is a sin. "
If he refused fire to someone half frozen, it is a sin. "
If he troubled the poor and cripples, it is a sin. "
If he misled strangers in the forest, it is a sin. "
If he blocked the way of a charitable person, it is a sin. "
If he spoke abusively to someone, it is a sin. "
If he threw thorns on the road, it is a sin. "
If he tore his dress angrily when it caught on thorns, it is a sin. "
If he told lies, it is a sin. "
If he drove away brothers and sisters, it is a sin. "
If he spat disrespectfully before someone, it is a sin. "
If he showed ingratitude to a priest, it is a sin. "

If he spat on Ganga [a stream or river], it is a sin. [*Response*: A Sin]
If he polluted Ganga with feces, it is a sin. "
If he crossed a river without paying respects to Ganga, it is a sin. "
If he broke the dam of another, it is a sin. "
If he let someone's water supply run away, it is a sin. "
If he urinated on burning embers, it is a sin. "
If he bared his rice-cakes [buttocks] in the sunshine, it is a sin. "
If he threw dirty water towards the sunshine, it is a sin. "
If he watched the snake swallowing the moon [an eclipse] and then slept, it is a sin. "
If he gnashed his teeth at innocent babes [in ridicule], it is a sin. "
If he committed adultery with a woman, it is a sin. "
If he raised his foot against his mother, it is a sin. "
If he laughed at a sister with evil in his heart, it is a sin. "
If he got on a cot while his father-in-law slept on the ground, it is a sin. "
If he sat on a raised veranda while his mother-in-law sat on the ground, it is a sin. "
[*Alternatively, for a dead female—*
If the daughter-in-law climbed up into the loft when her mother-in-law was in the house, it is a sin. "
If the daughter-in-law sat on the sleeping platform it is a sin. "
If he killed anything, snakes or cows, it is a sin. "
If he caught a bird and fed it to a cat, it is a sin. "
If he killed lizards and blood-suckers, it is a sin. "
If he poisoned someone's food, it is a sin. "
If he made false statements against someone, it is a sin. "
If he showed a wrong path, it is a sin. "
If he complained to the magistrate, it is a sin. "
If he went against natural instincts after reaching adulthood, it is a sin. "
If he committed even three hundred sins, may Lord Siva forgive his sins and take them from him! "
May all his good deeds open up the way,
Holding the feet of Brahma,
Holding the feet of Bassava set free today,
Holding the feet of six thousand godly saints,
Holding the feet of twelve thousand pious people,
Let the saints and the pious ones join together for him!
May he become one with them!
May he reach the pious group!
May the doors of heaven be open for him!
May the door of hell be closed to him!
May the hand of heaven be extended!
May the hand of hell be shortened!
May the ocean of death give way before the departed soul!
May his soul reach eternal bliss!
May he be reunited with the other life-partner!
May the door of heaven open suddenly!
May splendour appear everywhere!
May the burning pillars be cooled!
May the thread bridge become firm enough for his passage!
If his path is obstructed by thorn bushes, may he easily find a way through them!
May his path to the other world be clear of all obstructions!
May the house of wickedness be closed!
May the house of righteousness prevail!

May the mouth of the dragon be closed!
May the pit of worms be closed to him!
May the wicked hands of the deceased be prevented from sinning yet again!
May his hands be extended in charity towards others yet again!
May he pass on to the place of the golden pillars!
May he lean on the silver pillars after his journey!
May all his sins be forgiven!
May he seize the feet of a thousand priests!
May he seize the feet of three hundred priests!
May he seize Bassava's feet!
May he approach the feet of Brahma, the Deity who has given everything!
May he approach the face of Siva!
Thus may the soul of the Departed join Siva's generation!
So be it! [*Response*: So be it!]

(Based on Metz 1864: 80-1; Jagor 1876: 197-8; Thurston and Rangachari 1909, I: 114-15; Samikannu 1922: 36-8; Belli Gowder 1923-41: 1; Karl 1945: 22-4; Noble and Noble 1965: 262-72; Hockings 1988: no. 1125 a-d.)

Restrictions of space prevent us from exploring Badaga ethics more fully here: cf. Hockings 1988.

The Kuṟumba Tribes

DIETER B. KAPP and PAUL HOCKINGS

9

They are quite secluded from the rest of mankind, dwelling in holes and caverns in the sides of mountains, deriving a precarious and wretched subsistence from some ill-cultivated spots near their dens, from the animals which they may catch and destroy, and from presents received from the Todas and Burghers for assisting at their ceremonies.

—De B. Birch (1838:107).

Introduction

It is certainly ironic that while the Kuṟumbas were the first Nilgiri community to become known to scholars, they were the last to be studied at all closely.[1] In the year 1800, two decades prior to the first British settlement on the Hills, the peripatetic Dr Francis Buchanan had already encountered and reported on the Kuṟumbas of the eastern slopes. Admittedly his comments were both brief and inaccurate but they set the stage, so to speak. He identified two tribes south of the Kaveri River, namely the Cad (i.e. *ka:du* or 'jungle') Kuṟumbas and the Betta (i.e. *bëṭṭa* or 'hill') Kuṟumbas (Buchanan 1870: II, 126-9).

Confusion about divisions within the Kuṟumba tribe or tribes has characterized nearly all subsequent accounts of them. Metz (1864:106) noted three classes, 'Mullu Kuṟumbas, Naya Kurumbas, and Pania Kurumbas', and Shortt (1868:47) agreed with him on these three 'sub-divisions'. Breeks (1873:48-9, 52), expanding on the 1871 Census which counted 613 under six categories, produced a distinctly longer list of 'divisions', viẓ.: Kuṛumbas, Botta K., Eda K., Karmadiya K., K. Okkiliyan, Male K., Kambale K., Mullu K., Anda K., and also Pāl K., Naya K., Kurali K., Malsar K., Pania K., and Jain K. In the present century the Census of India (Nambiar 1965: 8) has identified 'at least five sects', viz.: Pal Kuṟumbas, Jen K., Urali K., Betta K., Mullu K.

Another dimension of the identity problem was touched upon by Thurston (1898) in his article 'Kurumba or Kuṟuba?' Here he had put his finger on a real economic and linguistic division within the group. Those who were called Kuṟubas were Kannada-speaking shepherds living in the Mysore Plain (P.K. Misra 1969). The Tamil form of the word, Kuṟumba, was applied however to inhabitants of the Nilgiris and borderland parts of the Coimbatore District, in both of which they were speaking a language belonging to the Tamil-

Malayalam sub-family of Dravidian, and subsisting by gardening and foraging. Since these Kuṟumbas tend to live in fairly thick, sloping forest, the tending of sheep is not a possible livelihood for them.

A more glaring dichotomy seems to separate Kuṟumbas of the present day—variously described by Europeans as of 'squalid and somewhat uncouth appearance' or 'a rude tribe...exceedingly poor and wretched'—from those of medieval times, then rulers and possessors of the territory of Tondamandalam, or Kurumbrana:du—in short, a fairly powerful and civilized people. Are they the 'same people', or is this a case of 'devolution?' History is unclear about the answer, so widely scattered are Kuṟumbas and Kurubas today and so various their occupational bases.

People of this name or something very close (Kuruba, Kuruman) have been reported across a wide swath of South India, from Pudukkoṭṭai District in the southeast, through Cuddappa, Tanjavur, Tiruchirappalli, Bellary, to Bangalore, Coimbatore, the Nilgiris, Mysore, South Canara, Coorg, and Malabar in the west (Macleane 1893:222). Undoubtedly they present considerable local variation.

Recently Kapp has examined just the Nilgiri groups and identified seven distinct Kuṟumba tribes there, to be distinguished not only by their names but also by their region of residence. This then is his categorization:

1. Ālu Kuṟumbas (SW, S, SE, and E slopes and glens)
2. Pālu Kuṟumbas (SW slopes and uplands)
3. Mudugas (SW foothills)
4. Bëṭṭa Kuṟumbas (NW and N foothills)
5. Jēnu Kuṟumbas (N foothills)
6. Muḻḻu Kuṟumbas (W foothills; cf. Misra's chapter, below)
7. Ūrāli Kuṟumbas (W. foothills).

Each of these is a distinct ethnic group[2], differing from the others in language, religious beliefs and other cultural features. The Ālu and Pālu Kuṟumbas and the Mudugas do intermarry with each other, as we shall see, but the remaining four Nilgiri tribes are endogamous. There are undoubtedly other related groups in Karnataka, such as the Kurubas and the Solagas, but they fall beyond the scope of this chapter because they live to the north of the Nilgiris.

Taking just the Ālu Kuṟumbas as an example, we find that this tribe is divided into two exogamous, non-totemic phratries (*ōriga*) and these in turn are further divided into clans. One contains three named, exogamous clans, while the other has four. The Ālu Kuṟumbas practise cross-cousin marriage between any of the clans forming each phratry. Within each phratry the clans share a fictive brotherhood, however, so that intermarriage is impermissible. Furthermore, there is some ranking within each phratry, since Nāgara and Bëllega are the two clans referred to as 'elder brothers' in each of the two phratries (Fig. 9.1). Children always belong to the clan and phratry of their father. It occasionally happens that Ālu Kuṟumbas enter into marriages with Pālu Kuṟumbas or Mudugas, two other Kuṟumba tribes with a similar clan organization. These latter two groups also regard each other as related, and they too intermarry with each other sometimes.

Despite the smallness of their settlements, Ālu Kuṟumbas recognize a village headman (*maniagara*), an assistant to him (*bandari*), a second head (*talevaru*) and an assistant for that man (*kuṟudale*); although the two assistants are not encountered in every village. All four offices are hereditary in the male line.

eldest clans, mainly resident in the Kundas:	NĀGARA			+	BËḶḶEGA	
younger brother clans, mainly on the northeast Nilgiri slopes:	KAYIGËṞU	GOBEĀḌA	ĪRAPAṆE		MĀCOLE/ BËḶḶAṞE	BËḶḶAKU

Fig. 9.1 The Two Ālu Kuṟumba Phratries and their Constituent Senior and Junior Clans.

There are six other sacred offices in the tribe. Foremost is the priest (*maṇṇugara*). Also important is another kind of priest, called *kāṇi-kuṟuma* , who operates as a kind of adjunct priest in a Badaga commune and receives compensation in kind for this ritual service (see below). Then there is a diviner (*kanigara*), an exorcist (*devvagara*); and a sorcerer (*odigara* or *odia*), who with the help of spells, herbs and roots can bring sickness or death to an enemy. Distinct from this last role is the therapist (*maddugara*), whose title literally means medicine-man, and who is of course a curer. Along with a priest and an exorcist, he will certainly be found in every village. One final role is also concerned with magic: this is the *pilligara* or wizard, who (like the *odigara*) supposedly can command such secrets as turning himself into an animal, something that the other Nilgiri communities mistakenly believe is possible for any Kuṟumba. Although little has been made known thus far about these various sacred and secular roles, Kapp has published (1980) a detailed account of the ordination of an Ālu Kuṟumba into the priesthood. In this, the central part of the ceremony is an elaborate offering to the seven ancestors and the seven ancestresses of the tribe.

Settlements

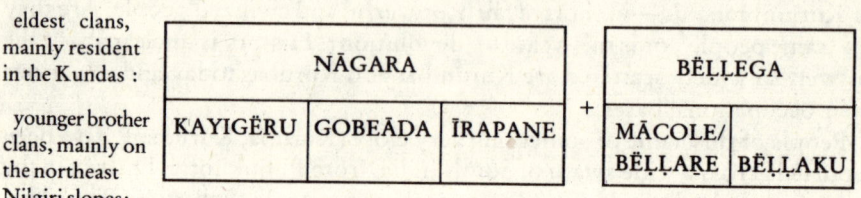

Noble (1968:105) has pointed out that various nineteenth century accounts of Kuṟumbas reported them to be dwelling 'in 1) caves or rockshelters, 2) dispersed dwellings near forest clearings, and 3) houses or huts in small hamlets'. Morgan, a forest officer, noted (1876:99) how isolated the lives of some Kuṟumbas were even as compared with the neighbouring Irulas, 'a couple frequently living alone, in the centre of dense forest and cultivating a small patch of raggy' (finger millet). Yet on the very next page Morgan showed that their social and economic isolation was by no means total: in addition to honey, 'These men collect almost all forest produce, such as soapnuts, myrobolans, dye barks, etc., which they sell for a trifle to the plain traders whose debtors they are, in return for salt, grain, chillies and other necessaries....' (1876:100).

The variability in Kurumba residence is still observable. While today the great majority of Nilgiri Kurumbas live in substantial huts, Hockings recently encountered a sizeable group of them, twelve families, living under a concrete road bridge at the foot of the hills. Clearly the geographical location of a group, more than its tribal affiliation and customs, determines whether its members will house themselves in a cave or a building. There are in fact relatively few good residential caves in the area; one very near Ootacamund which seemed quite convenient and habitable was tested by Hockings, who was surprised to discover not the slightest trace of even a brief occupation at any time in the past. Thurston and Rangachari (1909: IV, opp. 169) offer the only known photograph of Kurumbas living in a large rockshelter, although others were reported to have been occupied some fifty years ago (Noble 1968: 107). All the nineteenth century ethnographers agree, on the other hand, that the Nilgiri Kurumbas were usually to be found in hamlets of four or five houses scattered amongst poorly kept garden plots (Harkness 1832: 128-30; Shortt 1868: 49; Breeks 1873: 50). Bananas, mangoes, jackfruit, maize, yams, chillies and millets were the most usual garden produce, according to these and other authorities. Two writers, however, also mentioned that the Kurumbas were practising slash-and-burn cultivation in that period (Ward 1821: lxxvi; Cleghorn 1861: 140); they no longer do so. Breeks offers an excellent early photograph of one Kurumba settlement (1873: Pl. XXVI). Noble's study of Kurumba settlements (1968: 107-8, Pl. VIII and IX) emphasizes that the house plans vary greatly, and some are mere watch-huts beside their small fields. The dispersed settlement pattern is usually associated with sloping terrain. Building materials are invariably local timbers and grass thatch, unless a plantation or government agency has built a row-house for the tribal people of something more substantial, whether brick and mortar or corrugated iron sheets.

The Ālu Kurumbas and their Economic Setting

Space does not permit a detailed analysis of all of the varying cultures of all Kurumba tribes. Elsewhere in this book Rajalakshmi Misra examines the Muḷḷu Kurumbas, an agricultural tribe occupying comparatively low, rolling land in the Wainad Plateau. Here we will concentrate mainly on the Ālu Kurumbas, a group which contrasts in many ways with the Mullu Kurumbas. This tribe lives on the steep slopes and in the glens which border the Nilgiris and Coimbatore District to the east, at elevations ranging from 600 to 1600 m.

In 1981 the Nilgiris District had a 'census total of 4,874 Kuṟumbas', plus perhaps a few dozen Uralis. But there were in the mid-1970s approximately a thousand Ālu Kuṟumbas, living in about thirty-five villages through the long, narrow belt of sloping forest, as well as about 800 Pālu Kuṟumbas, 1,500 or so Muḍugas, and in addition the other groups noted. The Ālu Kuṟumbas still practise some shifting cultivation, set traps and snares for birds, mammals and fish, and gather wild foodstuffs in the jungle. From there they bring tubers, leafy vegetables, fruits, medicinal herbs and roots, honey, beeswax, resin, vines and timber. Their traditional crops (still cultivated) are common, Italian, and little millet, yams, maize, as well as numerous vegetables and spice plants, and plantain, jack, mango and orange. In their settlements may be found a few poultry, goats or occasional buffaloes. But although this may seem to offer a varied-enough diet and even a few saleable items, the persisting deforestation of their territory has in fact made it necessary for most people to work on tea plantations for a daily wage. This was true for some Kuṟumbas even a century ago. And in hard seasons the womenfolk have been known to go begging for food in Badaga villages too (Breeks 1873: 50-1).

The density of the Ālu Kuṟumba population has always been very low, as compared with other parts of South India, and even at the present time it is only about four persons per sq. km. Starvation and disease, especially malaria, have undoubtedly done their part over the generations to keep the Ālu Kuṟumbas at this low density. Probably too, in earlier centuries, the expansion of the Badaga settlement on the top of the massif forced Kuṟumbas settled there to seek homes in lower elevations, below 1400 m., where malaria was rife. This is certainly suggested by the existence of Badaga villages bearing Kuṟumba names (Hockings 1980a: 66). The deforestation of their area was first noticed in the later nineteenth century as tracts of sloping jungle were converted to coffee plantations. Then teak and some other timber species were removed by commercial interests from the area. By the mid-twentieth century a large dam-building programme, carried out in co-operation with the Canadian Government, was producing further ecological and economic effects. Most recently, in the Silent Valley and elsewhere, illegal commercial exploitation of the giant bamboos and other timber species, primarily for wood pulp, has done yet more to reduce the forest cover of the Ālu Kuṟumba territory and that of neighbouring tribes. Contrary to appearances, the soils underlying these forest areas are not abundant in plant nutrients, and the ecosystem has generally functioned to conserve and recycle those that are present. Thus the progressive deforestation of the area has caused much of the topsoil to be washed away towards the Bhavani River, with a consequent loss of plant nutrients. With loss of leafage and increased insolation, the soils have tended to bake hard.

In recent years the best these tribal people have been able to hope for has been a daily wage of Rs 3 when they work at plucking tea-leaves on the plantations. The shifting cultivation and hunting of wild animals have had to be abandoned nearly everywhere. An unrealistic state government once gave some Kuṟumba youths scholarships to learn basket-weaving (which they probably knew beforehand anyway), but then left them with no legal right to gather raw materials for this product in the Government Reserved Forests and, worse yet, no markets needing it! An adequate need for daily labour on the plantations proved more attractive to these young men.

No sketch of the Kuṟumba economy would be complete without a reference to their reputation for sorcery. Every other group living in the Nilgiri area has always believed that all Kuṟumbas are highly effective sorcerers, and the latter have done nothing to disabuse people of this notion. Sorcery perhaps fulfils different functions in Africa or Melanesia, but in this region it constitutes a complex and effective insurance system which leads to a modicum of satisfaction and peace-of-mind, for both the Kuṟumbas and their Badaga clients.

Exchange of Goods and Services with Badagas

Despite the tension-fraught nature of that relationship, the Ālu or Pālu Kuṟumba is no mere predator on the paranoid fears of Badagas he is associated with: he does engage in economic exchange with them too. Kuṟumbas are expected to supply the Badaga communes with three kinds of baskets, two types of winnow, and a large flat basket to hold grain while drying. They also provide 'their' Badaga communes with certain rare species of timber found on the jungle slopes, especially the hard wood needed for pounders. In the past honey, tamarind, beeswax and knotted nets regularly came from the Kuṟumbas, as also did some leaf umbrellas and sleeping mats made of split cane or reeds.

One duty a Kuṟumba might perform for Badagas under exceptional circumstances is to remedy illnesses that Badaga therapists have failed to cure. Certain Kuṟumbas become noted for their abilities to cure specific ailments, either by herbal medicine or by spells, and a Badaga sufferer may arrange for treatment by them through the Kuṟumba watchman whom he knows. Fees or costly presents would have to be given for this favour.

The Kuṟumba watchman (see below) may still play a necessary role in the ritual of the Badaga commune that he guards; but Badagas have no part whatever in Kuṟumba ritual. The sowing and harvest festivals of the Badagas in which the Kuṟumba must participate are the two most important in the annual cycle. In the former he must make certain ritual utterances and cut the first furrows with a plough—today with a mere ploughshare—at an early morning ceremony in January, and then he must sow the first seeds. In the harvest festival, which falls in July, he must sacrifice a buffalo calf early one morning in a Badaga's field. Two days later he leads a procession of Badagas to the temple where he assists their priest with preparations for the ritual. Two days after this he must make a wreath of millet sheaves, attach it to a stone arch (representing a temple gateway), and then reap the first grain of the harvest. Kuṟumbas now claim it was they who initially taught the Badagas about these two ceremonies. During their rare memorial ceremony for a generation that has passed away, the commune Kuṟumba also has an active part to play. On the second day (a Monday) he brings a bamboo ladder to use in building the catafalque, and he also cuts the first bamboo pole to be used in this construction. On the following Sunday and Monday he sacrifices two goats. His role throughout all of these ceremonies is that of an adjunct priest, and for this he receives about Rs 15 (or did so in 1936).

Although he does not help at any other Badaga ceremonies, it used to be considered a wise precaution on their part to invite his attendance at any auspicious ceremony, such as a wedding or naming, and then make him some small gift which might help avert the ever-threatening Kuṟumba malice. For a Badaga funeral he might also receive a payment, although Kuṟumba sorcery could not affect such an inauspicious event.

There are certain situations in which Kuṟumba musicians may play at a Badaga wedding or funeral in place of the more usual Kota band. Traditionally, it could happen either when a dispute had at least temporarily alienated a Badaga family from their Kota partner or when an epidemic was rampant in the Kota's village. Now, when most Badagas have stopped inviting Kotas to any of their ceremonies, some invite Kuṟumba musicians instead, while others dispense with the music altogether.

Until the 1930s when the institution of commune watchman began to lose currency the Kuṟumba got 25 paise each month for every pair of oxen in the commune, or more often its equivalent in grain. At harvest time he would also receive the same sum or else 6 litres of new grain collected from the villagers by the headman's assistant. In addition he and his family would be fed every time in the commune. The Kuṟumba could also rely on Badagas to satisfy all of his clothing needs; for which purpose the headman would first collect grain from the villagers with which to purchase the cloth by barter from a visiting Chetti trader.

When a Kuṟumba watchman dies the headman of the Badaga commune should collect some gifts from the villagers and take them, or have them sent, to the bereaved family. This used to be imperative, to secure the goodwill of the family from which the next watchman would also be drawn.

Finally, we may note that in the pre-British days the Kuṟumba watchmen only got their salt and crude sugar from the Badagas, who made annual expeditions to the plains for such needs. These had largely ceased by 1825, however. however.

Sorcery

No doubt the most interesting and socially important of the Ālu and Pālu Kuṟumba rituals are of that special kind which we label sorcery, witchcraft, or simply black magic. In comparative anthropological studies it has been found that each of these terms carries somewhat imprecise connotations and, for any culture where they are said to occur, it is necessary for the anthropologist to specify precisely what he means, and where in *that* culture he means to draw the line between, say, sorcery and witchcraft, or between the sorcerer, the therapist, and the priest. As for the popular phrase 'black magic', it will quickly become clear from the following paragraphs that what is 'black' (i.e. malevolent) to the Badaga villager is 'white' (i.e. economically beneficial) to the Kur̲umba who claims he can magically protect Badagas from this malevolence. As anthropologists we must report what our Badaga and Kur̲umba informants tell us about this sorcery, rather than attempt to divine whether it has any empirical reality.

Geographically the Kur̲umbas are much more isolated from the Badagas than are the Todas or Kotas; nevertheless certain Kur̲umbas have a crucial role in the Badaga community. A Badaga family does not have a specific Kur̲umba partner, but each commune, as we have seen, appoints a particular Kur̲umba to act as guardian and watchman to all the constituent villages. This is a lifelong appointment which passes from father to son, and involves warding off magical attacks from other Kur̲umbas with the watchman's own sorcery, as well as providing the goods and services we have already mentioned. As with Toda sorcery these 'magical attacks' were once a common way of interpreting crop failures and animal diseases as well as most human ills. A Kur̲umba watchman is expected to spend part of the time in his Badaga commune, where he generally lives on the veranda of the headman's house, for he may not enter any of the houses. If he does not appear there very often the Badagas nevertheless believe he is magically present each night or else knows by clairvoyance what is going on in the villages; yet his visits tend to be frequent, for he is fed by the commune. The standard remuneration should nowadays be 3.7 litres of wheat or millet from each household at each harvest.

Thus *vis-à-vis* their many Badaga clients, Kur̲umbas—*some* Kur̲umbas, called in Badaga *ka:ni Kur̲umba*—had four magico-religious roles as well as a complex economic one. The *ka:ni Kur̲umba* was a village watchman who, in his role as *sorcerer*, protected his particular Badaga commune from the supernatural malevolence of other Kur̲umbas whom we may term *wizards*, through the exercise of his own supernatural powers. Most Badagas still believe the Kur̲umba is the most effective of all south Indian sorcerers: he can kill people at a distance with a spell, can secretly remove internal organs from the living, can rape women without their knowledge, can enter a locked door, and can change into an insect or any sort of mammal.

The watchman position also entailed being an *adjunct priest*. At the sowing and harvest festivals of the Badagas, these watchmen had a distinct role to play in collaboration with a Badaga priest.

As with the distinction between sorcery and witchcraft among the Azande

(Evans-Pritchard 1937: 387-423) and other West African tribes, we find that Kuṟumba witchcraft is something that is merely alleged: people assert that it has happened, but they only see its consequences. Kuṟumba sorcery, on the other hand, is we believe something that is actually *done* by some Ālu and Pālu Kuṟumba sorcerers, usually males but occasionally females. (Fifty years ago one up-to-date Kuṟumba lady was reportedly 'experimenting with flying'; Hockings 1980a: 133, 21n.) Thus it is possible for an ill-intentioned Badaga to go to his Kuṟumba associate and purchase, for a high price, the performance of a spell that will magically produce the desired fate on that Badaga's enemy. Several of the epic Badaga poems detail both the means of performing Kuṟumba sorcery and its dire, usually fatal, effects. A fourth category of Kuṟumba associate was the *therapist*, usually a specialist in curing particular kinds of malady.

Until perhaps two decades ago Badagas often levelled charges against virtually defenceless Kuṟumbas; even today the suspicion directed towards them is strong. Formal charges of witchcraft, handled by Badaga commune councils, used formerly to unite the Badaga community against an external threat of (or target for) aggression; and the massacres which sometimes followed such charges were clearly a tension-reducing activity for the Badagas:

> In 1836, no less than fifty-eight Kurumbas were murdered, and a smaller number in 1875 and 1882. In 1891, the inmates of a single Kurumba hut were said to have been murdered, and the hut burnt to ashes, because one of the family had been treating a sick Badaga child, and failed to cure it.... Again, in 1900, a whole family of Kurumbas was murdered, of which the head, who had a reputation as a medicine man, was believed to have brought disease and death into a Badaga village. The sympathies of the whole countryside were so strongly with the murderers that detection was made very difficult, and the persons charged were acquitted (Thurston and Raṅgachari 1909: I, 86).

The techniques of killing varied from case to case, but usually involved burning down a hut or even a whole hamlet to make the slaughter seem accidental. In 1875 a head Kuṟumba had bewitched a Badaga woman into loving him and had supposedly caused a fever among Badagas. He was killed in a gun duel and altogether six households of Kuṟumbas were massacred with axes. The story was immortalized in the epic ballad of *Kadare Gauḍa*. In the gruesome case of 1882 the family of another sorcerer was impaled and burnt alive. Common to such killings was the fact that they followed closely on some catastrophe in the Badaga village responsible for the carnage, and also that one or two Todas were always involved. According to Badaga witnesses at the ensuing court trials Kuṟumba sorcerers had caused deaths or epidemics in the Badaga village, and this sorcery might be halted by the massacre only if a

Toda, himself considered a sorcerer, were to strike the first blow against the unfortunate Kuṟumbas.

Badagas claim that until quite recently a Kuṟumba council would sometimes authorize Kuṟumba youths to go out for a few weeks and practise the sorcery they had learned. At such times the Kuṟumba watchman would warn all the Badagas he was supposedly protecting that this was going on, and that they should not leave their homes after dark or take needless risks; nor should they blame him for any deaths, as each youth was permitted by the tribal council to kill one man. Since Kuṟumbas could also kill each other it was believed that no fledgling sorcerer would try his skill against members of a Badaga commune protected by a more powerful Kuṟumba, as the latter might well retaliate as his duty demanded.

Several of the remarkable epic ballads of the Badagas have as their moving force the envy or malice, not of a community of Kuṟumbas in concerted action, but of one isolated Kuṟumba individual whose sorcery must bring its tragic and inevitable result, the death of a Badaga hero or heroine. The mechanics of his magic are well described in these poems, but the motivation for it is less clear: one cannot fight against the psychopathology of a sorcerer, but must accept it as one's fate. Thus in *Ko:litippe* the heroine's mother-in-law, Ali Ma:di, was walking through the woods alone—in itself, a big mistake—when the Ka:ṇi Kuṟumba of Hadamotta, out looking for honey, noticed her. 'He turned himself into a black dog....He came round and round her leg... She was filled with fear. With that...he became a black cat...He looked at her, Ali Ma:di, and saw that she was like the rising sun...So he took out her kidney. He put grass in its place and thus made her die in five days...'

This was the standard operation, it was believed. Badaga informants explained that Kuṟumbas can only remove their victim's kidneys, never replace them; yet where, as in the above case, only one kidney is removed, another Kuṟumba can do magic to get rid of whatever junk had previously been put in the body in place of the removed kidney, and thus might save the victim. Otherwise, he or she dies within a couple of days if something very irritating like cactus or stones is inside; or may live as long as eight days if it is flowers and grass. Anyway, Ali Madi's fate was sealed.

Life-Cycle Rituals

Kuṟumba ritualization is perhaps elaborated to a lesser extent than with the Todas, Kotas or Badagas. Some writers have gone so far as to say that the Kuṟumbas have no wedding ceremony at all, and none to celebrate a birth either (Breeks 1873: 54; Grigg 1880: 212). That two such crucial events in people's lives should be marked in no way whatever would seem unlikely, and indeed ceremonial has been noted on these occasions. Kapp (1978b) discusses in considerable detail the birth and naming ceremonies of the Ālu Kuṟumbas;

and elsewhere (1978a) the head-shaving, ear- and nose-boring, tattooing, and puberty ceremonies. He has also studied the betrothal and wedding ceremonies.[3] With a death, the Ālu Kuṟumbas follow the custom of burial, using elaborate rituals which differ according to the sex, age, and rank of the deceased. In recent times they also have taken to cremation, but only in the case of a headman, supposedly out of respect for him. To the question, however, whether cremation has been adopted in imitation of the funeral practice of the Todas who are highly esteemed by the Ālu Kuṟumbas, or of the Hindus, a clear-cut answer cannot be given.

Religious Belief

Living as they do near the South Indian plains, it has not been altogether possible for the Ālu Kuṟumbas to preserve their ancestral religious beliefs and practices without the accretion of some Hindu ideas. Their religion is essentially an ancestor cult, and it is the ancestors, in particular their mythic ancestress, called Kuṟupade-Tāyi ('Mother of the 'multitude' or 'army' of the Ālu Kuṟumbas'), whose benevolence and protection they invoke by regular offerings. They live in constant fear of evil spirits which try to take possession of them; the fiercest of these spirits, known as Munirāvaḷa, being propitiated annually by a buffalo sacrifice. On the subject of spirits, we should also mention particularly the guardian spirit or tutelary deity of the Ālu Kuṟumbas, named Munīspura. He is considered a good spirit insofar as he controls the well-being of the tribe, and expects a yearly goat sacrifice as propitiation.

Noble has succinctly described the location of Ālu Kuṟumba sacred centres, and some variability in their form. He states (1968: 109):

> Away from Kurumba houses, close or far, there are burial-cremation centers. Combined with or some distance away from these centers there are memorial centers where water-worn stones are laid in dolmens, or upon especially erected stone platforms. Rockshelters have occasionally been used for worship...Kurumbas also maintain small sacred clearings, sometimes associated with pagoda trees, in which there may be a shallow altar, some...tridents, crude pottery images from the lowlands, and a large sacrificial knife.

The veneration and worship of Hindu gods is unknown to traditional Ālu Kuṟumbas, though two gods do play an important role in their mythological and religious concepts: viz. Brahmā (Buruma-Dēva), also called by some of the elders Āka, the Creator, who having created the world and living beings disappeared for ever; and Yama (Emme-Daruma-Rāja or Emme-Rāja), the 'Buffalo- (Justice-) King', the Lord of Death, the concepts associated with whom we shall now discuss. Besides Brahmā and Yama, other gods too, such as Indra (Indura), Sūrya (Cūria), Candra (Candura), and Śani (Cani-Bagava^{n}), figure in some of the legends told by the Ālu Kuṟumbas but not, however, in their religious beliefs. These stories are clearly rooted in Hindu traditions.

As far as the Hindu deities Śiva (Civa-Perumānu, Ispura), Pārvatī (Pāruvadi, Īspuri, Baddura-Kāli), Gaṇeśa/Vināyaka (Bināyagaru), Murukaṉ (Murugaru), Ayyappan (Ayyappa), Kṛṣṇa (Māyaⁿ-Kisuna), Lakṣmī (Laccimi) etc. are concerned, acquaintance with them and worship of them is of fairly recent date, and is due to the all-pervading Hindu influence which, in particular, the younger generation of tribals succumb to. So today the young Ālu Kuṟumbas occasionally visit Shaivite shrines and temples to offer *pūjā* although the elders strictly oppose this. Thus a tendency towards embracing Hinduism is much in evidence and is slowly but surely growing among young people. What has to be expressly noted, however, is the fact that the concept of reincarnation has not gained ground in the religion of the Ālu Kuṟumbas so far.

World View

The Ālu Kuṟumbas differentiate four main regions of the world: these are the region of men, the middle region, the upper region, and the nether region. The region of men (*nare-lōka*) is imagined as a huge disk floating in the centre of the world. Above the region of men, somewhere in the ether, there is said to be the middle region (*cagada-nādu*), which is visualized as a high and steep mountain. The top of this mountain forms the realm of the god Yama, his accountant, his messengers, and his helpers. A thread-bridge (*nūlu*) links the middle region with the entrance to the upper region (*mēlōka*) which is imagined as a still more elevated and steep mountain. On top of this mountain there is said to be the mansion of the guardians of the upper region, a huge demoness or Rākṣasī (*arakkaci*) and her daughter. Beyond this mountain there extends the upper region which is considered to have two parts, one (*dēva-lōka*) being the region of gods (*kadavuluga/dēvaruga*) and Ṛsis (*munivaruga*), the other being the region of the ancestors of the tribe, the paradise of the Ālu Kuṟumbas (*māyaⁿ-lōka*). At the entrance to the region of ancestors there is said to be a large 'refuse heap' (*gubi*) that plays an essential part in the religious beliefs of this tribe, as we shall see. The nether region (*pādala-lōka*) is thought to be located deep below the region of men, being demarcated by the roots of the two mountains which form the middle region and the entrance to the upper region respectively. It is the abode of Yama's host, the spirits (*gōlu*), as well as of the souls of animals.

Life on earth is characterized by deeds performed according to one's own will and wishes, though they are predestined. Life after death, i.e. life which starts with death and ends with the reaching of paradise, means purification from worldly sins. The span of life which has to be spent lying on the 'refuse heap' of the region of ancestors serves for purification from pollution sins. Life in paradise is marked by eternal bliss, i.e. eternal youth, love, abundance of vegetarian food, music and dance.

Concept of Soul

The Ālu Kuṟumbas hold a dualistic concept of soul, whereby each individual is endowed with two souls, the so-called Big Soul (*doḍḍ-ujuru*) and the Small Soul (*kill-ujuru*). Furthermore they distinguish between two shadows, the Visible Shadow (*tōṟō nālu*) and the Invisible Shadow (*tōṟāda nālu*). Thus the Big Soul goes always together with the Visible Shadow, whereas the Small Soul is closely connected with the Invisible Shadow. During life-time the two souls and the two shadows are inseparably united; when the after-life begins they are separated only to join finally and form an inseparable unit during the life in paradise.

Life-breath is not regarded as being associated with a soul. The Big Soul may also be termed the Dream Soul as it is said to leave the body at night and roam about at will during dreaming. It forms the more powerful aspect of soul. Those evil spirits which are desirous of taking possession of men are incapable of gaining control of the Big Soul, but they can take the less powerful Small Soul which they are said to carry off and keep prisoner if their attempt is successful.

At the occurrence of death the Big Soul along with the Visible Shadow leaves the body or, to be more precise, it is dragged out of the body by Yama, while the Small Soul along with the Invisible Shadow lingers in the dead body until the moment of interment or cremation. Then the Small Soul along with the Invisible Shadow is said to quit the body and take up its home in a long water-worn pebble which, after every death, is placed by the Ālu Kuṟumbas inside a dolmen (*nālu-padi*) located near the cemetery. Here the Small Soul dwells, but it may also leave its abode and hover around, although not however beyond the boundaries of the cemetery. The separation of the Small Soul, which is bound to the earth, from the Big Soul which sets out on its burdensome journey to the other world, lasts till the celebration of the so-called 'dry funeral' (*baṟa-cāvu*). This 'dry funeral' is a kind of memorial ceremony, being performed only once or twice in a century in memory of all fellow-tribesmen who have died since the celebration of the previous one. When the 'dry funeral' is completed, the Small Soul and the Invisible Shadow are said to unite themselves with the Big Soul and the Visible Shadow to form thenceforth an inseparable unit forever.

Yama and His Retinue

Yama is regarded as the Lord of Death and the Ruler over the Nether Region. He is feared for his function as the Taker of Life and the Ruler and Judge of Souls. His vehicle is a black buffalo (*emme*), and his ominous attributes are a noose (*pāca-kavuṟu*) and a net (*bīcu-bale*). His realm is the middle region which, as already mentioned, is visualized as being situated on top of an unimaginably high and steep mountain. There, in the midst of an impenetrable jungle, arises the large palace of Yama and his retinue. The entrance to Yama's abode is formed by another palace which is the residence of Citragupta (Cittura-Pattura), the accountant who holds the rank of a minister. By instruction of Yama, he keeps account of the deeds of men and puts everything down in a huge book. Besides Citragupta, Yama is assisted by two messengers (*emme-dūduva*^{n}). Only the name of one of them is given, namely Āgāca (cf. Toda *kosn*) which, most likely, is a corruption of the name of Agástya who, according to the Mahābhārata, is regarded as the foremost of the seven priests of Dharmarāja. Furthermore, the Ālu Kurumbas have the notion that there are three animals as helpers of Yama and Citragupta: a black dog considered as a true helper; a parrot; and a particular insect, called *uputi-kunni*, both regarded as mischievous helpers. The parrot and the insect may be characterized, so to speak, as 'over-zealous' in their help-mate function. Thus the parrot, by manipulating astrology, is forever trying to convince the overburdened Citragupta of the imminent death of some person. As for the insect, it pursues the same end, but resorts to another method: it is always in search of human hair which, once found, it rolls in ashes to make it white and then presents it to Citragupta, declaring that the particular person is obviously of old age and on the point of dying. (That is why the Ālu Kurumbas are always anxious to burn every single hair which falls out or gets stuck in the comb.) In both cases, however, the black dog prevents Citragupta from a rash decision by persuading him first to examine his records. With regard to the hair smeared with ashes, the black dog takes this and dips it into water to determine the truth.

Yama's subjects are represented by a host of spirits, called *gōlu*, which dwell in the Nether Region. On every night following a burial these spirits are said to come in crowds to the cemetery carrying along their Lord, who is seated in a palanquin. Once arrived there, they dig out the corpse, cut, fry, and feast on it and—while offering the liver to their Lord—sing and dance frantically until dawn.

After-Life Experiences of a Soul

Whenever the life-span destined for a person by Fate (*bidi*) comes to an end, Citragupta informs Yama of this and, accompanied by his two messengers, he, Yama sets forth on his buffalo to pull out the Big Soul (along with the Visible Shadow) of the doomed person by means of his noose or net. This can be done from a distance of about one mile. (This duty, the only action performed directly by Yama, may be entrusted to one of his messengers.) Having pulled out the soul he puts it into the hands of one of the messengers and returns to his abode. Then the messenger leads the soul for some distance and soon commits it to the care of the black dog, which then escorts it to a particular mountain peak on the Nilgiris, called Ammagallu, which is sacred to the Ālu Kuṟumbas. There is said to be a small pond here from which the soul is requested by the dog to drink some water by using a folded leaf of the *kurukatti* plant (*Hiptage madablota*). This is the first act of purification the soul has to undergo before leaving the Region of Men. Then the dog proceeds to lead the soul, which is clasping onto its tail, up to the Middle Region. Once there, it brings the soul before Citragupta who examines his records and announces the soul's further fate. This fate varies according to the deeds performed by the particular person during his life-time, or according to the cause of his death.

On the whole, five distinct fates are differentiated: the fate of those who have lived a life devoid of worldly and pollution sins; the fate of those who have killed animals; the fate of those who have committed worldly sins; the fate of those who have committed pollution sins; the fate of those who have died an unnatural death.

The soul of one who has lived a life devoid of any sin is straightway sent by Citragupta onto the thread-bridge that links the Middle Region with the entrance to the Upper Region, and required to cross it. This tightrope walking is rendered more difficult by the souls of two in-laws of that particular soul, females in the case of a male soul, males in the case of a female soul; they have been summoned by Citragupta from the region of ancestors in order to take a seat each at one end of the thread-bridge and shake it while the soul is crossing. Since, however, it is the soul of a sinless person it does not make a false step and reaches the other side safely. There it is immediately grabbed by the Rāksasi, the guardian demoness of the entrance to the upper region, and taken to her mansion where she places it on her large breasts and covers it. Now the soul is expected to lull the Rāksasī to sleep by telling her an interesting story. In the meantime, the Rāksasī's daughter is preparing some glowing charcoal which she spreads in front of the exit to the mansion. When the Rākṣasī has fallen asleep, her daughter pours water over the coals, and cautiously taking the soul from her mother's breasts leads it to the exit and requests it to walk over the bed of hot charcoals. Having performed this act of purification the soul can then ascend to the Region of Ancestors where it is well received by the souls of its relations, and a life of eternal bliss begins for it.

The soul of a person who has killed animals during his life-time is also sent onto the thread-bridge and requested to cross it. It is, however, unable to make the right steps. It tumbles when the thread is shaken, and falls down into the abyss of the Nether Region, where it is tormented by those animals which have been tormented or killed by it during its life-time. Thus it has to suffer physical pain for a certain period of time, which is fixed by Citragupta according to the quantity and gravity of its sins against animals. Once this period of purification is over, the soul is sent up to the Region of Ancestors and, if required, onto the 'refuse heap' of this region for a while.

The soul of a person who has committed worldly sins, such as harassment of fellow-creatures, molestation of women, theft, murder, black magic, etc., is sent back by Citragupta to the Region of Men as a spirit (*pijāci*) devoid of power; which means that it is not capable of possessing men, of eating, drinking, and sleeping, though being desirous of doing so. It is condemned to roam about restlessly within the boundaries of its ancestral village, constantly suffering from hunger, thirst, and fatigue, until the period of time fixed by Citragupta according to the quantity and gravity of its sins against fellow-creatures has passed. Then it is summoned by Citragupta and sent onto the thread-bridge, etc. as we described beforehand.

The soul of a person who has committed pollution sins, such as the physical contact of a man with a woman in labour or in menses, sexual intercourse during menses, sexual intercourse at the time of the celebration of religious ceremonies, or on a funeral day, or during the observance of fasts, extramarital relations, committing mistakes at the performance of rituals, etc., has to undergo purification at the entrance to the Region of Ancestors. Having arrived there it has to go to the large 'refuse heap' of the Region of Ancestors which is located near its entrance. There it has to dwell, feeding on waste, for a certain period of time fixed by Citragupta according to the quantity and gravity of the pollution sins committed by it during its life-time. After this expiation it is welcomed by the souls of its relations.

The soul of a person who has died an unnatural death, such as by continuous harassment, murder, black magic, animal bite, or otherwise is also sent back by Citragupta to the Region of Men as a spirit (*pijāci*) endowed with power, which means that it is capable of possessing people—females in the case of a male spirit, males in the case of a female spirit—and of eating, drinking, and sleeping. It is free to roam about at will, thereby harassing its murderer, possessing persons whom it takes a liking to, and supporting itself by extracting the flavour and essence from the dishes and beverages of the people whom it wishes to harass. Finally, it is allowed by direction of Citragupta to bring about the death of its murderer at the time this person was to die. Thus it is given permission to go on taking revenge until the life-span originally destined for it by Fate has come to an end. At that point, the soul is summoned by Citragupta and sent across the thread-bridge, and so on.

On the whole, the concepts of the Ālu Kuṟumbas which centre on Yama reveal a very refined and complex picture of the afterlife. On one hand, these concepts bear strong resemblances to those of the Hindus as depicted in the great epics Mahābhārata and Rāmāyaṇa, as well as in the Purāṇas; on the other hand, they disclose a number of features which are genuinely local and belong to the ancestral religious belief system of the Ālu Kuṟumbas themselves. At any rate, it is highly interesting to speculate on the extent to which concepts known to be a part of the Hindu literary tradition have taken hold, and on how these have been interwoven ingeniously with other traditional beliefs of the Kuṟumbas. It is questionable, however, whether these Hindu concepts were not perhaps originally a part of the indigenous folk and tribal religions of India, to be integrated only later with the Hindu cosmogony.

Notes and References

1. The information on the Ālu and Pālu Kuṟumbas was gathered by Dieter B. Kapp during a residence in the Nilgiris from May 1974 to April 1976. The fieldwork was sponsored by the German Research Association, to which organization this author owes his deep gratitude. Pages 237-41 are based mainly on fieldwork among the Badagas, done by Paul Hockings with financial assistance from the American Institute of Indian Studies, in 1962-3 and 1976-7.
2. The labels given to these Nilgiri groups, although applied by others, have been adopted by the Kuṟumbas themselves. A brief glossary may help with further reading:
   Ālu — milk
   Pālu — milk
   Kāḍu — forest, jungle
   Bëṭṭa — mountain
   Jēnu — honey (definitely not Jains!)
   Mullu — thorn (bush)
   Ūrāḷi — village head
   Ūr(u) — village
   Muduga — (?)
   When the term *Kuṟumba* is used by a Toda, Kota or Badaga speaker, it generally refers only to an Ālu or Pālu Kuṟumba.
3. A detailed account of the betrothal and wedding ceremonies, as well as of the funeral rites, will be published by Dieter B. Kapp in due course.

An Introduction to the Naikens: The People and the Ethnographic Myth

NURIT BIRD-DAVID

10

> *Jain Kurumbers and Kattu Naykans are a primitive race without a history and they are happy in their mountain slopes with means of subsistence always available in the shape of edible roots. Another decade, they will also be working for wages in the tea estates and earning their livelihood like their brother aborigines of Wynad.*
>
> —C. Gopalan Nair (1911:112-13).

Naiken people[1] live in the northern, northwestern, and western foothills of the Nilgiri region in what is usually referred to as the Nilgiri Wynaad or the Southeastern Wynaad. Their number is roughly estimated at 1,400 (1981). They have not previously been the subject of a participant observation study, but there are many references relevant to them. In general, in the various sources one will find references to up to seven 'Kur̲umba' groups, one of which is sometimes referred to under one of the following names: Jenu or Jain Kur̲umbas, Jen or Ten Kur̲umbas (all these epithets from *jēnu*, meaning 'honey'), Kattu Nayakas (*kāttu* meaning 'forest'), Nayakas, Naikens, Naikrs, Jenu Koyyo Shola Nayakas (meaning the 'honey-cutting lords of the woods'), or just Kur̲umbas. In some sources it is difficult to establish exactly what the connection is between the people referred to and the people who were studied. It is only clear that the studied group is associated with what can be called the 'Kur̲umba complex'.

Drawing on fieldwork conducted between October 1978 and October 1979, my main concern in this paper is to introduce the Naiken people. The Naikens are an integral part of the Nilgiris population: their study, therefore, needs to be tied in with the existing ethnography. Furthermore, the literary sources can be of use in the study of Naikens and, similarly, the study of Naikens may help in the understanding of the relevant literature. A sub-concern of this article is therefore the contrast between the people themselves (as observed during my fieldwork) and the references in the literature to Kur̲umbas. In the first part of the paper I examine the relevant literature, focusing in particular on the articulation of the confused, contradictory and almost mystical picture portrayed of the Kuṟumbas. In the second part, I consider the general characteristics of the Naikens, pointing in particular to the obstacles facing the observer who wishes to describe them in a categorical way. I follow with a description of one particular Naiken community. In the third and last part, I draw the literature and my description of the people together, considering the possibility that the confused portrait in the literature reflects to a great extent the nature of the people themselves, and not necessarily just the methodological handicaps of their commentators.

The Literature

In census reports of the late nineteenth and early twentieth centuries, and in collections of descriptions of tribes and castes, there are accounts of various groups referred to by a prefix followed by the name Kuṟumbas, or by some permutations of the name. Some of the groups are food-gatherers and shifting cultivators, others are agriculturists of sorts. They are spread throughout the Nilgiri Hills and the Mysore plains. The controversial issue in this literature concerns the possible common origin, if any, of this perplexing array of Kuṟumbas and Kuṟumba-like groups. (See summaries in Thurston and Rangachari 1909, IV: 155-77, Rooksby 1961: 26-51, and P.K. Misra 1969: 183-6.) Fürer-Haimendorf pointed out that the name Kuṟumbas is 'one of those tribal names which have done so much to obscure the ethnic picture of many Indian regions' (1952:19). Following his advice, as most other scholars since, I am only concerned here with Kuṟumbas of the Nilgiris region.

ACCOUNTS OF NINETEENTH CENTURY TRAVELLERS AND ADMINISTRATORS

Of the many relevant accounts from the nineteenth century and the first decade of the twentieth century, I refer here to seven sources which are useful for the following parts of the paper, and which convey the general nature of the body of references to Kuṟumbas. The earliest references to the Kuṟumbas of the Western Ghats and the Wynaad are found in the writings of Buchanan and Dubois at the outset of the nineteenth century. For their chronological primacy they are worth mentioning here. Buchanan, in his *Journey through Mysore, Canara, and Malabar* (1807), refers to the division of Kuṟumbas in *two* sections; the 'Cad Curubaru' and the 'Betta Curubaru'. He describes the former as follows:

> The Cad Curubaru are a rude tribe, who are exceedingly poor and wretched. In the fields near the villages, they build miserable low huts, have a few rags only for clothing, and the hair of both sexes stands out matted like a mop, and swarms with vermin...they work as daily labourers, or go into the woods, and collect the roots of wild yams..., part of which they eat, and part exchange with the farmers for grain.... These Curubaru have dogs, with which they catch deer, antelopes, and hares; and they have the

> art of taking in snares, peacocks, and other esculent birds (Quoted in Thurston and Rangachari 1909:IV, 1634).

Dubois refers only to the 'Kadu Kuṟumbas'. While those of Buchanan are seemingly 'more civilized', the Kadu Kuṟumbas described by Dubois are of the 'wild tribes which inhabit jungles and mountains':

> These savages live in the forests, but have no fixed abode. After staying a year or two in one place, they move on to another.... There they sow small seeds, and a great many pumpkins, cucumbers, and other vegetables; and on these they live for two or three months in the year.... During the rains these savages take shelter in miserable huts. Some find refuge in caves, or holes in the rocks, or in the hollow trunks of old trees. In fine weather they camp out in the open.... Roots and other natural products of the earth, snakes and animals that they can snare or catch, honey that they find on the rugged rocks or in the tops of trees, which they climb with the agility of monkeys; all these furnish them with the means of satisfying the cravings of hunger. Less intelligent even than the natives of Africa, these savages of India do not possess bows and arrows, which they do not know how to use (Dubois 1906: 76).

Dubois is the first, to my knowledge, to point out the fear of the Kuṟumba sorcery among the neighbouring people—a theme which is returned to in most subsequent references to Kuṟumbas:

> They have little or no intercourse with the more civilized inhabitants of the neighbourhood. The latter indeed prefer to keep them at a distance from their houses, as they stand in considerable dread of them, looking upon them as sorcerers or mischievous people, whom it is unlucky even to meet. If they suspect a *Kadu-Kurumbar* of having brought about illness or any other mishap by his spells, they punish him severely, sometimes even putting him to death (Dubois 1906:76).

The descriptions of Buchanan and Dubois relate to the Kuṟumbas of the lower slopes and the lowland. References which apply more to the Kuṟumbas of the Nilgiri plateau are found in accounts undertaken from Ootacamund after the 'discovery' of the plateau, and during the intensive British colonization there in the mid-nineteenth century. Breeks's work (1873) includes one such reference, which can be noted for its methodological comments on the circumstances in which information was obtained. Breeks mentions *four divisions* of the Kuṟumbas of the Nilgiris, given to him by the headman of a particular village: *Botta Kuṟumbas, Kambale Kuṟumbas, Mullu Kuṟumbas*, and *Anda Kuṟumbas*. Before that though he mentions: 'It is difficult to get a complete account of the tribal divisions recognized by them. One man will name you one (his own); another two divisions; another three, and so on' (Breeks 1873:48). In actually describing the Kuṟumbas Breeks refers to those on the upper slopes and those in the dense jungle further down at the base of the hills. Breeks is the first of the sources here mentioned to refer to the lack of birth, betrothal and marriage ceremonies among the Kuṟumbas—a theme which predominates in most other accounts, and which has been regarded with some incredulity by contemporary Nilgiri ethnographers: 'There seems to be no marriage ceremony amongst the Kuṟumbas; no early betrothals.... They have no birth ceremony' (Breeks 1873: 54).

In 1880, only seven years after the work of Breeks was published, in his massive handbook, *A Manual of the Nīlagiri District in the Madras Presidency* H.B. Grigg of the Indian Civil Service mentions *six* different 'caste divisions' in the Nilgiris: *Eda Kuṟumbas, Karmadiya Kuṟumbas, Kuṟumbas proper, Kuṟumbas Okkiliya, Male Kuṟumbas* and *Pal Kuṟumbas*, all of which differ by name from the four divisions mentioned by Breeks (see above). He takes the theme already mentioned by Breeks concerning ritual further and states: 'They are said to have no traditions of any kind' (Grigg 1880:213).

Morgan, a forest officer who wrote in 1876, contributes to the general portrait here being built up of the Kuṟumbas the theme concerning the barter of forest produce with villagers and traders of the plains. This is noted in all the references to the Kuṟumbas. But Morgan also highlights a factor which has been generally overlooked in other accounts, namely the important place of indebtedness in the regulation of this trade: 'These men collect almost all forest produce, such as soapnuts, myrobolams, dye barks etc., which they sell for a trifle to the plains traders whose debtors they are, in return for salt, grain, chillies, and other necessaries' (Morgan 1876:100).

A much revised version of the Nilgiri Manual of 1880 came out in 1908 in the volume edited by Francis. In the late nineteenth century there had been an increasing interest and a growing colonization of the jungle Wynaad, which is reflected in the volume compiled just after that period. While it was suggested by Breeks and Grigg that the Kuṟumbas of the low slopes are more 'primitive' even than those of the Nilgiri plateau, Francis asserts that the Kuṟumbas of the plateau are the 'more backward brethren' of the Kuṟumbas of the Wynaad. He distinguishes between *three* kinds of Kuṟumbas: The *Kuṟumbas proper* of the plateau; and the *Ur Kuṟumbas* and *Jen Kuṟumbas* (or Shola Nayakas), both residing in the Wynaad. But in general, Francis' descriptions can be noted for the many details contained in them, and some of the descriptions, of the funeral in particular, would appear to resemble quite closely what has been observed among the Naikens seventy years later. Referring like his predecessors to the lack of certain life-cycle rituals, Francis emphasizes that the people have 'religious ideas...of the vaguest'. But he notes nevertheless a particular feast that takes place annually—one which is mentioned also in other sources; 'A ceremony in their [the deities'] honour, subscribed for by the caste in general, is held in April every year, a cock or two being sacrificed, much rice cooked and eaten by the celebrants, and a dance being held' (Francis 1908: 156).

The last source to be mentioned here is a letter, quoted in Thurston's entry on the Kuṟumbas, from F.W.F. Fletcher, a planter in the Wynaad. The virtue of this is that Fletcher bases his comments on daily and regular contact with people he employs in his plantation. Fletcher makes a simple distinction between two groups, the Kuṟumbas and the Nāyakas:

> It may be that in some part of Wynaad there are people known indifferently as Kurumbas and Shola Nāyakas; but I have no hesitation in saying that the Nāyakas in my employ are entirely distinct from the Kurumbas.... The Kurumba of this part lives in comparatively open country, in the belt of deciduous forest lying between the ghāts proper and the foot of the Nīlgiri plateau. Here he has been brought into contact with European Planters, and is, comparatively speaking, civilised. The Nāyaka has his habitat in the dense jungle of the ghāts, and is essentially a forest nomad, living on honey, jungle fruits, and the tuberous roots of certain jungle creepers (Quoted in Thurston and Rangachari 1909:IV,176).

Fletcher mentions that the Nāyakas have those supernatural powers which are often also attributed to the Kuṟumbas, for example: 'Some Nāyakas are credited with the power of changing themselves at will into a tiger, and of wreaking vengeance on their enemies in that guise' (ibid.:177).

TWENTIETH CENTURY ANTHROPOLOGICAL LITERATURE

The confusion about what and who the Kuṟumbas are seems to have disappeared in the anthropological accounts of the early twentieth century. Reference is simply made to a food-gathering Kuṟumba tribe. The cultural map of the Nilgiris commonly depicted by anthropologists consisted of four tribes—the pastoralist Todas, the agriculturalist Badagas, the artisan Kotas and the food gatherer and sorcerer Kuṟumbas). A fifth one—the Irulas—was also often mentioned with the Kuṟumbas. The anthropological interest (particularly since Rivers' study of the Todas) focuses on the interrelationships between these four tribes (see detailed discussion of the 'traditional interchange', as Hockings refers to it, in Hockings 1980a: 99-131). Of these four tribes, only the first three have been extensively studied during this century (see references in Hockings 1978); the Kuṟumbas have not (until recently) been directly looked at. But there are references to the Kuṟumbas in all the other studies and in most of the anthropological literature where references to the Nilgiris are made. Or to put it more accurately, there are references to the *relationships* between the Kuṟumbas and the other three groups, while the Kuṟumbas themselves are quite literally just described as inhabitants of the jungles on the slopes of the Nilgiris, and as sorcerers and food-gatherers who are also engaged in selling forest produce. For example, the Kuṟumbas provide the Todas with the funeral post at which the buffalo is killed (Rivers 1906: 641). Or to take another example, each Badaga commune appoints for life a Kuṟumba man to act as guardian and watchman to all the constituent villages and to help the Badaga priest in certain agricultural rituals (Hockings 1980a: 122-3). Some aspects of these relationships, in particular those between the Badagas and the Kuṟumbas, are also mentioned in nineteenth century accounts (e.g. Breeks 1873: 50).

In those anthropological investigations which directly dealt with them, there emerge again the divisions among the Kuṟumbas. Fürer-Haimendorf (1952) during the summer of 1948 was the first to my knowledge[2] who briefly investigated some Kuṟumba settlements. He describes *three* 'types' of Kuṟumbas who are found in the Wynaad: the food-gatherer *Jen Kuṟumbas* (also named 'Naikr'), the shifting-cultivator *Betta Kuṟumbas* (also called 'Uralis') and the plough-cultivator *Mullu Kuṟumbas*. There followed two intensive studies of the plough cultivator Mullu Kuṟumbas, Rooksby (1961) on Mullu Kuṟumbas in Malabar, and R. Misra (1971) on the Mullu Kuṟumbas of Kappala in the Wynaad. Both see the Mullu Kuṟumbas as an endogamous and separate ethnic group, and Rooksby even suggests that they are more part of the caste system than part of a tribal one. Noble (1968) studied the relations between the five tribes of the Nilgiris and provides some useful information, especially about the settlements of some Kuṟumbas. The most recent study of Kuṟumbas, and at the same time perhaps the first intensive study of those Kuṟumbas on the upper slopes of the Nīlgiri Hills, was conducted by Kapp (1978a, b, 1980a, b) among three food gatherer, shifting cultivator and plantation labourer Kuṟumba tribes called *Ālu Kuṟumbas*, *Pālu Kuṟumbas* and *Muḍugas*. Kapp suggests that 'the Nilgiri Hills and the adjacent areas are inhabited by *seven* Kuṟumba tribes in all' (my italics): in addition to those studied by him, the Betta Kuṟumbas, the Jenu Kuṟumbas, the Mullu Kuṟumbas and the Urali Kuṟumbas. He further strongly argues that: 'All these tribes are distinct ethnic groups differing from each other in language, culture, religious beliefs, and customs and manners. With the exception of the Ālu and Pālu Kuṟumbas being sister tribes, and the Muḍugas, they are strictly endogamous' (Kapp 1978b: 168).

Finally, it is most important to add here that the notion of the four interrelated Nilgiri tribes is popularly held by many of the residents of the main Nilgiri settlements (whether they be Badagas, Todas, immigrant Indians such as Tamilians and Malayalis or Europeans). Many of the official tourist brochures, for example, describe the four tribes which inhabit the Nilgiris, and the interrelationships between them. There are in addition references to those still primitive Kuṟumbas away in the dense jungle on the lower slopes and at the foot of the hills who are called Jenu Kuṟumbas; who move fearlessly amidst wild animals; live in caves and on trees; and are fearful magicians.

The People

GENERAL CHARACTERISTICS

'Naikens'[3] is the name used by the people in question, but the terms 'Jēnu Kuṟumbas' and 'Kattu Naikens' are also used in this connection. It would appear that as the observer moves closer to the people, both in terms of spatial distance and social strata, so does the name change from Jenu Kuṟumbas to

Kattu Naikens and, most intimately, to Naikens. To take one example from the literature; the commentators whose base of investigation was in the Wynaad, where the people themselves reside, are in fact those who primarily, or at least additionally, use the name Naikens (e.g. Francis Fletcher). To take a concrete example: In the district Harijan Welfare Office I was informed that some Jenu Kuṟumbas live in the Government Game Reserve, about 50 km. away. The local officials in this game reserve did not know of any Jenu Kuṟumbas there. They spoke only about Kattu Naikens. A Malayali contractor who works closely with the people said that their name is Naikens, and speculated that travellers passing by saw the people collecting honey and called them Jēnu Kuṟumbas from that activity. Finally, the people in question did not know who the Jēnu Kuṟumbas were, and said they are Naikens themselves. It is not unlikely that, as Fürer-Haimendorf suggests (1952: 20), the name Naikens was given to them by Malayalam-speaking neighbours. But apart from the name Naikens they have no other name by which to refer to themselves; not even, as is often found in diverse societies, a name which literally means people, men or relatives. Following these people themselves I use the name 'Naikens'.

Gardner's description of the Paliyans fits well with the case of the Naikens, and it is perhaps useful to quote him here:

> 'Their various physical types fall within the range of South and South-East Asian Australoid types, formerly termed Negrito, Malid, Veddid and proto-Australoid. They are physically most similar to the Semang of Malaya and other Indian gatherers (see especially Evans 1937; Fürer-Haimendorf [and Fürer-Haimendorf] 1945: 35; Schebesta 1927; Skeat and Blagden 1906; see also Coon 1958: 29; Fürer-Haimendorf 1943: 17; Olivier 1961: 274-5; Sharma 1963)' (Gardner 1966: 390-1).

Fürer-Haimendorf in particular notes that the Naikens reminded him of the Chenchu he had written of (1952: 20).

There are small conglomerations of Naiken hamlets, each situated in a pocket amidst an area populated by other people, and nearly cut off from other similar conglomerations. Fürer-Haimendorf makes a similar point in his description of the Jēnu Kuṟumbas and the Betta Kuṟumbas where he mentions that the situation in the past is said to have been similar (1952: 25). Each conglomeration, which I shall hereafter refer to as *local community*, consists of 2-5 hamlets, at about 3-11 km. from each other. The hamlets move occasionally, but within the general area of the local community. In itself the local community does not necessarily occupy a contiguous territory, and there may be non-Naiken settlements interspersed between the Naiken hamlets. An intensive and regular contact is nevertheless maintained between the Naikens of the local community, and most if not all of them are related by kinship ties. Most marriages are contracted within this unit.

There may be a few fringe Naiken hamlets outside a given local community, usually located near market villages or places of employment. The Naikens of the local community may occasionally visit these fringe settlements and be visited by people thereof. There are usually at least a few kinship ties between the local community and the fringe settlements.

To refer to a hamlet of the local community, within intra-communal communication, Naikens use terms contextual to the position of the hamlet *vis-à-vis* the other hamlets of the local community. For example the term *male* ('up, mountain'), is used to refer to a hamlet situated on a hill above the other hamlets. To refer to fringe settlements Naikens of the local community use the name of a village near to it. For example, one fringe settlement consisting of two huts only is situated about 6 km. from the small market village of Devala, and is referred to as 'Devala'. Speaking to non-Naikens, Naikens refer also to hamlets in their local community by externally derived names. For example, in correspondence with the non-Naikens' familiarity with the area and the Naiken people thereof, they use the name of a nearby plantation; or a nearby prominent landmark (e.g. 'big rock'); or the name of an old, well-known resident there (e.g. 'Kungan's place'). Naikens have no other names for their hamlet than those described above.

In his description of Kuṟumbas, Breeks makes the point that 'Their villages... are so dispersed over the slopes and base of the hills, that the inhabitants of one locality know nothing of those at a distance' (1873: 50). This is also applicable to the Naikens today. By and large, they do not know of specific Naiken hamlets outside the boundaries of their local community and the fringe hamlets. Both within intra-communal and extra-communal communication they refer to all such hamlets by the term 'Waināḍu', whether or not the hamlets are in the geographical region so named.

It may infrequently occur that one or two of those single Naikens—who frequently move around—may go beyond the borders of the local community. They may then visit other Naikens, the connection with whom in certain cases may be established through favourably positioned fringe hamlets. Or they may meet such other Naikens through casual work taken up in plantations near Naiken hamlets, or even through accompanying ethnographers whose information about the other hamlets has been derived from external sources. Though the majority of marriages are endogamous to the local community, the occasional contact of single people with other external hamlets may lead to exogamous marriages. The couple is then free to choose where to live. If the two respective natal places are far from each other (i.e. generally speaking, further than a one-day's walking distance, or in some cases a cheap bus-fare distance), the newcomer to a given local community practically cuts links with his (her) natal place, and may never visit there again. Similarly the person who moves out of the local community is practically 'forgotten' by his relatives, and it is only sometimes mentioned that he married into and/or lives in 'Waināḍu'. In the less frequent cases of respective natal places being nearer to each other, such as for example marriages between members of the local community and of a fringe hamlet, mutual visits, mostly of single people, may gradually increase. This in turn will lead to an increasing number of intermarriages; and eventually, the external hamlet will practically become a part of what is here referred to as the local community.

Naikens use the term *nama sime* ('our place') to refer contextually to the place where one and one's nuclear family reside, to the place of one's hamlet, and to the place of one's forefathers (remembered back to three generations). When a Naiken is pushed to make a reference to something which corresponds to the enquirer's notion of a 'Naiken territory' he by inference uses the term *nama sime* to refer to all the places where Naikens live.

The Naiken language, which is often referred to by Naikens simply as *nama baśa* (meaning 'our language'), is basically of the Dravidian family, and contains elements of Kannada, Malayalam and Tamil. Kannada however is most predominant in this language, and many of the personal names and kinship terms are directly derived from Kannada. The dialect is distinct, and is hardly intelligible to Kannada speakers. All Naikens speak in addition some Kannada, Malayalam and Tamil. The decided majority of them speak Kannada fluently. But, in general, Naikens of a given community speak fluently that language which they use most frequently in their interaction with the non-Naiken neighbours; and the language which is used around them, and with which they communicate with their immediate neighbours, also affects their own dialect. That is, there are minor local variations between local communities and even between hamlets of the same local community as to the extent to which Malayalam and Tamil words are incorporated into predominantly Kannada-derived vocabulary. F.W.F. Fletcher made a similar observation for the Naikens he encountered in his plantation: 'although the two [Naiken] colonies are within five miles as the crow flies.... The low-country Nāyaka... speech is a patois of Malayālam. The Nāyaka on the hills above...speaks a dialect of his own...derived from Kanarese' (quoted in Thurston and Rangachari 1909, IV: 177). To take an example observed during fieldwork: the local community in question is located in the border area of Tamil Nadu and Kerala. Malayalam-speaking people moved to that area several decades ago to work in a plantation opened up there; they opened a tea shop from which Naikens purchase provisions, and they also supervise those Naikens who give casual labour in the plantation. While the older Naiken people, for example, use mostly Kannada-derived kinship terms, some of the younger people also use terms borrowed from Malayalam. The Malayalam influence is more apparent in those hamlets which are nearer to the plantation. The variations, it should be noted, did not cause any mutual intelligibility, and there remain clear differences between *nama baśa* (as the Naiken language with its different variations is referred to) and both Kannada and Malayalam.

Commenting on their dispersal (see above), Breeks (1873: 50) further continues his observation on the Kuṟumbas: 'They can hardly be said to have any tribal existence as it were, but are isolated scattered families.' Breeks would probably have described the Naikens in a similar way. Embedded in the local community (extending in certain contexts to include also the fringe hamlets) are the economic, social, ritual, and even the effective kinship domains. There are no pan-Naiken social and economic institutions. That is, there are no social aggregates such as council, age-set or club, which cut across the different local communities. The example of an annual feast celebrated by Naikens can illustrate the segregationist social configuration of the people. Naikens annually make offerings to the *devaru* (the 'gods'). Each Naiken makes an offering to the gods of the place where he lives, and to the spirits of his forefathers (if he has not moved too far away from where the forefathers lived). The congregation which annually gathers to make offerings to a particular set of *devaru* therefore consists of all the people of a given hamlet; of most of the people of the local community; and occasionally of a few individuals from the fringe hamlets. No one from 'Wainādu' attends or even knows of the event. There are significant variations of custom between the different celebrations, even though held in hamlets of the same local community.

Any economic, social, and even ritual extra-communal interaction which takes place is with non-Naiken neighbours and not with Naikens of other local communities. Thus, for example, while there is no exchange between Naiken local communities, Naikens are nowadays engaged in barter with their neighbours. Furthermore, while there is no ceremonial contact with other Naiken groups, Naikens of a given local group occasionally attend their neighbours' festivals. Abstracting further beyond the Naiken local community (and its frings hamlets), more often than not the Naiken does not refer to the Naiken people in general, but to the tribal people of the locality, namely, to the collective entity in the locality, of Naikens, Paniyas, and Kuṟumbas (the last mentioned often referred to by officials as Beṭṭa Kuṟumbas).

The Naiken knows that there are other Naikens than those in his locality, and very occasionally encounters them. But the social group he relates to, or to put it in other words, his practical Naiken universe, lies within the local community (extendable in certain cases to include the fringe hamlets).

Following emic perspectives I turn now to describe the Naiken group. I use the term *Naiken group* as alternative to the term 'local community', while reserving the term *Naiken people* for the more abstract category of all those people in the Nilgiris who are Naikens. The Naiken group, it perhaps needs to be recapitulated, recruits all those children born to resident couples and living in the local community. It recruits also those who marry into the local community, and who practically cut their ties with their natal places and adopt the local Naiken way of life: these can be Naiken people from 'Wainādu' or, occasionally, individuals from neighbouring ethnic groups such as Paniyans, Kuṟumbas, and, more rarely, Malayalis.

THE NAIKEN GROUP OF THE 'GIR VALLEY'

From January to September 1979 I was a participant observer among one Naiken local community.[4] which lives in a valley here referred to as the Gir

Valley. According to oral history, Wynaad Gauda people in the nineteenth century had patches of cultivation in part of the Gir Valley. These are the descendants of Badaga Gaudas who probably migrated to the Wynaad from villages in the Nilgiris proper during the eighteenth century (see Hockings 1980a: 31-2).

Plantations started in the Wynaad area in 1839 and, later, the upper part of the Gir Valley and the area towards Pandalur were planted with coffee. In 1874, however, rumours started spreading about the presence of gold in the South-East Wynaad, which culminated in a devastating gold boom between 1880 and 1882. During this short period, Pandalur—otherwise an unmemorable small Indian village—blossomed into a busy mining centre with a race course, a club, a saloon, rows of substantial bungalows and a post office. Most of the plantations in the area were dug up (by forty-one gold mining companies registered in London, and with prices for land reaching £2,600 an acre!). With the subsequent discovery that the mining of the low grade ores was not profitable, the whole area was taken back by the jungle, and only a few semi-wild coffee bushes remained amidst the forests (see Francis 1908: 16-19, 377 for a description of this period).

Around the turn of the century the Gir Valley was penetrated again, but from the Nilambur side. It was then part of the estate of the late Nelliyalam Raja, which was managed by the Nelliyalam Rani (Francis 1908: 365). A Scottish company opened a rubber plantation in Nilambur, and working its way up from there reached the bottom end of the valley, at the maximum elevation at which rubber can be grown. Five hundred acres of the valley were leased out, cleared and planted with rubber. A bungalow was built amidst it for the plantation management personnel. The plantation was named after the valley, and here is referred to as the Gir Rubber Plantation. During the First World War, as the result of difficulties in transportation, low prices for rubber, and diseases which affected the plants, the area was abandoned again. But before the jungle reclaimed the land, in 1928 a Malayalam-speaking Syrian Christian (who is still the owner of the plantation) took the lease over.

Unlike the tea and coffee plantations which predominate in the Nilgiris, the rubber plantation has less impact on its surroundings. Tapping work—a hereditary skill until recently—is what is mostly required and there is only a little demand for non-skilled weeding labour which the local tribal population can supply. Furthermore, tapping is required only during the dry season. Finally, the latex which is collected can be smoked and stored in this form throughout the season. In a plantation the size of the Gir Rubber Plantation the latex needs to be transported out of the plantation only once annually. Indeed until the 1950s Malayalam-speaking tappers came for work during the dry season, leaving their families back home. They were recruited, supervised and paid by contractors, who were the only persons to be directly engaged by the plantation. Workers and their contractors lived in temporary accommodation in the plantation area and had a little contact with the other inhabitants of the valley. A small jungle path, sometimes a jungle track, connected the plantation with the Gudalur-Calicut road. At the end of the dry season the workers climbed up the path, carrying with them the smoked latex, and left for their home villages.

Changes took place in the 1950s. The 1951 Plantation Labour Act abolished contract work and imposed on planters direct responsibility for their employees and detailed employment regulations. Most important, the Act forced planters to give permanent work to the immigrant workers. The Act, which came into effect in the later fifties, has been regularly implemented in the Gir Rubber Plantation. Accommodation (lines, as it is locally referred to) was built in the plantation area not far from the bungalow, and workers and their families settled there. The majority of these families were Mappilas (a poor Malayalam-speaking Muslim caste), but there were also several Syrian Christians, and one or two Hindu and Nayar families; coffee bushes, fruit trees and spices were planted in addition to the rubber in place of some aging rubber trees, to make the small, marginally-viable plantation more profitable, as well as to provide the immigrant tappers with work throughout the rainy season, and members of their family with work the whole year through.

A 'jeepworthy' track to the Gudalur-Calicut road was opened in the 1960s, but was (and is) only usable during the dry season. Because the track is long and twists through the jungle, all the valley residents still use only the steep and short jungle footpath. The plantation is still cut off from the outside, with no telephone line, no electricity connection, and no piped water.

While the Gir Plantation has changed little since the 1950s, in the sixties some changes took place in the forest surrounding it. The Gir Valley was then sold by the Nelliyalam Rani's son to a Tamil businessman. Only 'selection felling' was allowed in the region by that time and the Tamil needed to number, mark and measure the trees in particular areas, and only then could apply for permission to fell specific ones (e.g. old trees, or trees in very dense tracts). He carried out two major spells of felling—between 1962 and 1966, and between 1969 and 1974. Permission from the Nilgiri Collector was given for the felling of 900 trees in each spell, and it provides an idea—probably underestimated—of the extent of the felling. Permission was also given for the clearing of 600 acres in the border area of the valley for further plantation and the clearing took place from late in the 1960s to the mid-70s.

In the later seventies, while deforestation in the Wynaad accelerated and new plantations were opened to absorb the Tamil refugees from Sri Lanka, deforestation in the Gir Valley was stopped. Beyond the 600 acres cleared already, no further massive clearing took place. This has been due to particular administrative and legal circumstances (which need not be discussed here) resulting from the 1975 extension to the Gudalur *taluk* (until then exempted), of the 1948 Madras Estate (Abolition and Conversion into Ryotwari) Act. In response to this situation several Mappila workers have speculatively cleared and cultivated about 15 acres altogether. This area is in itself negligible in terms of the forest at large, but the activity had some impact upon the Naiken community in the valley.

Naikens do not provide a coherent representation of the past, but in general conversations informants occasionally make spontaneous and contextual references to it. The information thus provided (especially by those several informants in the age range of 65-90) is useful when collated for outlining the history of the Naiken group against the background of the general developments of the Gir Valley. The veterans among the Mappila plantation workers have been for thirty years in close contact with the Naikens, and accounts provided by them add further to the information given by Naikens. Since they acutely remember past events, Mappilas' accounts also are useful for cross-checking and putting in context the Naikens' information. Finally, the estate has kept detailed records of its individual workers and those from 1948 onwards were made available to me.

According to Naikens, about 130 to 150 years ago three Naikens came to the valley from what is now the Wild Life Sanctuary of Mudumalai, some 24 km. away: they were a husband and a wife and the husband's brother. They were the first Naikens in the valley and they settled on the rocky side in the bamboo forest. One informant recounts that one of the brothers dreamt that this place would be good to live in; another, that the trio just wandered here and there and then happened to come to the valley. In any case, it is said that the trio liked the place because there was no plantation there, that they 'became familiar with mountains, rocks and roots' and they ate *ganazu* (wild yam). The trio were the grandparents of the oldest living Naikens in the group. This account is not implausible. From early sources (e.g. Francis 1908: 219-20, Grigg 1880: 448-9) it is clear that in the second half of the nineteenth century intensive felling operations took place, and restrictive measures upon the tribal inhabitants were laid down, in the Mudumalai and Benne forests. This perhaps did cause the migration of some Naikens. Furthermore, though each Naiken provides only a shallow genealogy, when the particles of information are drawn together, there does emerge one genealogical tree linking in consanguine ties the majority of living Naikens to the trio, while the remaining few are those who moved into the local group after marrying the descendants of the trio.

Naikens mention the Wynaad Gaudas who lived in the valley, but speak of no particular relationship with them. There are a few references to occasional agricultural labour in return for tobacco, salt, etc. and perhaps it was the Gaudas in whose fields some Naikens worked occasionally.

There is no mention of the 'gold' period, and it is most likely that, isolated in the rocky bamboo forest, the Naikens were little affected, or even had no contact at all with the gold miners.

During the first phase of the rubber plantation (1900-22) some of the Naikens were occasionally engaged for initial clearing work. After the plantation was established, the Naikens had little contact with it and remained undisturbed in their place. Some casual work was taken up by Naikens also in the 1940s during the second phase of the rubber plantation.

The picture portrayed by the veteran Mappila workers of the Naikens in the early 'fifties is consistent and clear. One theme in the descriptions naturally concerns the extent to which Naikens adapted to life in the forest: 'They did not bother about rain and sun, just wandered all around', 'like animals', 'just brought wild yam and fruits and ate'. Another theme concerns the initial contact between the Naikens and the newcomers: the linguistic difficulties of communication, and the shyness and reserve of Naikens when Mappilas came to their hamlets even after some Naikens started to work in the plantation: 'They did not speak with other people... if someone went to a village the children ran into the forest; the rest sat with stooped head and just said they would come tomorrow to work'. Another interesting theme concerns the striking lack of concern which Naikens showed for their kin and even family members: 'Nothing was a problem to these people... they did not bother about anybody...even a father did not bother about his children or about his own father... if there was food they ate it, if there was not they did not.'

Following the developments undergone by the Gir Plantation in the 1950s, Naiken involvement with it increased in the late 1950s and the first half of the 'sixties. This was, to a large extent, in response to the demand placed upon the Naiken labour by the plantation. The 1951 Plantation Labour Act implied that it was economical for the plantation to employ local labourers—namely Naikens. Furthermore, following the changes in land use, there was an increased need for non-skilled labour (mainly for weeding) throughout the year, and the Naikens, accustomed to the jungle terrain, were more efficient at these tasks than the other labour force locally available to the plantation, mostly tappers' wives. In 1958 the first three Naikens were employed as permanent workers. They were all from the Gir hamlet. Nine more Naikens joined the plantation in 1960 and an additional ten took up regular work in 1964. These were from all three hamlets, but the largest number of workers was recruited from the Gir hamlet. Throughout this period, however, even those who worked regularly in the plantation did not completely forsake the forest way of life. Plantation records further confirm accounts of Mappilas on this issue. For example, the plantation personnel brought down to the valley bulk quantities of provisions, and workers could by way of payment take provisions and/or money. Records show that the Mappilas largely took grain, while the Naikens took comparatively little grain and preferred tobacco, salt, etc., or money with which they bought such commodities.

In an attempt to protect the local labourers who were not covered by the 1951 Plantation Labour Act a third category of employment was enforced upon South Indian plantations in 1964, in addition to the permanent and the temporary categories. The casual category, as it was termed, applied to workers who work regularly in the estate, but not exceeding 15 consecutive days, and with no obligation to report for work daily. Though they do not attend work regularly, and have no obligation to do so, the casual workers nevertheless are entitled to most of the benefits endowed on the permanent workers such as minimal wage, paid leave and bonus (calculated in proportion to the number of days in a year which they did attend) etc. A clear trend of resignation of Naikens from permanent status in favour of the casual status took place in the late 1960s and early' seventies.

Employment in the plantation was by far the largest element of change in Naikens' occupations. But other changes were experienced both by those Naikens who did work in the plantation and those who did not. These involved the intensification of those customary occupations prior to the involvement with plantation work, and have resulted from the increasing local demand both by the plantation and its non-Naiken workers for minor forest produce, such as bamboo, grass for thatching, leaves which are used for wrapping provisions, firewood, and so on. Naikens directly exchanged these commodities for some grain, tobacco, and tea and pasties, which became popular with them. The recent encroachment on the forest by some Mappila workers, small as it is in terms of the total forest area, created further demand for Naiken labour and skills: especially for clearing the plots, for bringing from the forest building materials, and for constructing huts 'Naiken-style' in the plots for the Mappilas who left their lines accommodation. Naikens also worked occasionally as watchmen for these Mappilas, as well as for the plantation itself.

Following the recent developments in the Gir Valley, from 1975 onwards, the movement of Naiken hamlets has changed somewhat in nature. Their huts are still being rebuilt regularly (every 6-18 months), and the flux of people between huts has not changed much. But huts have been rebuilt within the same general locations, so hamlets remain more or less in the same places. The restrictive movement is related to the status quo on land use enforced on the Gir Valley. Naikens were asked not to move into new areas, and at the same time were advised by their Mappila associates to establish rights over their territories by continuous occupation there.

Finally, it is important to note that throughout this century, perhaps unlike the majority of the Nilgiri tribal settlements, the Gir Valley local group has not been affected by any Government welfare programme, and has not benefitted from any help given to Scheduled Tribes. Only one indirect and short-lived enterprise could have, but hardly did, advance the Naikens. For several years during the early 1970s a school existed in the valley for the children of plantation workers, and some Naiken children did attend it. But then it was closed for insufficient number of pupils and because of the reluctance of teachers to remain there. None of these Naiken children could read or write, and all recall only the meals provided in the school.

THE AREA AND THE LOCAL COMMUNITY

The Naikens studied live in an area of about 2,000 ha. (5,000 acres) lying at the edge of the Wynaad tableland and at the higher elevations of the steep drop from this table into the Nilambur valley. The area is about 15 km. away from Pandalur (which is a small cluster of huts, with some brick houses and a few shops), included administratively in the Nelliyalam Panchayat, which itself had a total of 25,446 people, including 1191 Scheduled Tribes (1981 census). Pandalur is about 20 km. away from the *taluk* centre of Gudalur, and about 70 km. of steep and twisting road from the district capital, Ootacamund. The area which these Naikens occupy consists of a narrow and steep valley dropping from about 1,000 to 300 meters above sea level. It is here referred to as the Gir Valley[5]. The upper elevations of one side of the valley are partly rocky and covered by thick bamboo forest. The other side and part of the valley bed are covered by a deciduous evergreen forest (Fig. 10.1).

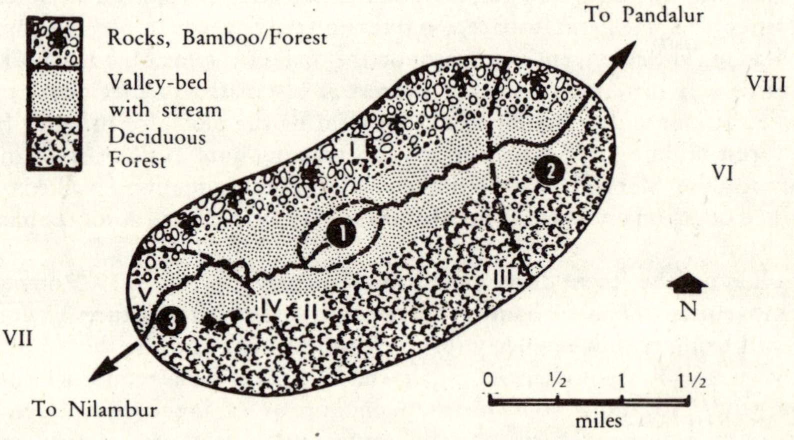

FIG. 10.1 Schematized Map of a Local Community

Key: Hamlets of the local community—
I-Taravad
II-Gir
III-Male
IV-Bridge
V-Upandgap

Fringe hamlets—
VI-N.P.
VII-M.P.
VIII-Devala

Historical sequence of occupation—
1-Wynaad Gauda people (?)
2-first phase of coffee plantation
3-second phase of rubber plantation

bungalow and lines of huts

Around the turn of the century, as we have seen, 500 acres at the bottom of the valley were cleared and planted with rubber. Until the late 1940s mostly Mappila seasonal workers were employed by the plantation. Restricted clearing and felling also took place during the late sixties and early seventies. The deforestation altered the balance of work required to obtain a livelihood from the forest: with the felling of trees the undergrowth became thicker and more difficult to move through. Some trees which held bee-nests were felled, and some creepers of particular importance in the forest economy (also the soapnut *Sapindus,* and the wild yam *Dioascorea)* were destroyed. However, areas amounting only to about one-eighth of the valley in total were cleared; (or about one-fifth if the original 200 ha. of the rubber plantation are included), and this has not been detrimental to the viable pursuit of a forest-based mode of subsistence.

According to Naikens, they first lived in the rocky and bamboo-forested side of the valley, a place which was not easily accessible from the bed of the valley and from the rubber plantation. The first fission in the Naiken settlement took place sometime in the thirties: one Naiken party moved away from the valley to live near the Nelliyalam Palace, where one (or perhaps two) men were engaged in work with the Rani's elephants. Some of the remaining Naikens started in the forties to take occasional weeding work in the plantation. They only worked during the rainy season, however, when the utilizable forest produce was least available. The second fission then took place: one party of those who began to work in the plantation moved down to the valley bed and built a hamlet near the plantation. At the same period, sometime during the fifties, the party who went to Nelliyalam returned to the Gir Valley and camped on the opposite side of the valley. I refer to the first, second and third settlements by the respective names: Taravad, Gir, and Male. Each of the three hamlets moved location every 6-18 months. Individual families then moved independently. There was a continuous flux and many intermarriages between the hamlets. Each Naiken could gather and hunt in any part of the Gir Valley. But a triplicate territorial division—conceptual and not on the ground—which corresponds with those three hamlets, underlies the spatial distribution of Naiken hamlets in the area and, mostly, certain aspects of the present Naikens' ritual and cosmology. I refer to subsequent hamlets positioned within these three general tracts respectively by the same names: Taravad, Gir and Male.

At present (i.e. middle fieldwork time) there are 69 Naikens in the Gir Valley, including 2[illegible] male and 24 female adults. The total Naiken population constitutes about one-third of the total population in the valley, while the Naiken adults account for two-fifths of the adult population of the valley. The Naikens to date are distributed in five hamlets. Three correspond to Male, Gir and Taravad. Two others have recently branched respectively from Taravad and Gir and are referred to as Upandgap and Bridge. The Gir hamlet taken as the central point, the distance to the other hamlets ranges between 3 and 6 km. There are three fringe hamlets to this local group: two are referred to by the name of nearby plantations and here are called M.p. and N.p., and the third is referred to as Devala. N.p. and M.p. are about 6 km. away from the Gir hamlet, while Devala is about 24 km. away. The hamlets of M.p. and N.p. are more closely related to the Gir Valley local group than the hamlet of Devala, and it is possible, and even likely, that in the near future they may become part of the local community of the Gir Valley. Here all three are still considered as fringe hamlets to the local community on which the discussion focuses (Fig. 10.1) The distribution of the Naiken population between the five hamlets is given in Table 10.1.

TABLE 10.1
The Distribution of Naiken Population

Hamlet:	Adults[6] Male	Adults[6] Female	Children
Taravad	4	6	3
Male	6	6	5
Gir	8	6	12
Upandgap	1	2	3
Bridge	3	4	0
Total	22	24	23

Two-thirds of the employable Naiken people in the valley are involved with plantation work, while the remaining third by and large subsist on gathering and trading in minor forest produce. Of those who work in the plantation only six are permanent employees. In general, the cohesiveness of the local group as a whole has been retained. Away from the plantation, in their hamlets, the observer cannot differentiate between the two Naiken sub-groups (see Bird-David 1983).

MATERIAL CULTURE

Each of the Naiken hamlets is positioned in a small forest clearing from which a footpath leads to a nearby water point, and others spread into the forest in various directions. The hamlets, on the valley slopes and at a relatively high elevation, possess open views to picturesque evergreen slopes. There are some plants randomly spaced around each hamlet: wild flowers and creepers brought from the forest; plants obtained from neighbouring people, some of which are just ornamental flowers; and semi-wild coffee bushes from the jungle. There are usually one or two teak trees and a few jack-fruit and mango trees. For the last 5-8 years Naikens have also planted plantains and arecanut which they obtained from the Gir Plantation. Plants are thought to be associated with the individual person[7] who planted them—whether a man, a woman, or a child. Since huts continuously move within the hamlet's area, and individuals frequently move between hamlets, the plants—associated with individuals and planted intermittently—are intermingled and imply no pattern of land division between individuals.

Most of the early accounts suggest that the Kurumbas lived in hamlets with up to four or five huts each (e.g. Breeks 1873: 50, Grigg 1880: 211,

Francis 1908: 15). The Naiken hamlets, not unlike those of Kuṟumbas consist of one to five huts. The huts (where there are more than one) are randomly positioned in the forest clearing with no apparent pattern, with no central focal place, and usually with their openings facing away from each other. The walls of the huts are woven from strips of flattened bamboo sections. The roofs are thatched and made of grass collected in the forest. The huts vary greatly in their details—a point made by Noble (1968: 107-8) for Kuṟumba house plans in general. For example, in some of the huts half of the wall is made of mud, while the rest is made from bamboo; others are only roofed by palm leaves; and one or two huts are built onto rocks with only one or two walls perhaps added to enclose the living space. During collection expeditions (and in some reported cases in the past), Naikens live in caves or in the open, each family near its own fire.

The huts contain one small living space (or a 'room'); or in some cases up to three separate spaces, each with its own opening. Usually husband and wife construct the hut. They can complete it within one to two weeks, but mostly, working at their leisure, they take a month or even several months. A second or third 'room' is built either by the couple or by those who move to live in it. A second, or a third, 'room' is created in a one-'room' hut either by making an internal division in it, or by adding an annexe, often only a rough lean-to shelter. Such an additional 'room' is created in a variety of circumstances. For example, for a young couple who still live with parents and may still eat with them; or for an unattached adult who stays with a family and may cooperate with it in certain collection activities. When an existing second 'room' is available an unrelated family may move into it instead of building, or until they build a new hut. Each nuclear family or an unattached adult lives in a clearly separate living space.

In the centre of each living space is the fire, around which the members of the family eat, spend their leisure time, and sleep. The importance of the fire is in particular apparent where two families, or a family and a single adult, share the same hut and even share the cooking. I hereafter refer to either a nuclear family or a single person by the term *social unit.* Though food then may be cooked jointly each social unit eats its share separately near its own fire. Even an aging single parent who lives with his child's family has his own fire, perhaps only a few metres away from that of his child.

Naikens work skilfully with forest materials and are often requested by outsiders to make such items as grinding posts and pestles, walking sticks, or baskets. The range of items they themselves use is very limited however, and a large part of it they actually obtain from the outside. Starting with what they make themselves: each social unit has a bamboo basket which is carried on the back by strips tied around the shoulders. Each family has a few bamboo containers which are used to keep salt, chilli, water, and grain (if any). Some Naikens also have mats, made of grass, while others sleep on flattened bamboos or rags. Among the items which are obtained from the outside, the billhook is of paramount importance both for domestic purposes (e.g. preparing food and cutting firewood) and for earning a living from, even moving through, the thick jungle. The majority of Naikens in addition have axes. Naikens use mostly metal digging sticks today, which they either pick up from old constructions in the area or purchase from shops. Only a few old Naikens still follow past practices and prepare a digging stick each day from a branch. Naikens buy simple cloths from shops in Pandalur, which when torn they use as sleeping mats or to carry bundles, etc. They usually wear one cloth until it is torn, though they may have in store another one, which they bought previously when they had more money than necessary for their basic requirements. Finally, each nuclear family (and some single people) have a few metal cooking pots and plates bought from Pandalur. In addition, Naikens have assorted odds and ends such as wooden boxes, beads, broken mirrors, small bags made of cloth, empty glass bottles and other *bric-à-brac*. Most important of these, in the eyes of Naikens at least, is the small cloth bag in which they keep the constantly used tobacco, lime, and arecanut.

It is important to note that usually each social unit has one billhook (though some couples who both work in the plantation and get their billhooks there may have two). Similarly, each social unit has one digging stick. Members of each family not infrequently exchange between themselves the cloth they wear. Finally, and perhaps most significant, though both males and females use it to the same extent, each couple has only one cloth bag for the tobacco, arecanut etc.

In addition to the above items, which can always be seen in any Naiken hut, there are a few others which may be seen occasionally. These include various trapping appliances and fishing baskets made of bamboo, and bows and arrows. Only some of the adult males possess the knowledge to make and use this equipment. In general these are the individuals who were interested and so learnt it in childhood by observation and by joining adults who were engaged in trapping and hunting activites.

Music plays an important part in the Naiken annual celebration, and for that purpose there are two drums and a few flutes in the local community. The drums and the flutes appear to be similar to those used by Kota musicians (and occasionally by Kuṟumbas who replace them) in Badaga and Toda ceremonies. Naikens provide the skin required for the drum, but the body of the drum is made by outside artisans. They prepare the wooden base of the flute, but metal bits are added to this base by blacksmiths in one of the villages around the Gir Valley. The drum is kept by the person who is responsible for the celebration, and is usually played only at the celebration. The flutes on the other hand are kept by those individuals who are fond of playing music, and who often play them at their leisure near their huts. In the festival, music can be played by any Naiken. At the end of the celebration, in return, when the offerings to the gods are distributed between those gathered there, the 'flute' and the 'drum' (namely the people who played them) get their own share. Significantly, and with relevance to the general social configuration, in one case it has been observed that a man from a fringe hamlet who was playing his flute received in addition to his share of the offering also a gift of money.

A few sacred items are stored from one annual celebration to another in a box usually kept in a corner of one of the huts: cheap jewellery and a cloth (often with bright colours and a flowery pattern) which the shaman wears during the ceremony, a large knife which he holds when in a trance, and a variety of objects which are displayed during the festival. The last mentioned include small stones of elongated and smooth shape (a human shape according to Naikens); crude pottery images; metal images from the lowland, some of which are of Hindu deities; and a few pots and beads which were used by particular ancestors. (A comparable description of the sacred objects of Kuṟumbas is provided by Noble 1968: 109.) These objects are displayed in a hut newly built, or in the dilapidated hut from the previous year which has been repaired for the festival. In the past, Naikens say, the objects were displayed on certain rocks or under certain trees in the forests where the festival took place. Finally in the forest near their hamlets Naikens to this day have some stones which are positioned on a rock platform, and which are worshipped together with the other objects during the annual feast. The objects of each celebrating hamlet, the huts constructed in each for the festival, and even the way in which the objects are displayed, vary from one hamlet to another. Between one annual festival and another no reverence is shown for the hut of the previous feast, for the box with its sacred objects, or, by and large, for the stones.

ECONOMIC ACTIVITIES

Naikens are engaged in a variety of activities: for example, gathering for consumption; gathering for trade; fishing; honey collecting; some hunting; labour in the plantation; occasional day labour in the fields of the Mappilas; making household items from bamboo and timber; constructing huts and fences for Mappilas; occasional work with officials who visit the estate; acting as watchmen for the plantation and the forest's owner, etc.

In general, Naikens do not set out upon any one particular forest-based activity. Having got up between half-past-six and half-past-seven; and having idled around their fires until about nine a.m.—eating perhaps left-overs from the previous evening, tea and pastries when obtainable from the tea shop, or nothing at all; leisurely they then go into the forest.[8] Walking there they forage for whatever may come their way. Someone may notice holes in a rock and leave the path to find out if there are any birds' eggs there. Another may later see bees flying above some bamboo and the party would go there looking for honey. A large wild yam creeper may be identified and the party would stay there for a few hours to dig it up. In the majority of cases each couple, or a nuclear family, goes to the forest on its own. They always have their billhook with them, and they usually have their digging-stick and bamboo basket as well. The last two mentioned, however, as well as all the other possibly required equipment, can in any case be improvised quite easily. Each party in the forest can therefore act on any opportunity which may occur. The family returns usually in the late afternoon, but sometimes a few days later. Certain forest activities require some prior intent, decision and organization. For example, the collection of a certain type of honey (the 'big' honey) has to be carried out on a dark night by three or four men. In such cases what takes place is a *de facto* organization of the working party, and a *de facto* decision about the timing of the operation.

Most of the 'non-forest' activities, with the exception of plantation work, are embarked upon in a similar manner. A Mappila may encounter a Naiken near the tea shop or at the Naiken hamlet and ask him to do some work that day. Naikens may pass by a working party in the forest and there and then be engaged to help. A trader will come to the hamlet and Naikens give him some minor forest produce which is kept near the huts, and which they have gradually collected as and when they noticed it while in the forest.

To a certain extent work in the plantation is also marked by this unpurposeful conduct. People go to the plantation when they wish to, without *a priori* planning, and furthermore combine such 'traditional activities' as described above, with work in the plantation. The workers have to register for work at seven a.m. They are then told what work they are to do that day and where, and walk to the location in question. At the plantation they are mostly engaged in tasks similar to those traditionally carried out in the forest, e.g. clearing paths, weeding, or making fences from bamboo. They end work between two and three p.m. On their way to and from the plantation, and even on their way from the registration place to the site of the work, and during work, Naikens are on the look-out for any usable forest produce. Back from plantation work, they go to the forest primarily to collect firewood or building materials, and they may then forage for any other usable produce. Furthermore, on average Naikens attend plantation work four days a week. In the bountiful months of January and February they may be absent from work for weeks. On those days, and on the paid weekly rest day, they often go into the forest. Finally, in season, even within the framework of the plantation work, and in return for their usual daily wage, Naikens are occasionally delegated the tasks of collecting minor forest produce. In general, it can be concluded, the Naiken is quite flexible as to how he obtains those minimal commodities he requires. Acceptable to him are gathering in the forest for consumption or for trade, or any casual, daily-based or intermittent work engagement. He forages, so to say, from the world at large.

Morgan (1876: 100) has pointed out the important place of indebtedness in the regulation of barter between the Kurumbas and the neighbouring villagers. Indebtedness, it can be said for the Naikens, regulates not only the barter in minor forest produce, but also all other economic transactions with outsiders (*e.g.* with the plantation, with the Mappilas who employ them, and so on). Usually, having received payment in advance or an independent loan of cash or goods, Naikens are linked by debt to certain outsiders, and are constantly exhorted to repay by labour or by minor forest produce. But, upon doing so, Naikens request, and usually receive, some money or goods which put them again into a position of indebtedness, and so the 'barter' relationships are maintained.

SOCIAL ACTIVITIES

Rarely does the observer see the adult residents of one hamlet gathered together. The spatial distribution of huts in the hamlet does not call for it, nor do those subsistence activities which Naikens are engaged in. During the evening, having eaten their evening meal, couples most often sit near their own separate fires. Their children sit with them or, occasionally, visit other fires, or play in between the huts with their mates. During the night, the members of each nuclear family sleep all huddled together on one mat near their fire. In the morning, couples independently leave for the forest or the plantation, and in the former case, their young children accompany them. Those remaining in the hamlet—e.g. old people, new mothers, or couples who do not go to work that day—may spend the latter part of the morning sitting all in one space somewhere in between the huts. Their spatial distribution then may well accord with that of the huts in the hamlet, namely, individuals and couples sit at a distance from each other, facing in different directions. They do not gossip with each other, and usually the intermittent verbal exchange between them consists of infrequent reference to here-and-now happenings (e.g. a Jeep passing by, flower buds which have appeared during the night, etc.). Purposeful conversation, when necessary, usually takes place in the evenings when individuals leave their fires for a short while and visit other fires.

Outsiders who come into even superficial contact with Naikens comment on the close tie between the conjugal couple and the weak links between lineally and laterally related families, namely parents and their married children, and married siblings. Indeed, the Naiken child growing up gradually leaves his parents and moves around quite independently. When he establishes his own family of procreation, and members of his family of orientation are in different hamlets, the child rarely, if at all, visits them. The link between spouses, on the other hand, is particularly warm—they are close companions and depend only on each other for help.

A prominent nineteenth century theme in accounts concerned the lack of marriage ceremonies among Kuṟumbas, and the formation of marital unions which principally depended upon the respective choice and consent of both spouses. Grigg wrote of the Kuṟumbas: 'It sometimes happens that after a couple have cohabited for some time they agree to live together for life' (1880: 213; also Breeks 1873: 54, Francis, 1908: 153). A similar situation prevails for the Naikens. There are no marriage ceremonies to speak of. Marriages are often merely *de facto* arrangements whereby the couples start to live together. Even when this is not the case, and some vague suggestions for matches are made, marriages only follows on from the mutual final choice and consent of both spouses. When the 'bridegroom' or the parents of a young 'bride' can afford it, a meal may be offered to a small gathering to celebrate the event.

Hierarchy in the community is minimal, but old people in general are more respected and consulted than young ones. In explaining personal views concerning the primacy of certain individuals, informants often provided criteria such as 'he listens to you when you talk' and 'he knows best how to deal with outsiders.' The orders of primacy provided by adults in one hamlet, for example, were significantly at variance from each other. Certain succession principles and social offices emerge during, and have meaning in connection with, the annual celebration (see below), but in general they are dormant and with little manifestation throughout the year.

RITUAL ACTIVITIES

Nineteenth century accounts emphasize that the Kuṟumbas have 'religious ideas...of the vaguest' (Francis 1908: 155), and that 'They are said to have no traditions of any kind' (Grigg 1880: 213). One annual celebration only is mentioned in many of these accounts. Breeks for example writes that 'They knew of no god peculiar to the Kuṟumbas, nor had they any temple, but at a certain season they took offerings of plantains to the *Pujāri*...who attended on Maleswara (lord of the mountain), the god who lived on a hill known by that name' (1873: 55; see also Francis 1908: 156.) Broadly speaking, the descriptions in these accounts fit the Naikens well enough.

There are only a very few rituals to celebrate life cycle events. It was ascertained not only from informants' verbal accounts, but also from observation of births which took place during my fieldwork, that there are no birth ceremonies. It was similarly ascertained that marriages are by and large not celebrated. At the onset of puberty the girl is kept separate, and she only washes in the river before she joins the community again. But during subsequent menstrual periods she remains in her hut. The only life cycle event which is marked more elaborately by ritual is burial. To take one example: after the burial, oil drops are poured onto water. Should the drops touch each other in the water, it implies a good omen, and kin or the old people among those gathered wet their heads with the mixture of oil and water. (Francis described a very similar ritual for the Jenu Kuṟumbas, 1908: 156.)

Only one celebration takes place in certain hamlets annually, some time during April or early May. The residents of the hamlet and most of the people in the local community gather together for 24 hours, from late one afternoon to the next one. Throughout the 24 hours the sacred objects are displayed, music is played, and those gathered there dance in two sex-differentiated circles. One to three shamans go into trance, and through them the other

Naikens speak with the gods. Mostly each of the speakers speaks on behalf of the community at large. The main and continuous topic of the conversation between the living and the gods concerns the promise that the living would follow 'the ways of previous generations' and would make offerings to the gods if the gods in return would protect the living from mortal illnesses. The souls of people who died during the preceding year are brought into the celebration hut during this annual festival, to be joined with the gods. In some celebrations, one or two couples act out *tamasa* (funny play-acting). Finally, a meal of cooked rice is eaten by all those gathered there, at the end of the 24-hour festival, and the offerings to the gods are distributed among them.

Within the context of the annual celebration there emerge two offices. The first is the office of shaman, which is non-hereditary, and is open to any of the gathered who may go into trance. The second is the *modale* ('the first'). A distinction is made between those who first settled in the particular territory and their descendants—the 'firsts'—and those who joined the hamlet from other places. The *modale* is the oldest man of the 'firsts,' and his role will pass on to his oldest son or, failing that, to a young brother, or the oldest daughter's husband. It is the *modale's* responsibility to see that the festival is celebrated annually, and he has to supply the resources required. The other members of the 'firsts' are expected to contribute money as well, but are not in any way pressured to do so. Any other person is welcome to contribute towards the expenses. In general a sort of bilateral principle of descent underlies which feast one has to contribute to. It is said that half of the children are affiliated to the gods the mother makes offerings to, and half to those of the father. But the final choice, it is stated, lies with the children themselves, when they depart from their natal families (see also Fürer-Haimendorf 1952: 21). Such a choice is expressed simply by a *de facto* contribution of money to one or the other feasts. Celebrations in the Gir Valley are held in three hamlets: Taravad, Gir and Male. Between one annual feast and another, the offices of the shaman and the *modale* are hardly activated. Furthermore, the shaman and the *modale* cannot be differentiated from other people by any social or economic criteria. They tend, however, to be the oldest male members of the hamlet, and as such may be more respected. The shaman in addition is also infrequently sought after to intervene in certain eventualities which are described below.

Naikens, in general, attribute to their gods little but the power to protect from and inflict certain mortal diseases. The illnesses which Naikens associate with the gods can be described as mainly those which involve sudden high fever, complete lethargy, etc., and which are not easily explained in Naiken terms by observable natural phenomena. Examples given to me by Naikens for some naturally explained complaints are: backache, through carrying too heavy a load; and stomachache caused by eating something bad.

Naikens say that these illnesses are caused by the gods because the person in question failed to follow the 'way of the previous generations,' or because some other person induced the gods to do so by making offerings. While for 'naturally explained' illnesses Naikens take a variety of medicinal jungle plants, for those which are not they sometimes appeal to the shaman. There is no concern with the identification of the person who may have brought the illness about; the shaman only finds out if the illness is indeed supernaturally caused. He does so either through divination—in accordance with whether there is an even or an odd number in the handful of certain seeds taken randomly from a pile; or, rarely, through trance. Should the answer be positive, the sick person either promises to make offerings to the gods in the following annual celebration, or purchases, for example, a cloth, or a statue, to be added to the sacred objects displayed during the festival.

Naikens are noted among their neighbours for their supernatural and medicinal skills. The shaman may be consulted by outsiders on such matters, and is paid with a gift, usually money or cloth. While Naikens only seek the shaman's advice in the case of certain unexplained illnesses, outsiders seek his help for a wide range of causes—prosperity, cures, and even success in jobs. Outsiders ask for medicines and cures also from other Naikens, in particular those with whom they are more familiar (e.g. the Naiken watchmen they employ).

The People and the Ethnographic Myth

From the first section of this paper it is clear that a problem of identity, underlies in different ways the literature relevant to the Kuṟumbas in general and the Naikens in particular. In the early literature there emerges a confusion as to how many separate divisions there are among the Kuṟumbas. A puzzling point also concerns the dispersal of Kuṟumba hamlets (even those of the same division), and the lack of communication between them. The information suggests varied, non-unified, and localized Kuṟumba culture and occupations. Furthermore, the culture is in general portrayed as *scanty in its ritual*; with 'religious ideas of the vaguest,' and with no 'traditions of any kind.' In the main stream of the twentieth century Nilgiri anthropological literature the problem, one could say, is almost reversed. There is a clear and simple formulation of the food-gatherer Kuṟumba tribe. Furthermore, this Kuṟumba tribe is depicted to be customarily involved in highly patterned interrelationships with three other Nilgiri tribes—the Todas, the Badagas, and the Kotas. Not only that, but the most important component in these interrelationships is perhaps that the Kuṟumbas provide both *ritual* services and objects to the other tribes. Further scanty details are provided of the Kuṟumba society (and culture) itself. Recently it has been argued that the prevailing notion of *a* Kuṟumba tribe is erroneous (Kapp and Hockings, above). But multiplying this notion by seven as it were, it was suggested that there are seven Kuṟumba tribes—each a separate and well defined ethnic group with its own language, social structure, body of customs and religious ideas, and so on. Clearly, there seems to be some dissonance between the two pictures portrayed: a dissonance 1980a: 99). Or, in other words, the picture portrayed of the Kur̲umbas in the any radical changes undergone by the Kur̲umbas, since—a point often overlooked—the traditional interchange between the Nilgiri tribes flourished in the nineteenth century, and was declining after the First World War (Hockings 1980a: 99). Or, in other words, the picture portrayed of the Kur̲umbas in the more recent literature applies in many ways also, if not more so to the nineteenth century and the first decade of the twentieth century. The early accounts of Kur̲umbas, one could summarize, are problematic for the details provided in them; while the twentieth century references which also concern, to some extent, the nineteenth century Kur̲umbas are problematic for the details that are missing.

From the second section of this paper it is seen that Naikens do not conform to the model of a closed and bounded tribal society. Many traditional societies have no clearly defined boundaries, but the issue with the Naikens seems to be of a different order. There is here a formation of small pockets of Naikens dispersed amidst other populations, with no institutionalized communication between them, and no pan-Naiken structure (whether ritual, social or economic) linking them. A nucleated pattern of social organization characterizes each segregated local community. The local community in general is distinguishable for the variability within it, more than for its homogeneity. Occupations and, to some extent, even certain cultural practices, seem to be characterized by flexibility and by contextuality rather than by customary cultural prescriptions.

Broadly speaking then, there is a certain affinity between the Naikens as described here and the Kur̲umbas as portrayed in the nineteenth century literature. As is clearly seen from the above paragraphs, such affinity extends to the general formation—namely the nucleated, locality-centred, and segregated sub-local communities. But it also extends, as is pointed out throughout the description of Naikens, to other important aspects of the religious, social and economic activities.

From the description of Naikens it seems that the early references to 'Kur̲umbas' cannot be criticised (as they often are) merely because they are vague and even contradictory. The Naikens, who clearly were a part of the 'Kur̲umba complex', show that such vagueness and contradiction may well indicate, or at least be explained by, the very nature of the people themselves. A great deal of speculation is involved in any inference from the Naikens of today back to the Kur̲umbas of the last century. But such an exercise would nevertheless seem worthy of consideration because of the affinity between Naikens and Kur̲umbas; because of the problematics of the literature which otherwise cannot be assessed; because the Kur̲umbas of the nineteenth century cannot any more be reconstructed independently from the available sources; and finally (and least) because of the *de facto* explanatory relevance which such inference may suggest. Judging by the Naikens it can be hypothesised that the individual Kuṟumbas did not provide a uniform, overall picture of Kuṟumbas division and their names, because they could not do so; because the entity of 'Kuṟumba' people was not so structured. Kuṟumba informants, like the Naikens perhaps, had each a highly ego-centred and, in this sense, personalised view of the Kuṟumba people. They each knew their own group, and in addition a few other groups, the number of which depended upon the very circumstances of the location of their own place. Or to put it in different words, using Breeks' careful information (1873: 48), it is not that it was 'difficult to get a complete account of the tribal divisions recognised by them'—an account which Breeks presumed to exist. But indeed, as Breeks continued, 'one man will name you one (his own); another two divisions; another three, and so on.' Should the Kuṟumbas resemble the Naikens, descriptions concerning the lack of marriage have not necessarily resulted, as some scholars politely suggested, from the reserve of Kuṟumbas and the difficulties of communication with them. Perhaps, though brief and therefore crude, such descriptions did reflect upon the relative skeptical concern of these people with religious ideas and upon the relatively small importance of ritual in their lives. Perhaps it could be said then, such descriptions did result from the problem of communication with the Kuṟumbas, but problems which hinge less on the Kuṟumbas themselves and more on the expectations, or presumptions, of the writers (and readers) of these accounts. Finally, judging from the Naikens, the wide range of economic occupations attributed to the Kuṟumbas are perhaps all connected by the simple common denominator, namely, the readiness to be engaged in any casual, daily labour, the returns for which are those simple commodities required by the Kuṟumbas.

Inferring on the other hand from the nineteenth century accounts to the study of Naikens, it seems that in its broad lines the Naiken social configuration has probably persisted for some considerable time. Fürer-Haimendorf's description (1952) further supports such a possibility.

Since even observations are somewhat affected by theoretical pre-suppositions of the observers, whether explicitly or implicitly, it is important that previous accounts generally back, or at least do not negate, the picture here given of the Naikens. It seems necessary then to explore how to describe analytically the entity of the Naiken people; and following on from that, perhaps how to describe analytically the various Kuṟumba peoples and the relations between them. These problems are difficult and lie outside the scope of this paper. But one tentative thought can perhaps here be briefly articulated. Each Naiken local community can perhaps be viewed as one variation on the themes which correspond with slightly different Kuṟumba peoples. The Kuṟumba groups being perhaps, like the Naikens, characteristically open, informal and adjustable—the differentiation between clusters is likely to be related to the adjustments made to different economic set-ups, and different cultural surroundings.

The problem of the 'Kuṟumbas' in the recent anthropological literature is more difficult to understand than that in the early literature, because the portraits depicted in the two bodies of literature seem in discord with each other. Secondly, because from the many intensive studies conducted among the three tribes and from references to them in the nineteenth century accounts, it cannot be doubted that Kuṟumba persons did fulfil the functions for the other tribes as described in the literature. From the mere perspective of geographical proximity, it is highly likely that the Kuṟumbas involved in the inter-tribal traditional interchange are those who resided near to the Badaga, Kota and Toda settlements, namely, at the higher elevations of the Nilgiri slopes. At the same time it is not implausible that Kuṟumbas of the lower slopes had contact with those other three tribes. Rivers for example mentions that 'Poles of the proper length [for the second funeral ceremonies of Todas] are said to grow only on the Malabar side of the Nilgiris.' (Rivers 1906: 641). Similarly Breeks reports information obtained from 'Two Kuṟumbas who came to the Kūndah *Kotagiri* (bringing their hoes to be sharpened by the Kotas) from a *Motta* [hamlet] in Malabar...' (1873: 55). The Kuṟumbas at the somewhat higher elevations of the Nilgiri slopes are those studied and referred to by Kapp as the Ālu and Pālu Kuṟumbas. The role played by the Kuṟumbas in the traditional interchange is described in the work on those groups, but mainly on the basis of Badaga accounts, and on the base of the existing literature (see for example Kapp and Hockings above). It could not be done otherwise nowadays since (it needs here to be repeated) the period concerned is mainly the nineteenth and early twentieth centuries. Kapp describes the Ālu Kuṟumbas and the Pālu Kuṟumbas (of 1976) as a highly structured society with an elaborate social structure, and with both elaborate ritual and complex cosmology. Indeed Kapp himself argues that 'most of the information to be gathered from articles and studies which have been written on this tribe during the past 150 years is scanty, vague, and often wrong. Not even the name by which they call themselves to be differentiated from other Kuṟumba tribes, namely, Ālu Kuṟumbas, has been mentioned in any of the available sources on the subject' (1978b: 167). It is an even more speculative exercise than the one taken above, to infer from the Naikens of today at the low elevation of the Nilgiri slopes, to those nineteenth century Kuṟumbas probably at the high elevations, who are portrayed in the early anthropological literature. But while accounts contemporary with the nineteenth century 'Kuṟumbas' do not well describe the present Ālu and Pālu Kuṟumbas who, nearer to the plateau and its hill stations, perhaps did undergo radical changes in recent decades; such accounts suggest a picture closer to the present Naikens. A speculative hypothesis based on the Naikens is therefore perhaps called for, to shed some light on the issues of the role of Kuṟumbas in the traditional interchange *as seen from the Kuṟumbas' point of view*, and the issue of the discord between the portraits of Kuṟumbas in the nineteenth and twentieth century literature. Needless to say, such a hypothesis requires scrutinizing consideration by those who have worked with the other tribes and in the light of their work.

The actual customary interchange of both ritual services and objects between the Kurumbas and the other Nilgiri tribes can be summarized in three points (other components of the interchange such as the provision of minor forest produce and the fear of their sorcery are not relevant for the current purpose): first, the Kurumbas provided ritual objects procured from the forest and important to the other tribes. Second, the Kurumbas offered supernatural curing sought after in certain cases by the other Nilgiri people. Third, individual Kurumbas acted as watchmen, each for one Badaga commune. The appointment was for life and was hereditary; and the watchman played an important part in certain agricultural rituals of the Badaga commune, helping the Badaga priest.

With regard to the first point: it has been suggested that nineteenth century Kurumbas (as well as Naikens) provided any forest item needed by outsiders in return for which they could obtain those simple commodities which they required. It is not unlikely that while for the recipient Nilgiri tribes certain items provided by the Kurumbas were of particular ritual importance (e.g. the funerary pole for the Todas); for the provider Kurumbas these were merely one among many other types of forest produce for which there was external demand (e.g. poles cut from certain trees).

With regard to the second point above: judging from the Naikens, the Kurumbas could have acted as healers and as sorcerers mainly within external contexts and, less importantly, within internal ones.

With regard to the third point, the most dominant of the three points: leaving aside for the time being what the Kurumba watchman gave to the Badagas, it is worth noting that he himself received in return some cloth, some money, and simple commodities such as grain, tobacco, etc. In itself the engagement as watchman seems not to have been uncommon among the nineteenth century Kurumbas (as indeed it is not among present Naikens), and Kurumbas could have been employed as watchmen by Badagas as by any other outsiders. That the Kurumba watchman was employed by the Badaga commune as a whole and not by individual Badagas was perhaps more related to the nature of the Badaga society than to the nature of the Kurumba society. From Hockings' work it is indeed clear that the commune was of particular importance in the Badaga social organization, and indeed in certain cases acted collectively under its headman and priest. For the Badagas the appointment was for life, but the Kurumba was not present in the village all the time: 'If he [the Kurumba watchman] does not appear there very often the Badagas nevertheless believe he is magically present each night or else knows by clairvoyance what is going on in the villages' (Hockings 1980a: 123). Judging by the Naiken, it is not unlikely that the Kurumba, as he saw it, went occasionally to the Badaga villages where, on performing certain duties, he received those simple commodities he required. The quote above in Hockings' careful and detailed description continues: 'yet his [the Kuṟumba's] visits tend to be frequent, for he is fed by the commune' (ibid.). Judging by the Naiken, and by the nineteenth century literature, the Kuṟumbas were not patrilineal and perhaps did not even have hereditary offices. But the Badagas were patrilineal and did have hereditary offices. Perhaps then it was the Badagas who upon the death of their Kuṟumba watchman enrolled his son. Hockings' description seems to indicate that such an interpretation on the Kuṟumba's behalf is not unlikely: 'When a Kurumba watchman dies the headman of the Badaga commune should collect some gifts from the villagers and take them, or have them sent, to the bereaved family. This used to be imperative to *secure the goodwill* of the family from which the next watchman would also be drawn' (ibid.: 126; my italics). So the meeting-point between the cultural requirements of Badagas for a hereditary watchman office and the egalitarian and almost mercenary Kuṟumbas lay precisely perhaps in this gift from the Badaga village to the family of the deceased Kuṟumba. For the Badagas it was perhaps merely a matter of securing goodwill; but for the Kuṟumbas it may have been the establishment of a new 'barter' relationship. (See above, on the importance of indebtedness or payment in advance in the economic interactions with outsiders for the Kuṟumbas and the Naikens). Finally there is the issue of the essential participation of the Kuṟumba watchman in Badaga ritual; essential, that is, probably only for the Badaga. The Kuṟumba, one could almost prophesy by now, indeed received for his participation some money and those invariable commodities he was after—the tobacco, some grain, and so on. Inferring from the Naiken, an adjustable and contextual code of behaviour was not foreign to the Kuṟumbas, who could operate quite satisfactorily within the framework of the Badaga ritual.

It seems that from the Kuṟumba point of view, his involvement in the Nilgiri's traditional interchange may not have been as regulated by customary norms as it is presented by the other tribes, and that the form which this interchange took may not necessarily have reflected upon a structured 'traditional' Kuṟumba society. A case of a similar type of interrelationship between the foodgatherer Mbuti pygmies of the Congo and their neighbouring Bantu-speaking cultivators has been reported in Turnbull's book, subtitled 'The two worlds of the African Pygmies' (1965). Turnbull clearly shows that this interrelationship (which extends to the ritual domain) is interpreted in different ways by the different sides, and is entered upon for different purposes—by the Pygmies mainly for economic returns. Hockings summarizes the traditional interchange between the four Nilgiri tribes in two diagrams of social and economic relations by respective arrows from givers to receivers (Hockings 1980a: 100). It is perhaps not insignificant that in the diagram of economic relations, arrows go from and to the Kuṟumbas, as is the case with the other three tribes; while in the diagram of social relations, arrows only go away from the Kuṟumbas, and are only balanced by returning arrows in the diagram of economic relations. Or to quote Hockings who elsewhere puts it clearly for the Kuṟumba-Badaga relationship: 'The Kurumba watchman...play[s] a necessary role in the ritual of his Badaga commune; but Badagas have no part whatever in Kurumba ritual' (1980a: 125). The simple formulation of a Kuṟumba tribe may similarly have followed on from the views expressed by the other Nilgiri tribes, which anthropologists studied directly. Such a view was perhaps embraced by scholars because not unlike their subjects of study, and indeed understandably, they presumed the Kuṟumbas to be a bounded highly structured 'traditional' tribe. If the Kuṟumba place in the Nilgiri traditional interchange was indeed as it is here hypothesised, there is not necessarily any clash between the portraits of Kuṟumbas in early literature and the Kuṟumbas in the twentieth century literature.

NOTES AND REFERENCES

1. This paper is a version of the introductory chapter of my doctoral dissertation, submitted to the Department of Social Anthropology, University of Cambridge. The research and fieldwork have been supported by a College Bursary from Trinity College, Cambridge; the 1978 A. Wilkin Studentship; the 1979 H.M. Chadwick Studentship; and research grants from the Smuts Memorial Fund and the Wyse Fund. I am indebted to all of the above for their financial support. I am also grateful to my supervisor Dr Alan Macfarlane, to Mr K. Lakshmanan, my assistant and interpreter during fieldwork, and to Mr Phillip of the Plantation for his hospitality. An abridged version of this paper appeared in *Modern Asian Studies* (21: 173-89) in 1987.
2. A short article on the Jain Kurumbers of Wynaad was published in 1929 (Raghavan 1929), which in a way intermediates between nineteenth century accounts and the twentieth century literature. I have not seen Scherman's article (1942).
3. I spell the name 'Naikens' and other words in the Naiken language as they sound to me. The language is an unwritten one, and I am not qualified to provide a phonetically accurate spelling.
4. In addition to observations, I drew my information from on-going interactions and conversations with the Naikens in the community studied. There was no 'acculturated' Naiken there who might act as the key informant. Although among those I regularly interacted and talked with were shamans, *modales* and old people, there was no Naiken in the community who was inclined to reflect upon his society and semi-process it, as it were, for the anthropologist.
5. The intensive study conducted, and the small size of the population studied, would enable easy identification of individual Naikens. To protect both the people and the plantation in question from intrusion, fictitious names are given here to the area and the plantation, but the locational details I provide should allow identification accurate enough for the purpose of comparative regional studies.
6. By 'adults', I refer to those married or over twenty to twenty-two years of age.
7. Any Naiken can collect the produce of any plant for his own consumption. But the right over what is obtained through the barter (if any) of the produce of these plants is reserved for the person associated with the plant.
8. Taking Naiken perspectives, though the hamlet is positioned in the forest, any move away from the hamlet is referred to as 'into the forest'.

The Iruḷas

—A. WILLIAM JEBADHAS AND WILLIAM A. NOBLE

11

These are... worshippers of Vishnoo, a remarkable circumstance considering the almost universal Sheivism of the aboriginal tribes of S. India. Near Rungasawmy's peak, and scattered about the slopes and base of the hills to the south and south-east, there are several Iroola mottas. These villages consist of seven or eight split bamboo huts plastered with mud, and generally built round a square.... Near their villages they have large gardens of plantain and lime trees, and cultivate the neighbouring ground in the Cautcaud fashion, changing the field every year....

—C. Macleane (ed.; 1893: 373)

Visitors to the lowlands stretching eastward from the Nilgiri massif will, if looking upward and westward, notice a dominating peak. This is the one called Rangaswami Betta (Pl. 21A). In its vicinity, mainly within the Nilgiris District, live the Irula farmers and labourers to whom this chapter refers (Fig. 11.1). Because we did not have adequate time to work among some lowland Iruḷa colonizers who use ploughs and cultivate wet rice, only those Irulas who cultivate millets and use hoes are covered here. Gathering activities for income are mainly part of a long-standing relationship between the Irulas and the lowlanders. From the first census of 1871 until 1981, the total Iruḷa population within the Nilgiris District rose from 1,470 to an estimated 5,900 (Grigg 1880: 30). As the Iruḷas were among the first of the traditional Nilgiri groups to be visited by a Scotsman (in 1800), changes or lack of change since that time will be dealt with here. The hamlet of Koppayūr, lying next to Kil-Kotagiri Estate and the Iruḷas residing in it will serve to illuminate certain points. Iruḷa agriculture demonstrates the significance of tropical gardening among tribals in South India, and of the retention of millet cultivation (in Southeast Asia, rice tended to replace millets everywhere). While the modern spread of plantation agriculture led to Iruḷa employment, this spread also produced much conflict. Despite Iruḷas being classed as tribal, there are men among them who serve as Hindu priests. These priests, who come traditionally from families residing in the lowland colony of Kallampalaiyam, minister to Iruḷas as well as to caste devotees who come annually to the isolated Rangaswami Betta ceremonial centre.

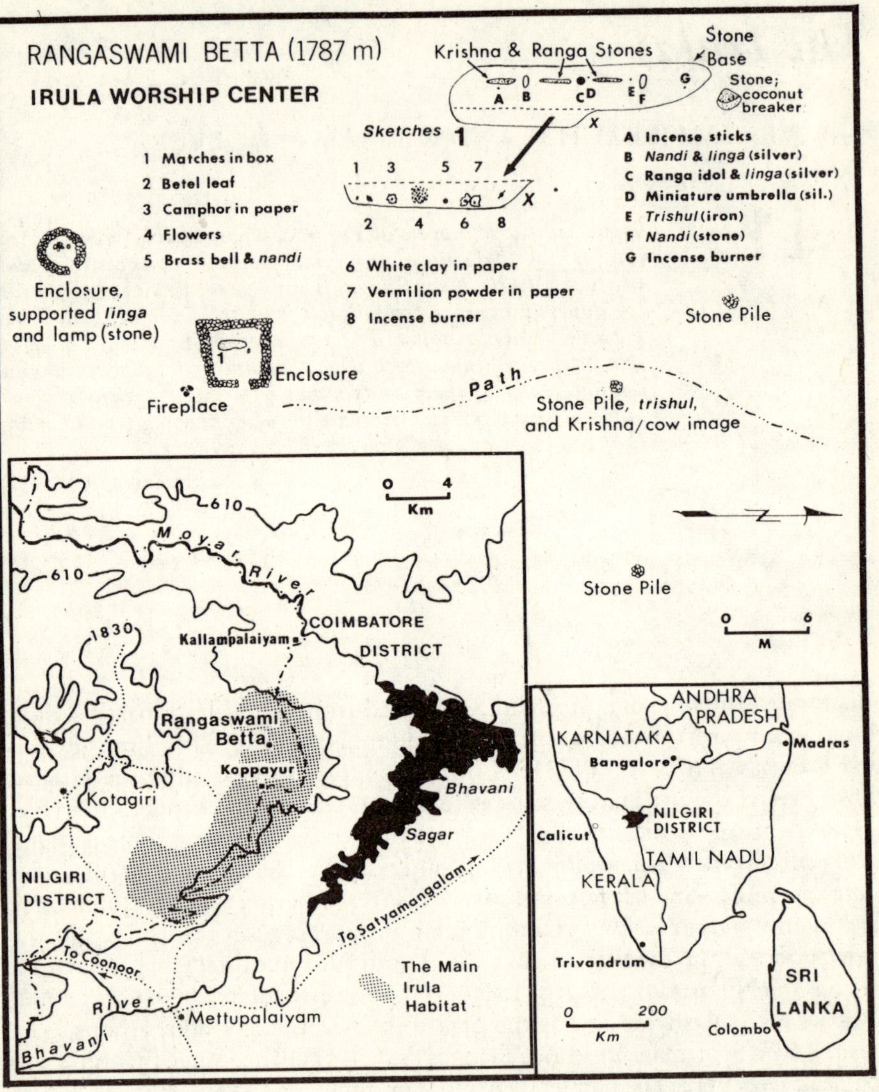

FIG. 11.1 Rangaswami Betta: Irula Worship Centre.

The Earlier Iruḷa Economy

If we broadly interpret economy as a survival system, earlier accounts provide us with some glimpses of how the Iruḷas survived prior to change that resulted from the British imperial presence. Francis Buchanan, a superb observer and the first Briton to record a visit to the Nilgiris, managed in AD 1800 to encompass pertinent ingredients of the Iruḷa economy in this brief description:

> *On the hills the Eriligaru* [Iruḷa] have small villages [really hamlets]. That which I visited contained seven or eight huts, with some pens for the goats; the whole built round a square, in which they burn a fire all night to keep away the tigers. The huts were very small, but tolerably neat, and constructed of *Bamboos* interwoven like basketwork, and plastered on the inside with clay. These people have abundance of poultry, a few goats, and in some villages a few cows, which are only used for giving milk, as the *Eriligaru* never use the plough. They possess the art of taking wild-fowl in nets…; and sometimes they kill the tigers in spring traps, loaded with stones, and baited with a kid. Near their villages they have large gardens of plantain and lime trees, and they cultivate the neighbouring ground after the *Cotucadu* fashion, changing the fields every year…. Besides cultivating their gardens and fields, the *Eriligaru* gather wild *Yams [Dioscorea]*, and cut timber and *Bamboos* for the people of the low country… (Buchanan 1870, I: 462).

We learn that the farming Iruḷas lived in nucleated settlements, just like the majority of Indian farmers. Also typical, a few houses in a hamlet were built next to a courtyard. Interwoven bamboos plastered with clay were an essential ingredient in their house construction. That Buchanan should mention the setting of fires at night to keep tigers away is indicative of how wild and dominated by forest was the region in which the Iruḷas then lived. Gardens adjacent to hamlets and shifted field plots cut out of forest characterized Iruḷa agriculture. In them, hoes were used; ploughs, never. Dry field shifting was then so much a part of Iruḷa existence that they, like the Kannikars of Kerala, used to calculate a person's age by the number of shifts which had taken place during that individual's life. Livestock, ranging from chickens, to goats, to cows, were reared. There is some indication that the Iruḷas used nets when hunting. They made deadfall traps, some large enough even to kill tigers. Wild yams were gathered. We may infer that cut timber and bamboos were gathered up and sold to lowlanders, or exchanged for produce.

With regard to Iruḷa year-round economic activities, Harkness suggested a model in which the single annual grain harvest yielded food for a comparatively short time. Iruḷas afterwards depended on garden produce and food obtained through hunting and gathering. As their food became harder to obtain, there was a shift by some from sedentary group-living in hamlets to nomadic roaming in ever smaller bands, farther away from hamlets, and with a dependence upon temporary shelters. Because Harkness' observations were so accurate, his model can now be outlined and supplemented. According to

Harkness (1832: 93-4), Iruḷas partially cleared the forest, and turned up earth with hoes or used dibbles to scratch the earth's surface into furrows. Grain was broadcast over the clearings, but no attention was given to a growing crop. If grain was planted at some distance from a community, families would move to their grain fields at harvest time. Family members ate grain from their field, but would also invite neighbours and passers-by to eat the grain. Grain to be eaten in one day was gathered according to the day's need. When grain in a field cultivated by one family had been consumed, its members moved on to a grain field cultivated by another family. Thus, grain in each field was eaten in turn until all grain raised by the community members was finished. Limited grain production, when combined with crop sharing and no desire whatsoever to store grain for the future, resulted in rapid grain consumption. Community members thereafter had to depend upon forest and garden products for their sustenance. Of their gathering, Harkness (1832: 94-5) went on to say:

> ...Many of them live, for the remainder of the year, on a sort of yam, which here grows wild, and which, after the name of these people, is called the Erular root. To the use of this root they accustom their children from infancy, and when it fails them, which is sometimes the case, they have hardly any resource from starvation. As it becomes scarce in the vicinity of their village, they wander through the forests in search of it.... It is...while they are wandering about the forests in search of food, that, driven by hunger, the families or parties separate one from another, each eager only to satisfy his own cravings.

The model provided by Harkness is soon afterwards corroborated somewhat by Ouchterlony (1848: 62). He stated that the Iruḷas grew a limited acreage of grain, ate field produce immediately, made no effort to store grain, and then turned in the rainy season (easterly Monsoon) to their garden products. When Iruḷas became hard pressed for food, hunting and the sale of beeswax to lowlanders became important. How far can we accept portions of the model? If ideal weather conditions prevail and pests remain inactive, consumption of grain from one field after another is possible. This practice, however, seems unlikely. Farmers tend to value short-term harvesting soon after grain ripens. It is also highly unlikely that Iruḷas failed to store any harvested grain. Due to inefficient cereal production and a high degree of cooperative sharing, it is probable that grain supplies were in any case soon exhausted. It seems likely that Iruḷas did then depend upon garden products, hunting, and gathering. Gathered yams (51-6)[1] may have provided a large portion of the staples eaten in this period. It is improbable that hamlets would be totally unoccupied at any time of the year. The nomadic wanderings of some Iruḷas, farther and farther away from hamlets, and a corresponding division of the population into smaller groups, are not difficult to accept. The sale or exchange of gathered products with neighbours and lowlanders probably aided also in obtaining food.

In addition to the products gathered for barter or sale, which could also be utilized by the Iruḷas, these are examples of food sources available to them. At higher elevations in the main Iruḷa habitat, there were then large tracts of montane subtropical forest. From these and plants on related savanna the Iruḷas could obtain the fruit of the 'wild olive' or rahoo seed (61), the hill gooseberry (155), and the Nilgiri barberry (25). Covering much of the eastern Nilgiri slopes at lower elevations, and particularly over areas below most Iruḷa hamlets, there was much shrub savanna. Scattered savanna trees like dhavam (16), red sanders (148), and teak (185) are in the *Anogeissus*, *Pterocarpus*, and *Tectona* series (Gaussen *et al.* 1962). Shrubby growth is represented by bracken (147), lantana (99), and a profusion of small palms (129). In the shrub savanna, the following wild yams (51-6) existed in large enough quantities normally to provide high yields in relatively short periods of gathering input: common yam, *D. hispida* Dennst., betel yam, prickly yam, and thorny yam. It is worth noting, too, that lowlanders may actually have identified Iruḷas by the betel yam (53), the main one collected by them. With Tamil approximated into English, this translates at *erilai kilangu* (two-leaved (opposite) bulbous-rooted plant). With the addition of the Irula plural suffix of *garu* for 'people', the term *Eriligaru*—meaning the person or people who eat this root—would have developed. This is precisely the term which Francis Buchanan (see quotation on p. 283) used for them. If the less respectful Tamil term of *ār* was substituted for *garu*, the name would become *Erular*. (And later, with the additional tendency of the English to create words euphonious to their speech, this could have become abbreviated to Iruḷa.) Roots of three wild yam-like smilax plants (166-8), the china root or wild sarsaparilla, are also edible. The young leaves of rusty mimosa (6) and the more tender leaf and stalk portions of wild taro (42) may be cooked and eaten. For preparing the latter, commonly growing alongside streams and in swamps, the Iruḷas learned to add Indian sorrel leaves and stems (125). These neutralized the effects of acidic rods within the taro.

For the earlier economy, there are these additional gleanings. Harkness (1832: 90,130) observed banana, chili, edible root, jack, lime, and orange plants growing in Iruḷa gardens. Bananas, edible roots, and jacks may serve as staples. Limes and oranges would compensate for vitamin deficiencies in the other garden staples. Keys (1812: xlix) and Hough (1829: 109) noted the use of hoes in Iruḷa dry field preparation. Ouchterlony's account (1848:62) is particularly useful for its specific identification of finger, Italian, and little millet as the cultivated grains. Mustard was mixed in with these. By seeing and collecting grain amaranth, Buchanan (1870, 1: 462) proved the spread in cultivation of a New World plant (from Malabar, where it was just possibly introduced by the Portuguese). From Cleghorn (1861: 140), who spoke against and helped end shifting slash-burn agriculture, we have proof that Iruḷas were still shifting their fields in the 1850s.

Shortt (1868:62) wrote that Iruḷas collected wild fruits, herbs, and roots to appease hunger along with beeswax, drugs, dyes, gum, honey, and medicinal herbs. The gathered products were exchanged with lowlanders for cloth or food. While searching for honey, Iruḷas sometimes suffered severely from encounters with sloth bears (*Melursus ursinus* Shaw). Earlier, Keys (1812: xlix) had stated that Iruḷas were expert in collecting honey from rocks and cliffs. Harkness (1832: 95) mentions them as ensnarers and hunters of wild animals. Ouchterlony (1848: 62) listed muntjac (or barking deer, *Muntiacus muntjak* Zimm.), sambar (*Rusa unicolor* Kerr.), spotted deer (*Axis axis* Erxl.), and other game as being hunted with much skill. Use of bows and arrows by Iruḷas is not mentioned in the literature: probably they used hunting nets and spears. Breeks (1873: Plate 78) obtained a photograph of a net and spear used mainly to catch and dispatch muntjacs. The spear had an iron head manufactured by the Kotas. The same photograph also shows a small net in which jungle fowl were caught.

Morgan (1876: 100-1) outlined two methods used by the Iruḷas, Kuṟumbas, and Sholigas in hunting (or gathering?) flying-squirrels (the large brown *Petaurista philippensis* Elliot and small Travancore *Petinomys fuscocapillus* Jerdon), mouse-deers (Indian chevrotain, *Tragulus meminna* Erxleben), and other small game. Flying-squirrels are best sought in the day, when they sleep in holes within tree trunks. As one man climbs up to such a hole, others groan and hiss loudly while beating nearby bushes. The vibrating sounds produced by a man climbing a tree would normally cause the flying-squirrel to take flight, but the sounds produced by men on the ground are terrifying enough to prevent it from leaving. Thus the man who climbs the tree will probably be able to grab the flying-squirrel and twist its neck. In the dry hot season, when high savanna grasses around scattered trees are burned down, a few grassy patches escape being burned. Mouse-deer and other game take refuge in such patches; but while a strong wind is blowing, grass along the edge of a single patch is ignited. Then flames race through the vegetation, and animals which flee are clubbed to death by men surrounding the burn. If some manage to escape the men, they may be killed by dogs. After a burn has cooled sufficiently, men cross it to collect the carcasses of animals caught in the flames. Morgan also mentioned fish gathered after being drugged with crushed poison nut seeds (175) thrown in water. To serve the same purpose, we can add the use of the fruit from *Xeromphis spinosa* (Thunb.) Keay (199) and honey cauray (30). Iruḷas may eat the ripe fruit and leaves from the latter, and its stem yields a fiber.

Iruḷas in the early 1800s appeared to have had some caste-affiliated beliefs and Hindu-inspired rituals. The first positive mention of Iruḷas serving as priests to Ranga (Vishnu) dates to 1812 (Keys 1812: xlix). Although the Abbé Dubois (1906: 196) made earlier reference to a sacred Nilgiri peak which was probably Rangaswami Betta, it was not until 1848 that this peak was identified in a published source (Ouchterlony 1848: 62). There, as in Dubois, it is mentioned that Hindu pilgrims flocked to the peak to offer money and produce.

Breeks (1873: 68) noted that Iruḷas did not eat buffalo or cow flesh, which would be in keeping with Hindu caste position. Apart from being a food source, chickens and goats served as the sacrificial animals. Harkness (1832: 88) witnessed the throat-slitting of cocks and a goat by Iruḷas. Shortt (1868: 64) mentions cockerel and goat sacrifice to Māri, the Hindu goddess of smallpox. It is, however, probable that this was not a fully accurate recording. Iruḷas do not directly worship Hindu deities, so much as make sacrifices to appease spirits; this, an aspect of their animism, and the basic worship of the male and female principle (found also among the Kotas) are important to the Iruḷas.

Persistence and Change

Most elements of the early nineteenth century Iruḷas' economy have persisted. Iruḷas continue to live in nucleated communities. Their houses are built on comparatively flat terrain, or on terraces fashioned by digging into slopes and then piling up earth on the outward sides. Houses on flatter ground often surround courtyards on two or three sides, but houses built upon terraces on steep slopes tend to be aligned into single rows next to elongated courtyards (Fig. 11.2; Pl. 21B). A feature not mentioned by Buchanan ia the confinement room where menstruating women and mothers undergoing a postnatal pollution period must stay (Pl. 21C). That Iruḷas associate spirits with natural objects (trees for example), may in part explain the general absence of temples in their communities. A recent decline in wild predators, mainly panthers (*Panthera pardus* L.) and tigers (*Panthera tigris* L.), and the preference of elephants (*Elephas maximus* L.) for the wilder lower elevations have almost eliminated any need for Iruḷas to keep nocturnal fires going in courtyards. The Iruḷas themselves are partially responsible for the decline in predators. After their livestock have been killed by predators, they have followed a widespread practice of sprinkling poison—often an insecticide, such as Folidol—onto the meat. Any animal coming later to feed on the flesh might then die from poisoning. Courtyards serve mainly as work areas, basking areas, flats to dry agricultural produce, or arenas for dancing, merry-making, and ceremonial events. As keepers of chickens, goats, and sheep, Iruḷas construct assorted livestock huts with floors usually on stilts. Large, specially-made baskets are sometimes placed at night over chickens at ground level. Keeping cows for milk is the exception rather than the rule, and such cows spend their nights in huts with or without walls (Pl.21D). Iruḷa communities are also likely to have various store rooms, sheds, huts, and platforms to store firewood, grain, and household objects. At Koppayūr, in 1963, there was even a special hut for storing drums played during leisure hours. Water to supply domestic needs in most Iruḷa hamlets is carried from nearby streams, but some Iruḷas who live in house-rows constructed for them by the Government of Tamil Nadu now

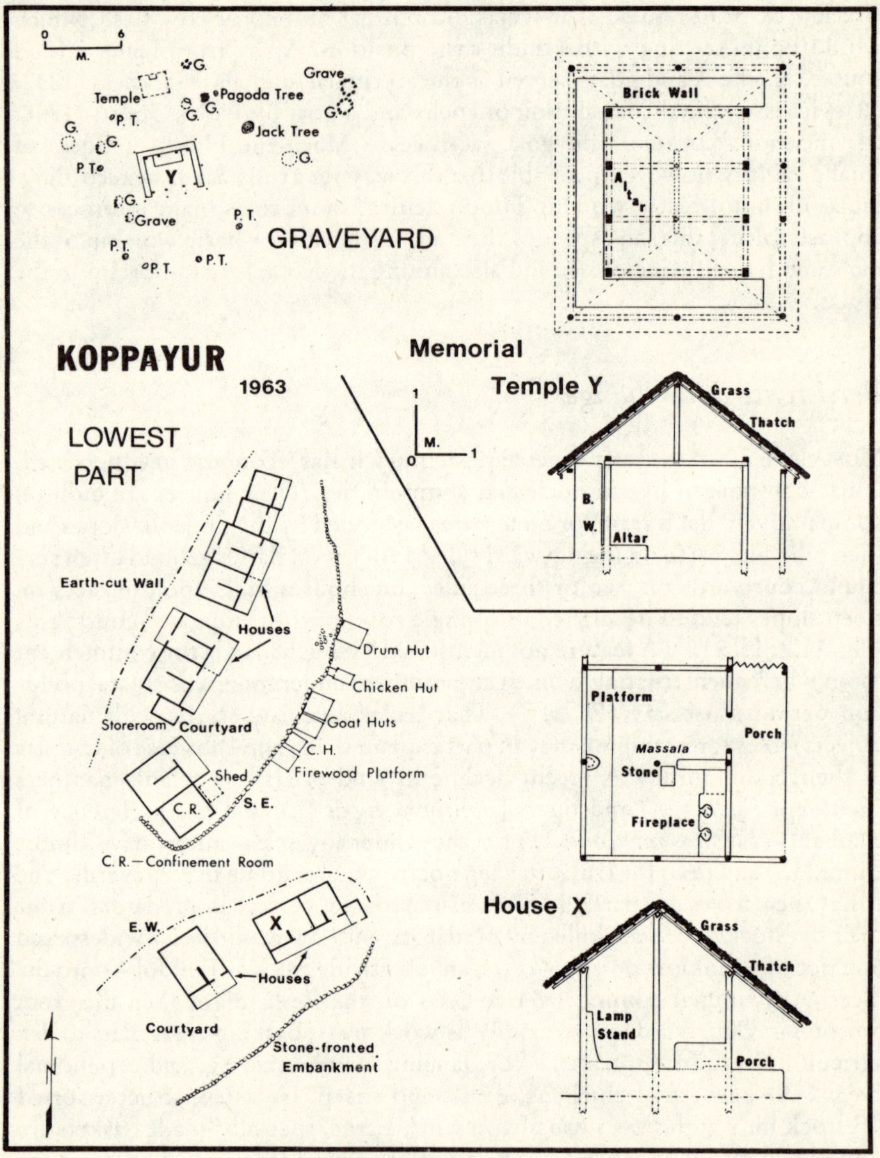

FIG. 11.2 Koppayur.

use piped water. Some communities are close to roads, but others are reached only by narrow paths.

Outside Iruḷa hamlets, either close or some distance away, there are graveyards with memorial temples. As the Iruḷas bury their dead in a seated posture, each grave is demarcated by a small, circular earth mound which is sometimes encircled with stones (Fig. 11.2). Following a practice observed by other South Indian tribals (the Kundavadians of Kerala, for example), the Iruḷas provide sacred identity to a graveyard by planting temple (or frangipanni, 140) trees there. Of the few sacred centres which Iruḷas do visit elsewhere, the shrine on top of Rangaswami Betta (Fig. 11.1) continues as the one most hallowed to them. Gardens are still conspicuous features around Iruḷa communities, and their composition varies greatly. Apart from multilayer gardens having earlier species mixed with those introduced by Europeans, there are also single layer gardens and kitchen gardens. The Final Land Settlement (1881-4) and Madras Forest Act (1882) officially prohibited the use of forested land for agricultural purposes, and thus Iruḷa shifting slash-burn agriculture eventually came to an end. It is noteworthy that the early nineteenth-century millets still dominate in dry field cultivation near hamlets, and Iruḷas still tend to shift cultivation on land originally allotted them or since purchased. In contrast to many northern Indian inhabitants who stall-feed their livestock, Iruḷas follow the widespread southern Indian practice of letting their livestock graze or browse on nearby tracts with grasses and shrubby regenerate growth.

The montane subtropical forest tracts in the Iruḷa habitat proved to be ideal for tea growth at the higher elevations and for coffee growth at lower elevations. As the plantations spread, Iruḷas lost usufruct rights over considerable tracts and were forced to contend with large, capitalistic, efficiently run enterprises close to their own hamlets. Their main benefit of easy employment on plantations brought with it the termination forever of any seasonal dependence upon gathering, trapping, or hunting. Iruḷas did also benefit from the modernizing amenities associated with plantations. When eventually most of them worked on plantations, they led a dual existence, one related to daily work rounds largely controlled by foreigners, and another still tied to many of their own traditional ways.

Elimination of any reliance upon the products of gathering, trapping, or hunting was certainly related in part to an increased liking for rice which was now obtainable through wages. Wild yams and other edible roots may still, however, be easily gathered with the help of dibbles or crowbars. And the Iruḷas may at any time use the products which they gather for barter or sale with others. After 1900, produce collected by Iruḷas was being taken into special collection centres established by the Madras Forest Department. Lushington (1902: 147) listed the main gathered products as beeswax, deer antlers, gum, honey, avaram bark (33), red creeper (196), myrobolam fruit (185), tamarind fruit (183), and soapnut (160). That these products continue to be gathered is evident from Table 11.1, listing mainly the plant species that are economically advantageous to Irulas. Both medicinal and non-medicinal products are listed there. It should also be kept in mind that the many ramifications in plant use cannot be explained in this brief presentation. For example, fruits from jack (18), mango (108), and tamarind (183) trees, although also growing in gardens, are among the leading products collected outside them. From gardens, human and other agents have widely dispersed viable seeds of these three species. From the margosa or neem tree (22), which yields oil and oilcake from its seeds, the bark, flowers, leaves, and seeds are all used for medicinal purposes. Apart from the more specific gathered products, another major source of income for Iruḷas is the firewood they cut and remove from forests. Until recently, Iruḷas were supplying forest products to the Forest Department and to contractors. Contractors all too frequently exploited them, and the Forest Department tended to provide only the minimal acceptable return. The Government of Tamil Nadu is now promoting co-operative societies, so that Iruḷas may obtain more reasonable returns for their labour input.

There is persistence in some of the old traditional ties between the Iruḷas and their neighbours. Some Iruḷas weave baskets, winnowers, and winnowing-fans of split bamboo. Banana pseudostem strips are also employed in basketry. Leaves from the small date-palm (129) are made into brooms. A few men fashion a wind instrument called the *köylu* out of bamboo, wild jack (18) wood, and porcupine (*Hystrix leucura* Gray & Hard.) quills. These artifacts, along with some gathered products, are bartered or sold to neighbours. The Iruḷas are especially known as suppliers of honey collected from the hives of forest and rock bees (*Apis indica* L. and *A. dorsata* L.). An example of bartering trade is offered by the Kinar (otherwise Kil-Kotagiri) Kotas and nearby Iruḷas. These Kotas obtain brooms, bamboo artifacts, honey, punk used by priests to light fires (priests should not use matches), and resin incense or black dammer (29) from Irulas. In return, the latter get both field and garden implements from Kotas.

By the early nineteen-hundreds the Iruḷas had started using guns (Thurston and Rangachari 1909, II: 373, 379). Owing to strict game controls, however, Irulas generally no longer hunt big game: if they do, guns are still used. Deadfalls, small nets, and snares are occasionally employed too to entrap or kill animals and jungle fowl.

We have thus far outlined—with an historical perspective—some basic cultural characteristics which in recent decades have either ended, persisted, or changed. In order to consider further four significant aspects of the Iruḷa cultural complex, we shall now examine these topics: houses and memorial temples, fields and gardens, the shift from plantation work, and the Iruḷa priesthood.

HOUSES AND MEMORIAL TEMPLES

Members of a South Asian caste or tribe often construct and functionally organize each home and each temple in a similar manner. Among modern South Asian scholars, Fürer-Haimendorf and others have most clearly demonstrated this truism—and in relation to people as diverse as the Apa Tani, Konyak Naga, Raj Gond, and Reddi (1962: 14-15, 70; 1969: 23-9; 1948: 87-90, 95; 1945: 49-60).

Although Iruḷa houses vary in size, each tends to conform to an accepted norm. Some houses stand separate, but many are built contiguous to each other. The common side-gabled house has a rectangular earthen base, a front porch, upright Y-forked posts supporting cross-poles, wattle and daub walls running between posts, and a thatched roof (Fig. 11.2). Walls with split bamboos, sticks, or banana pseudostem strips woven onto embedded vertical poles are not always coated with mud. Sometimes a final wall covering of vertical banana pseudostem strips is held in position by horizontal braces lashed into position (Pl.21E). A right doorway is made by lashing a horizontal pole onto two posts, and a door formed by woven split bamboos is tied to one doorpost. Some houses have a doorway in the left rear. Roof thatching is laid over split bamboos lashed onto rafters overlying the crosspoles. Citronella grass (48) is commonly used for thatching, while the stems of Johnson grass (170) are ideal for holding thatch in position. Roofs may also be made with banana pseudostem strips held in position by poles that are lashed on. Each house interior is divided into two roughly equal parts, one more secular behind the front entrance and the other, with fireplace, more sacred. The two parts in larger houses are separated by an earthen partition extending inward from the front wall. The Iruḷas are Hindu-like and Badaga-like in their strict exclusion of all outsiders from their kitchens. Next to the rear wall, a shallow earthen platform is constructed for household utensils. Over this platform and next to the central rear wall post, an earthen lampstand is erected. An interior fireplace is built next to the left front wall (Pl. 21F). Lastly, there may be a storage platform above a high central earthen partition and the left end wall top. With changing ways Iruḷas live increasingly in non-traditional structures: the State Government and plantation managements have started to furnish them with substantial row dwellings (Pl. 21G). Iruḷa children now receive free board, lodging, and education while away in boarding schools (Pl. 21H).

Memorial temples were the most characteristic structures in the original Iruḷa religious system when unaffected by Hinduism. As an extension of their belief in spirits, the Iruḷas worshipped the spirits of their departed. An ancestral spirit may be associated with a water-worn stone placed on an altar within a memorial temple. Typically, immediate family members who can afford to do so will bring a stone from a stream bed at one to several years after the death of their relative. Once the stone is emplaced, a feast has to be provided for the kin and others in attendance. Among the Irulas there are twelve exogamous lineages (they are named Chambai, Devennan, Kalkaṭṭi, Kuduvan, Kuppan, Kurunāga, Ollaga, Pērava, Porigan, Punga, Uppikan, and Vellaikan). A deceased lineage member is normally always buried in the graveyard surrounding the memorial temple of his or her lineage. If a dead person must be buried far away, it is desirable for some portion of the corpse eventually to be brought to and buried near the associated memorial temple. Even though the members of a lineage will always use their own ancestral memorial temple, this structure need not be located away from other such temples. Since the members of three lineages dwell in Kallampalaiyam, for example, there is a cluster of three corresponding memorial temples. Before any major undertaking, a devout Iruḷa will visit the memorial temple of his or her lineage to offer incense and a traditional food offering (see the last section below) to the ancestral spirits. In addition, new clothing may still be offered first to the departed before being worn by living Iruḷas.

The memorial temples are constructed amidst the small earthen burial mounds (Fig. 11.2). The one at Koppayūr (named after a *koppa*, a graveyard) and others near Kil-Kotagiri Tea Estate are virtually identical in architecture and their functional organization. Each is a rectangular structure with an open front, a U-shaped brick wall surrounding the other three sides, and a thatched roof with short crest. The roof rafters extend over a ridgepole above two kingposts, ultimately supported by posts within the wall and pole plates surmounting posts outside the wall. Overlapping rows of grass thatch are tied onto horizontal split bamboos lashed over the rafters. The grass species used for thatching and for tie-downs are liable to be the same as the already cited two that are often used in house roofs (48, 170). Within the memorial temples, centred against the back walls, there are shallow earthen altars upon which are lain the water-worn memorial stones (Noble 1976: Pl. 13H), and an occasional substitute sculptured stone or horseshoe-shaped piece of metal (horn-like, symbolically related to a deceased owner of cattle).

FIELDS AND GARDENS

Even though we have listed lower elevation cereals and have included lower elevation plants in the general list of garden plants (see the Table 11.1), the ensuing discussion primarily pertains to Iruḷa dry field and garden agriculture at higher elevations. Cultivation of the lower elevation cereals involves a somewhat different system, and lower elevation garden plants do not thrive in the higher elevations. Until some scholar provides a viewpoint of the Irulas from the plains, conceptual gaps will exist in our knowledge of them.

Excluding those communities inhabiting that portion of the Nilgiris District that lies within the northwestern Wynaad Plateau, Badaga, Iruḷa, Kota, and Kuṟumba farmers grew millets but no hill rice when the British first came. It is suggested that the complex related to millet cultivation, harvest, and post-harvest operations represents a remnant tradition harking back to Neolithic times. The Iruḷas usually sow Italian and little millet together (Pl.

22A), but finger millet is cultivated separately. Because amaranth and mustard are mixed in with the three cereals, they too grow scattered through the grain fields. Excluding the sowing of seed by men, both adult sexes perform all the agricultural tasks. However, such monotonous tasks as weeding or reaping are often relegated to the women. Hoes with narrow blades and hoe-forks are the most frequently used implements in soil preparation. Although Iruḷas may no longer cultivate crops wherever they desire, a few of them still own enough land to permit the periodic shifting of grain plots. The shifts appear to be prompted by topsoil erosion on steep slopes, non-application of fertilizers, and weedy plants obtaining an increasingly firm foothold. Terracing would slow down erosion, but Irulas do not terrace their fields. The plots are cultivated preferably for about three years before being permitted to regenerate with natural vegetation. Now, because some individual landholdings have decreased in size as a result of land being inherited equally through subdivision by sons, longer cultivation periods may be necessary.

To obtain their annual crops, Iruḷas clear and prepare fields by May. Ashes from plants burned after their removal are spread, and the earth is tilled with hoes and hoe-forks. Seed is sown mainly in late April and early May, after the 'mango showers' have fallen. However, if the showers have not brought enough rain, sowing may be postponed until June. Broadcast grain is worked into the earth with the aid of hoes and hoe-forks. The fields with growing plants are weeded to varying degrees. Excepting the little millet, grains are harvested mainly in August and September. This means that Italian millet is harvested from fields in which far from mature little millet still grows. An occasional finger millet crop sown after the first easterly Monsoon rains is reaped in December and January, when the later maturing little millet is also harvested. Irulas use very small sickles for removing the individual grain heads, but amaranth stems are cut off with knives. In the harvesting of amaranth and finger millet, the plants are visited periodically to permit the removal of grain as it ripens. Stubble left on fields provides forage for livestock. Harvested Italian millet grain heads may be stored away until needed. Irulas use threshing sticks to thresh all their grain. When thoroughly sundried, the grain is stored in large woven bamboo baskets and grain chests kept inside or on platforms within houses. For storage of up to a year or more, maize cobs and beans from kitchen gardens, or grains—especially finger millet—from fields, may be placed in a clay pot. This then has its lid thoroughly sealed on with clay before being put away in a warm section within a dwelling. That Iruḷas use a technique which enables finger millet to be stored for years suggests, too, that they were grain storers prior to the first visits by Englishmen.

The role which gardens play in tropical agriculture remains poorly understood by many. Gardens are generally related to a continued vegetative ground cover and are therefore ecologically stable. For the dwellings of man, they provide shade and serve as a shatter belt to reduce the effects of wind and rain (Pl. 22B). Although some of the garden plants produce carbohydrate-rich staples, others do provide foods with protein and vitamins. Many gardens are also related to year-round production. This is due to particular species which yield all the year, but there are also plants which yield in different seasons. All these factors are demonstrated in Iruḷa gardens. Until the early 1800s and in each annual season following grain consumption, Iruḷas must have gathered a wide assortment of foods from their gardens.

When thinking of Iruḷa gardens, the classificatory system which follows will enable us to focus upon aspects of them. We should not forget, however, that in reality there usually exists a continuum generally related to plants which cover the land close to Iruḷa hamlets on a permanent basis. This continuum may actually extend into fields as well, for it is not uncommon for example to find perennial bananas gaining a foothold in plots with millets. The presence of at least eighty plant species within Iruḷa gardens (Table 11.1) is indicative of the diversity in combinations which may exist in garden continuums. Some portions of Iruḷa garden continuums are long enduring, like some forests, but other portions undergo frequent change. Some plants are removed when their roots are dug up, others die naturally and are replaced with different species. Changes in Iruḷa dietary preference may lead to replacement with the more desired species, while the wish to grow a commercially productive plant may lead to its spread and the consequent decline of several other species (Pl. 22C).

Some Iruḷa gardens exemplify tropical gardens *par excellence*, dominated by perennials, but with a glorious hodgepodge of annuals and perennials yielding produce on a vertical dimension extending from beneath the surface of the soil up to branches twelve or more metres above the ground. Other multilayer gardens have both horizontal and vertical conformity. Thus, a garden with scattered perennial jack, mandarin orange, and coffee plants (typically pruned, and kept to two metres or less in height) will have three general crown zones. Then there are single layer gardens, each with a single perennial species forming a single production zone. Stands containing only bananas continue to be the conspicuous example of this type of garden. Coffee stands shaded by wild tree species, having a continued growth permitted by the Irulas, are another example. Additional kitchen gardens are often close to kitchens, and are usually adjacent to or within other gardens. They are typically small patches especially benefitting from deposits of hearth ash and settlement sweepings. Most kitchen gardens have annuals which are planted or sown after the first 'mango showers' in May. In contrast to the year-round production in multilayer gardens with perennials, produce in kitchen gardens sometimes tends to be harvested at the same time as dry field crops. Kitchen garden plants are sometimes mixed together and are often planted in mounded rows. However, species such as amaranth, maize, or turmeric may be grown individually over microplots.

TABLE 11.1

Gathered Products (Mostly from Plants)

Mainly for Medicinal Purposes

Cumbi gum, or Dikamali, *Gardenia gumnifera* L. (resin)

*Indian beech, or pongam oil tree, *Pongamia pinnata* (L.) Pierre (seeds)

Indian laburnum, *Cassia fistula* L. (mainly the pod)

*Indian sarsaparilla, *Hemidesmus indicus* (L.) Schult. (roots)

*Jambolan, or Java plum, *Syzygium cumini* (L.) Skeels (bark, fruit, seeds)

Kantakari, *Solanum surattense* Burm. (roots)

Lichens

*Mahua, *Madhuca longifolia* (Koenig) Macbride (bark, leaves, flowers, seeds)

Mukia maderaspatana (L.) Roemer (shoots, roots, seeds)

*Myrobolam—belleric, *Terminalia bellerica* (Gaertn.) Roxb. (fruit)

*Myrobolam—chebulic, *Terminalia chebula* Retz. (fruit)

*Myrobolam—emblic, *Emblica officinalis* Gaertn. (fruit)

Pergularia daemia (Forsk.) Chiov (entire plant)

Queensland hemp, *Sida rhombifolia* L. (entire plant)

Sida, *Sida cordifolia* L. (entire plant)

Sweet flag, *Acorus calamus* L. (roots, leaves)

Wild asparagus, *Asparagus racemosus* Willd. (roots)

Wild snake gourd, *Trichosanthes cucumerina* L. (entire plant)

Mainly for Non-medicinal Purposes

*Honey and beeswax

Gaur horns

Sambar and spotted deer antlers

Acacia pennata (L.) Willd. (bark tannin)

*Acacia—distiller's, *Acacia leucophloea* (Roxb.) Willd. (bark clears and flavours arrack, bark tannin)

*Acacia—distiller's, *Acacia leucophloea* (Roxb.) Willd. (bark-clears and flavorus arrack, bark tannin)

Acacia—shikai, *Acacia concinna* DC (pod detergent and bark tannin)

Acacia—black wattle, *Acacia melanoxylon* R. Br. (bark tannin)

Acacia—silver wattle, *Acacia dealbata* F.V.M. (bark tannin)

Albizzia amara (Roxb.) Boiv. (powdered leaves for detergent)

*Bamboo, *Dendrocalamus strictus* Nees (basket making)

*Bamboo—spiny, *Bambusa arundinacea* (Retz.) Willd. (poles)

Bowstring hemp, *Sanseveria roxburghiana* Schult. (fibre)

Ficus racemosa L. (edible fruit, pickled when green)

*Indian jujube, *Ziziphus mauritiana* Lam. (edible fruit, pickled when green)

*Indian wood apple, *Feronia limonia* (L.) Swing. (fruit)

*Karaunda, *Carissa carandas* L. (edible fruit, pickled when green)

Lemon grass, *Cymbopogon flexuosus* (Nees ex Steud.) Wats. (oil in leaves)

*Margosa, or neem, *Azadirachta indica* A. Juss. (oilseeds)

Mauritius hemp, *Furcraea gigantea* Vent. (fibre)

*Small wild date, *Phoenix humilis* Royle (edible fruit, leaves for brooms)

*Soapnut tree, *Sapindus trifoliata* L. (dried, powdered fruit as a detergent)

Solanum torvum Sw. (fruits for pickling)

*Tanner's cassia, *Cassia auriculata* L. (bark tannin, powdered leaves for detergent)

Vempadam, *Ventilago maderaspatana* Gaertn. (root bark for dye)

DRY FIELD CEREALS

Mainly for Non-medicinal Purposes

Mainly at higher elevations

Finger millet, *Eleusine corocana* Gaertn.

Italian millet, *Setaria italica* Beauv.

Little millet, *Panicum sumatrense* Roth & Schult

Mainly at lower elevations

Barnyard millet, *Echinochloa colona* (L.) Link

Bullrush millet, *Pennisetum typhoides* Stapf & Hubb.

Common millet, *Panicum miliaceum* L.

Sorghum, *Sorghum bicolor* (L.) Moench

GARDEN PLANTS

Grain

**Amaranth, *Amaranthus caudatus* L., but mostly *A. hypochondriacus* L.

**Maize, *Zea mays* L.

Fruit

Banana, *Musa sapientum* L.

**Custard apple, cherimoyar, *Annona cherimolia* Mill.

Citron, *Citrus medica* L.

**Guava, *Psidium guajava* L.

**Grapefruit, *Citrus paradisi* Macf.

Jack, *Artocarpus heterophyllus* Lam.

Lime, *Citrus aurantifolia* Swingle

Loquat, *Eriobotrya, japonica* Lind!.

Mandarin orange, *Citrus reticulata* Blanco

Mango, *Mangifera indica* L.

**Papaya, *Carica papaya* L.

**Passion fruit, *Passiflora edulis* Sims.

Peach, *Prunus persica* Benth. & Hook.

**Pineapple, *Ananas comosus* Merr.

Pomegranate, *Punica granatum* L.

Pummelo, *Citrus grandis* Osbeck.

Star gooseberry, *Cicca disticha* L.

Tamarind, *Tamarindus indica* L.

**Tree tomato, *Cyphomandra betacea* Miers

Watermelon, *Citrullus vulgaris* Schrad.

Vegetable

Agati, *Sesbania grandiflora* (L.) Poir.

Bitter gourd, *Momordica charantia* L.

Brinjal, *Solanum melongena* L.

Broad bean, *Vicia faba* L.

Cabbage, *Brassica oleracea* L. var. *capitata*L.

Catjang cow pea, *Vigna unguiculata* (L.) Walp.

**Chow-chow, *Sechium edule* L.

Cluster bean, *Cyamopsis tetragonoloba* (L.) Taub.

Cucumber, *Cucumis sativus* L.

Gourd, *Lagenaria siceraria* (Molina) Standl.

Horse gram, *Dolichos uniflorus* Lam.

Indian nightshade, *Solanum nigrum* L.

Indian spinach, *Basella alba* L.

**Kidney bean, *Phaseolus vulgaris* L.

Lablab bean, *Dolichos lablab* L.

**Lima, or sieva bean, *Phaseolus lunatus* L.

Loofah—angled, *Luffa acutangula* (L.) Roxb.

Loofah—smooth, or sponge gourd, *Luffa cylindrica* (L.) M.J. Roem.

**Manila tamarind, *Pithecellobium dulce* (Roxb.) Benth.

Pea, *Pisum sativum* L.

Pigeon pea, *Cajanus cajan* Millsp.

**Pumpkin, *Cucurbita pepo* DC

**Tomato, *Lycopersicum esculentum* Miller

Root

**Canna, *Canna edulis* Ker.

Elephant foot yam, *Amorphophallus campanulatus* (Roxb.) Blume

Giant taro, *Alocasia indica* Schott

**Potato, *Solanum tuberosum* L.

**Sweet potato, *Ipomea batatas* Lamk.

**Tapioca, *Manihot esculenta* Crantz

Taro, *Colocasia esculenta* (L.) Schott var. *esculenta*

Turmeric, *Curcuma longa* L.
Yam, *Dioscorea alata*. L.

SPICE
Cardamon, *Elettaria cardamomum* Maton
**Chili, *Capsicum frutescens* L., *C. annuum* L.
Curry-leaf tree, *Murraya koenigii* Spr.
Mustard, *Brassica juncea* Hk. & T.
Pepper, *Piper nigrum* L. (medicinal fruits)

FLOWER
Bachelor's button, *Gomphrena globosa* L.
**Canna, inedible *Canna indica* L.
Changeable rose, *Hibiscus mutabilis* L.
Chrysanthemum, *Chrysanthemum* sp.
Crossandra, *Crossandra undulaefolia* Salisb.
**Four o'clock plant, *Mirabilis jalapa* L.
Indian oleander, *Nerium odorum* Soland
Jasmine, *Jasminum flexile* Vahl
Marigold, *Tagetes patula* L.
**Periwinkle, or vinca, *Catharanthus roseus* (L.) G. Don. (medicinal roots)
Rose, *Rosa* sp.
Rose of China, *Hibiscus rosa-sinensis* L.
**Slipper plant, *Pedilanthus tithymaloides* Poit

SWEETENING
Sugarcane, *Saccharum officinarum* L.
Wild date palm, *Phoenix sylvestris* Roxb.

BEVERAGE
Coffee, *Coffea arabica* L., *canephora* Pierre

OIL
Castor, *Ricinus communis* L.
Sesame, *Sesamum indicum* L.

NARCOTIC
Hemp, *Cannabis sativa* L. (grown by few, for personal use)
**Tobacco, *Nicotiana tabacum* L.

FIBRE
Kapok, *Ceiba pentandra* (L.) Gaertn. var. *pentandra* (syn. var. *indica* Bakh.)

*most frequently collected
**New World domesticates

(Source: Scientific nomenclature for domesticated plants is always difficult. This table contains our best effort to define scientific names adequately. For the reader who wishes to go further, in 1961 an international code of nomenclature (30 pages) was published in *Regnum Vegetabile*, Vol. 22.)

THE SHIFT FROM PLANTATION WORK

As mentioned previously, plantations in the Nilgiris had a profound effect on the Irulas. During 1838 Dawson started the first coffee plantation near Coonoor, and two other plantations were opened during 1840 in the Kotagiri area (Athrey 1953: 11). A few plants grown from Chinese tea seeds were planted in 1835 at Ketti and elsewhere (United Planters' Association of Southern India 1960: xxxiii). Fortune, in 1854, brought some tea seeds from bushes recently imported from China, and gave the seeds to Henry Mann. The planted seeds formed the nucleus of the first tea plantation near Coonoor, and in 1858 Cleghorn (1861: 18) noted the presence of about 2000 vigorous tea plants on the plantation. In 1979 there were 7,165 ha of coffee and 18,186 ha of tea on the Nilgiri massif (United Planters' Association of Southern India 1979: 635). The areal spread of coffee by then appears to have stabilized. In the decade 1966-75 there was a dramatic increase of 4,132 ha of tea, and a further 608 ha of tea were established in the following four years. Plantations for long were speading over the eastern slopes where the Irulas lived. Tea came close to Koppayūr, while recently planted coffee now extends right up to this hamlet on three sides.

By 1847 Iruḷas had started to work occasionally as coolies on what must have been coffee plantations. In their plantation work then, Ouchterlony (1848: 62) considered them expert tree-fellers and hewers of planks or rafters. By 1868 some had been acculturated considerably by working alongside imported South Indian labourers on plantations (Shortt 1868: 63). After a post-1900 visit to an Iruḷa hamlet near a coffee plantation, Thurston and Rangachari (1909, II: 376) recorded that it was 'in the possession of pariah dogs and nude children, the elder children and adults being away at work'. They also refer (1909, II: 377) to this statement made by another writer in the early years of this century:

> The Irulas...generally possess a small plot of ground near their villages, which they assiduously cultivate with grain, although they depend more upon the wages earned by working on estates [plantations]....The Irula women are as useful as the men in weeding, and all estate work. In fact, planters find both men and women far more industrious and reliable than the Tamil coolies.

Many Iruḷas were working on plantations in 1963. Remembering that some fluctuations in employment have always existed, as a corollary of measures enabling Iruḷas to meet their private needs for seasonal work input, over 500 Iruḷas were sometimes employed at Kil-Kotagiri Estate. In the daytime Koppayūr was nearly deserted.

The desire to obtain income from coffee is leading to change and a new developmental phase, with the Iruḷas ultimately benefitting. The change is reflected in the experience of the Iruḷas at Koppayūr. In 1963 several Iruḷas already owned about 3.8 ha of yielding coffee next to the hamlet. In 1974 the Iruḷas at Koppayūr received about 20.5 ha of land from the Government of Tamil Nadu. This land is, unfortunately, on a steep slope beneath Kolikuttai; also it was equally distributed among 53 individuals. Up till 1978 these Iruḷas received no financial assistance from the government—despite their evident need for capital to prepare the land, grow coffee plants in nurseries, and then transplant the young plants. Their desire to be self-reliant and their ill-feeling toward a management which was responsible for the planting of coffee right up to Koppayūr, and on land used in 1962-3 by Iruḷas for cereal cultivation, contributed towards a decline in the number of Iruḷas employed at Kil-Kotagiri Estate. When one of us recently inquired of the management, only 48 Iruḷas (six from Koppayūr) were then employed. The management also confirmed that, because the Iruḷas were turning to the establishment of small coffee gardens, a far smaller number worked for wages than in 1963. When the same writer visited Koppayūr, there seemed to be a landscape reflection of the embittered desperation felt by some of its inhabitants. Population increase had led to the construction of more houses, including some on new terraces, and the memorial temple was in an advanced state of disrepair. From 1979 the Co-operative Land Development Bank (an agency of the Tamil Nadu Government) at Kotagiri has financed the Iruḷas at Koppayūr. They are now

(1980) growing coffee plants in nurseries and plan to plant these under the most favourable conditions during the next westerly Monsoon. Because the Kil-Kotagiri Estate management has become so aggressive, and the local village headman and clerk side with the management's drive to obtain disputed land, the Iruḷas now fear that they will be evicted from Koppayūr—and forced to settle on their new land, which is generally unsuited to human occupancy.

THE IRULA PRIESTHOOD

Some South Indian tribals perform ritual for Hindus. Thus, in some mountains south of the Nilgiris, Peter Gardner found two examples among Paliyans who tend to return periodically to a nomadic existence based upon gathering (Gardner: 1982: 465-7). In one example forestry officers employ an elderly Paliyan to serve as officiating priest in ritual honouring Sandana Māriammāl, 'Goddess of the Sandal Tree'. The forestry officers participate in the ceremonies which precede sandalwood extraction from reserved forests. In the other example a father and son serve as hereditary priests in the Mahalingam and Sundaralingam temples honouring Siva, near Saptūr (Gardner 1976: 307). Pilgrims by the thousand, mainly Taṃilian lowlanders from different castes, come annually to worship at the two temples. Gardner suggests that the deviation from the norm exhibited by tribals serving as priests is derived from Hindus regarding the tribals as being pure 'by analogy with peripheral members of Hindu society who live a life of austerity' (Gardner: 1982: 462). A sacred and isolated Nilgiri peak such as Rangaswami Betta posed a problem when it came to the need for ritual to honour Ranga on only eight days of each year, spread over the duration of two months. In their willingness to serve as priests, tribal Iruḷas provided a solution to the problem. Since Hindus will devise a rationalization for a departure from the norm, that suggested by Gardner may equally well apply to the Iruḷas. We wish to stress, however, that the Iruḷas are promoting a dualistic system of worship. A participant observer at Rangaswami Betta will see an Iruḷa priest performing typically Hindu ritual before stones, on one of which Lord Krişṇa's symbolic image is clearly sculptured. Ranga, in the Hindu's perception, is equated with Krişṇa and Viṣṇu. The Iruḷas themselves are worshipping the male principle.

An Iruḷa priest officiates on the top of Rangaswami Betta during every Saturday in the Tamil months of Puracṭāṭi and Aippaci, beginning from the middle of August. On the morning of 13 October, 1962, the last Saturday in the sequence, some Chettiars came from the small local town of Sholurmattam. By the time they reached the top of Rangaswami Betta, at 11.15 a.m., the Iruḷa priest had prepared everything for worship. A sketch of the altar and ritual paraphernalia reveals all that is needed for the typical Hindu ritual (Fig. 11.1). In the actual ritual, the priest first placed marigold and rose petals on top of all the altar stones and metal statues. He placed fingertip spots of vermilion on all the stones. Smashed coconuts and their milk collected in vessels, peeled bananas, yellow hands of bananas, and burning incense sticks were all placed as offerings on the altar (Pl. 22D). Bell ringing and the waving of camphor incense before each image then climaxed the ritual. Accompanied by the incense of burning camphor, a mixture of flower petals from the altar, white clay, and vermilion powder in a flat brass dish were offered by the priest to the worshippers. They in turn placed coins on the dish (offerings to the priest) and wrapped their shares of the mixture in scraps of newspaper.

Another ritual phase was ushered in with bell ringing, and the people responding with a brief invocation to Ranga. This phase involves the belief that prayer is answered when flower petals fall (a procedure also used by Badagas). The first to come was a male, who knelt and then bowed his head to the ground. He prayed long and passionately, while the other Chettiars watched intensely for flower petals to fall. None did, and so the petitioner left after the priest had again rung the bell and the others had once again chanted the brief invocation. A mother brought her baby to the altar, and left shortly after lowering the infant three times while making her brief petition. No petals fell. Food offerings were then divided. Worshippers and priest first ate the bananas, and then the fruits of coconuts after further smashing were consumed. Thus refreshed, the Chettiars left for their homes.

The same general sequence occurred when a group of Iruḷas came from Kolangiri. In the phase of the ritual involving petitioners, flower petals fell soon after the first two females came forward. When no petals fell after the third female came, the priest added more petals and arranged them further (Pl. 22E). When still no petals fell, that third person left. A girl came to pray, and some petals immediately fell. A more elderly female prayed intensely for a long time, went into a trance, lost her balance, and sprang back into the crowd of watchers. They helped her into a kneeling position, and some petals soon fell. They also fell right after the sixth and last woman came to the altar. Most of the women prayed, stood up to see if any petals would fall, and then again went into a kneeling or low bowing stance with prayer if no petals had fallen by then. After the Iruḷas had departed, the priest had his lunch and cooked a preparation of rice and crude sugar for ritual presentation. Four packets of this mixture, wrapped in banana leaves, were placed on the aforementioned altar, but a fifth was placed in a nearby stone circle (probably not one of the distinctive prehistoric sites).

When we examine more carefully the ritual paraphernalia related to the 1962 Ranga worship, there seems to have been a remarkable catering to the god Siva as well (Fig. 11.1). One altar stone has a sculpture of Lord Krisṇa playing a flute before a cow. As Krisṇa is the most popular *avatar* of Viṣṇu, this stone is in no way unusual. The other two plain altar stones honour Raṅga, who is an aspect of Viṣṇu. On the altar itself there was a stone bull (the Nandi, Lord Siva's vehicle) and a silver bull with a silver *linga* (of the type worn by Lingayats) around its neck. The intent of any Nandi, including these, is to indicate attendance on Lord Siva. This indication is substantiated by the fact that a silver image of Ranga had a silver *linga* hung around the neck. When the Irulas were asked why the *lingas* were present, they unanimously agreed that the presence of *lingas* indicated the prohibition of beef eating: out of respect for Siva and his vehicle, the Nandi, the great majority of Hindus will refuse ever to eat beef. When the Irula priest made his rice and crude sugar offerings, that within the stone circle was placed before a crude stone *linga* held vertical by supporting stones.

The *liṅga* and bull are probably the symbols most commonly related to Shaivite worship in India. Because one deity is usually honoured at one time in specifically Hindu ritual, and because a Hindu will normally worship either Siva or Viṣṇu (often in *avatar* form), what happens at Rangaswami Betta seems to be an odd departure from the norm. Breeks, in noting corresponding discrepancies at Kallampalaiyam about 1870, rashly suggested that the Irulas 'don't know the difference between Siva and Vishnu' (Thurston and Rangachari 1909: II, 374). If, however, we accept the viewpoint of Irulas personally worshipping the male principle only, the discrepancies disappear. When one of us interviewed the priest (Ūmai Rangan) who officiates at Rangaswami Betta, he insisted that such was the nature of Irula worship there. As the Irulas can readily appreciate the symbolic significance of the *linga* as representative of the male principle, their personal symbolic use of the *linga* and associated bull may easily be incorporated into the ritual which they perform for the benefit of Hindus as well.

When the Iruḷas' relationship with Kriṣṇa and Raṅga worship is examined in the light of folk legends, the intertwined duality of the Irula and Hindu belief systems becomes even more complex. Irulas in the Kalkatti lineage, who migrated downward to colonize Kallampalaiyam, once resided at a place called Kurinji. This site, now leased to the Irulas by the Forest Department, where Pukkuni Goundan still resides, is located just below Rangaswami Betta and about 9.6 km. from Kallampalaiyam. A family in the Kalkaṭṭi ('stone-offering') lineage traditionally provides the priests who serve at Rangaswami Betta, and the officiating priest spends each Friday night at Kurinji before performing ritual on the peak's summit. While Kurinji was still a thriving hamlet, Vaishnavite Iyengar Brahmins from a place called Betikuttai on the plains used to provide the ritual. The relationship between Rangaswami Betta and the Hindu worship of Kriṣṇa and Viṣṇu as Raṅga thus became firmly established. established.

One Iruḷa legend seems to be inspired by many others related to the Kriṣṇa Cult. Another legend explains how the Irulas became priests. According to the first legend, Irulas at Kurinji were tending the cattle of an Okkiliyan (Tamil synonym for Vokkaliga or Okkaliga) Gaundan living on the plains. At night the cattle were kept in a shed. Every morning afterwards the Irulas observed that one cow had no milk. Upon close examination, the Irulas discovered that the cow escaped each night and went to the top of Rangaswami Betta. There she dropped her milk over the stone dedicated to Lord Kriṣṇa.

Because the Iruḷas were convinced that the stone was related to the supernatural and the male principle, they started worshipping at the site. An explanation of why Iruḷas do not eat beef or buffalo flesh is also derived from the belief that they are too pure—and especially so in the case of cow's flesh, because of what that one cow once did at Rangaswami Betta. Later, but still some generations ago, an Iyengar Brahmin was performing the ritual at Rangaswami Betta. An Iruḷa of the Kalkaṭṭi lineage went to the altar. The Brahmin priest was offended, because he felt that an unclean Iruḷa had polluted the area most ritually pure. The Iruḷa then, with his hand, took some of the hot food being prepared as an offering to Ranga. After eating this, the Iruḷa became entranced and worshipped Ranga (inferred, as the male principle). After observing this, the Brahmin priest did not take offence any more. He declared that Iruḷas should serve as priests and that he had no rightful place there. From that time the same family in the Kalkaṭṭi lineage has provided the priests who officiate at Rangaswami Betta.

Despite the intricacies of a dual belief system, and evidence against the Iruḷas being orthodox Hindus, these tribals enjoy major Hindu festivals along with the general populace. Events related to Māṭu Poṅgal in 1962 provide an example. An Iruḷa priest, the father of the man who served on top of Rangaswami Betta, came to officiate during the Māṭu Poṅgal festival on 14 and 15 January, 1962 (Pl. 22F). Festivities varied in different hamlets and among individual families. Those who decorated their houses placed margosa flowers from the plains on their roofs. Strings with mango leaves and marigold flowers were festooned in doorways. On the porches, designs were drawn with hearth ash and chalk. At Mettukal, starting on the fourteenth, a long drama was performed late into the night. On the fifteenth, the horns of cows were painted with green and red. Then with banana, coconut, and incense offerings, unpolished rice boiled in milk was offered on banana leaves to cows and calves. Later, to the accompaniment of drums and a clarinet (*köylu*), Iruḷa men and women (Pl. 22G) danced in turn. Kuṟumba dancers and musicians wandered from one Iruḷa hamlet to another. One musician played a harmonium, and the other sang with cymbals, and was joined by the three dancers and all sang responsively. The dancers had whitened faces: one, depicting Sita, was dressed in a *sari*; another, depicting Rama, wore headgear and a mask with cowrie shells representing teeth over his mouth (Pl. 22H); the third dancer was in western dress. The performers were paid with food and coins.

Conclusions

The earlier literary references prove that Iruḷas resided in hamlets and practised shifting cultivation when some British visitors first contacted them. It may be argued that Iruḷa agriculture then had a Neolithic-like base, represented by the cultivation of millets in dry fields and by gardens with perennials. The

Irulas, whether for their own consumption or for barter and sale with neighbouring people, were seasonally dependent upon products gathered in the wild. Lowlanders may actually have named the Iruḷas after the roots of a particular wild yam species that constituted a primary staple in the Iruḷa diet. The Iruḷas also ate the flesh of animals trapped in deadfalls or hunted down. In their hunting, nets and spears were most probably used, not bows and arrows.

The original Iruḷa settlement pattern and most agricultural practices remain intact. After intervention by the British, general usufruct land use and related shifting agriculture were ended; hunting was greatly restricted. A few Iruḷas who owned sufficient land still tended to shift their dry field plots. Plants introduced to India by Europeans have been freely incorporated into Iruḷa gardens. With the spread of capitalist-inspired plantations in their area, the Iruḷas increasingly became labourers on nearby estates—until most of them were eventually working as labourers. Altered by the amenities in their dual existence, the Iruḷas never again needed to depend seasonally upon gathered products and hunting. However, the gathering of both medicinal and non-medicinal products continues still to provide income. The former roles played in this by the Forest Department and contractors are now being assumed by Iruḷa-run co-operatives.

In focusing upon four aspects of Iruḷa existence, these factors become clear: Iruḷas, like many other caste and tribal people in India, construct their houses and memorial temples in architecturally and functionally identical manners. Among Iruḷas, memorial temples are closely allied to lineages. On the altars within these temples honouring ancestral spirits are placed smooth, water-worn stones dedicated to the departed. In dry field agriculture only one crop of finger, Italian and little millet is generally obtained annually. Hoes aid in field preparation, and small sickles or knives are used to harvest the individual heads of grain. Gardens are conceived of as being parts of vegetative continuums, and are typed as (1) a glorious hodgepodge or (2) having multi-level conformity or (3) being single-layered or as (4) kitchen gardens. When the Iruḷas serve as priests at Rangaswami Betta and elsewhere, they provide ritual that is satisfactory to Hindus. Yet simultaneously they are in a dualistic manner satisfying their own religious requirements. Thus they continue basically to be non-Hindus.

NOTES AND REFERENCES

1. Full botanical identifications for the plants referred to in this chapter may be found in the Appendix to Chapter 2 (pp. 63-78) and in Table 11.1 above (pp. 296-8).

The Muḷḷu Kuṟumbas

RAJALAKSHMI MISRA

12

The Mullu and Naya Kurumbas are believed to possess the power of killing men by sorcery, and so greatly are they feared that, if a Badaga meet a Kurumba in a jungle alone, death from sheer terror is not unfrequently the consequence.

—J.F. Metz (1864: 116)

Mullukurumbas[1] are members of a distinct tribe, living on the western slopes and Wynaad plateau of the Nilgiris. They live in uni-ethnic settlements, locally known as *vīṭu*, and their community is strictly endogamous. Till a few decades ago they had a very effective tribal council. Though many other tribal groups also lived in the same region, the Mullukurumbas had no commensal relations with others and enjoyed a special social status in the region, superior to all except the Chettis and Kurichiyas (Misra 1972).

Mullukurumbas speak the Malayalam language. They are mentioned in the works of several of the pioneers in Indian anthropology like Thurston and Rangachari (1909), Ananthakrishna Iyer (1930), and Aiyappan (1948); but they have somehow been missed in the recent lists of Scheduled Tribes. In the 1971 Census they were grouped together with the Kuruman, whose population in Kerala and Tamil Nadu has been recorded respectively as 15,116 and 11,269. According to Luiz (1962), the Mullukurumbas numbered about 7,000 a couple of decades ago. Though my study of the Mullukurumbas was only in one settlement of theirs, which had a population of 159 in 1965, from discussions with them and visits to the other regions where Mullukurumbas live, I feel Luiz's estimate can be accepted.

History

In referring to the pre-British period, Aiyappan (1948: 91) describes the Wynaad region where the Mullukurumbas live in the following words:

> The eastern half of Wynaad contiguous with the Nilgiris and Coorg is inhabited by Kanarese-speaking tribes and the western half, by Malayalam-

I am indebted to the Director of the Anthropological Survey of India for his kindness in sanctioning the preparation of this chapter, and for permission to publish it here.

speaking tribes. When the Nilgiri-Wynaad Plateau began to be opened up by European planters, they found the country in the possession of a few landlords, chiefly Malayalees, for the whole of the Wynaad area, until its cession to the East India Company by Tippu Sultan was part of the dominions of the Raja of Kottayam. The actual cultivators were the tribes, such as the Kurichiyas, Mulla Kurumbar and Chettis.

Aiyappan adds that '...They [Mullukurumbas] are experts in hunting, and are good shots with their bows and arrows. During the Kottayam Raja's rebellion against the English, the Mulla Kurumbar fought very valiantly with the Raja' (1948: 95).

Thurston and Rangachari quote from the Madras Census Report, 1891: The final overthrow of the Kurumba sovereignty was effected by the Chōla king Adondai about the seventh or eighth century AD, and the Kurumbas were scattered far and wide. Many fled to the hills, and in the Nīlgiris and the Wynād, in Coorg and Mysore, representatives of this ancient race are now found as wild and uncivilized tribes (1909, IV: 156).

Luiz (1962: 198) puts aside the possibility of the Mullukurumbas being the descendants of these ancient Kuṟumbas or of the Pallavas, but agrees with the claim of the Mullukurumbas that they are the Vedas (*Vētans*) of South India, who after coming into Malabar were called Kurumans.

The Mullukurumbas believe they are one of the native groups of the region, and consider themselves the descendants of the band of hunters created by the God Siva and Goddess Parvati. This claim brings them closer to the Hindu fold. Of late they have been worshipping Hindu Gods more and celebrating Hindu festivals in the style by which local caste Hindus do, as we shall see below.

Inter-Group Relations

Till four decades ago Wynaad was mostly inhabitated by tribes such as the Chettis, Kuṛichiyas, Mullukurumbas, Uralis or Vettu Kuṟumbas, Kunduvatians, Karimpalans, Kadars, Patians, Uridavans, Tachanad Muppans, Kanaladis, Adiyans, Paniyans, Pulayans, Jenu Kurubas or Ten Kurubas and Kattu Naikens (Aiyappan 1948: 92). But since independence, members of various other castes and communities such as the Nairs, Āsaris, Tiyas, Izhavas, Christians, Mappilas (Muslims), Harijans and others have come to live in this region, and this has affected the life of local tribes.

Of these groups, it is generally the Mappilas who have opened shops and who sell provisions, pottery, agricultural tools and other necessary items for daily use; they allow the Mullukurumbas and other tribes also to buy things on credit. The Christian immigrant population, coming with their agricultural background, have their eyes on land possessed by the tribals. They have encouraged the tribals to seek credit from them, particularly by mortgaging land to them. Slowly the Christian immigrants have thus usurped tribal land. The ties the Mullukurumbas have with Mappilas and Christians, the two major immigrant communities, are thus mainly economic but exploitative in nature.

In certain pockets of this region we find that Harijan ex-servicemen have been settled by the state government. These Harijans have received grants of lands from the government in recognition of their service to the country. In comparison with the richer Mappilas and Christians, these Harijans are financially poor, have a low ritual and social status, and politically are not very powerful. The Mullukurumbas have a hostile attitude towards these Harijans because they feel that the Harijans are the usurpers of the lands on which they had been practising shifting cultivation till three decades ago. The advancement of some money in the form of a loan by Christians and then their taking away land have been viewed as equivalent to purchase of their lands, while the Harijans received their land as a free grant from the government. The Mullukurumbas obviously never liked this and it has caused hostility against the Harijans. Members of other communities are quite few and there is no specific pattern of relationship between these people and the Mullukurumbas or other tribes.

As for the relationship between the tribes of the region, it is interesting to observe that there is a tribal hierarchy among them like the caste hierarchy among castes anywhere else. Each tribe knows its position in the hierarchy and, in acceptance of it, plays its role according to the requirements of the situation. According to this tribal hierarchy, only two tribes in this region, namely the Kurichiyas and the Chettis, are recognized as superior by the Mullukurumbas. As mentioned earlier, there are no commensal relations among them. A tribe of a higher status will not accept cooked food or water from a tribe belonging to a lower order in the hierarchy.

Intra-Tribal Relations

The Mullukurumbas are distributed in an area of about 20 miles radius. The land on which their population is spread is divided into four regions, namely thorny scrub land (*kāra nāṭu*), areca-nut land (*pākka nāṭu*), stony land (*kallu nāṭu*) and fine land (*nēriya nāṭu*). *Kulam* in the Mullukurumba dialect refers to clan. There are four exogamous *kulams* among them, namely Vadakku, Villippa, Katippa and Veṅgada. Though they are unable to trace the origin of these clans, they continue to identify themselves as members of one or the other clan. Clan descent is determined through the mother and it has only ritual importance at present.

Lineage membership is traced through one's father for the purpose of inheritance of property and succession to office. Each Mullukurumba settlement is believed to consist of particular households that all belong to one lineage. Hence all the members of a settlement share birth and death pollution and extend mutual services on occasions like marriage and death. Thus the Mullukurumba are organized on one side into four matrilineal clans and on the other into several patrilineal lineages.

Among the Mullukurumbas, nuclear households (parents with children), patrilocal extended households consisting of three generations (grandparents, parents and children), and fraternal joint households (brothers, married or unmarried, with their spouses and children) are the common types found. Though these appear as different types when seen at synchronic level, seen diachronically, these are only various stages in the developmental cycle of a household. As with the Badagas (see p. 211), a nuclear household consisting of parents and children grows into a patrilocal extended household consisting of three generations (grandparents, parents and children). When the grandparents die, their sons who are brothers, married or unmarried, live together as a fraternal joint household until such time as there is some friction among them or their spouses. Then the fraternal joint household breaks into nuclear households once more. Thus as a result of recurrent processes of augmentation and depletion by natural events like birth and death and social events like adoption, marriage and partition, the genealogical composition of a Mullukurumba household varies.

A Mullukurumba Household

Irrespective of the type of household, the authority to take the final decision on any issue is in the hands of its eldest male member. His wife the mother in the household, is feared and respected next only to the father but, after the death of the father, she is not given so much importance. After the parents, the eldest son of the family is given much responsibility and importance and accepted as the head of the fraternal joint household. A daughter-in-law is in charge of all household work—washing vessels and clothes, sweeping, bringing water, husking the grains, cooking, etc. During the peak season, she works on the fields also, if the family possesses land and cultivates it. If there are more than one daughter-in-law in the household, the work is shared by all of them; hence the Mullukurumba girls prefer to marry into a household where there are several grown-up sons.

Married daughters are mere guests in their natal homes, and that too for only a short time, as they must continue to play the role of daughter-in-law in the homes of their husbands, where their absence is not long tolerated. Because of their prolonged stay with the paternal relations, the children are more fond of their paternal relations, particularly paternal grandmothers. The relationship between the daughter-in-law and unmarried daughters in the household is not very stable. The former commands the latter to do household work during the absence of mother-in-law, while in her presence the daughter-in-law accepts the authority of the mother-in-law and submits to it.

The relationship between brothers and sisters is not always very intimate, perhaps because of the age groupings among them. Girls of each of the following age groups spend their leisure together and also sleep together in a separate room in any one of the bigger houses in the settlement:

* 7 to 11 years
* 12 to 16 years
* 17 and above (till married)

There are similar age groups among the boys too. The expectation among the Mullukurumbas that the junior siblings should respect their senior siblings has perhaps created a gap between them. On the other hand there is an intimate relationship between girls of the same age group, and so also among the boys.

Avoidance and Joking Relationships

The relationship between the daughter-in-law and her real or classificatory father-in-law or her husband's real or classificatory elder brothers is one of avoidance. She does not look at or talk to them. On the other hand, she is friendly to her husband's younger brothers, and they crack jokes with each other. She is friendly to her elder sister's husband but avoids the husband of her younger sister. People who stand in an avoidance relationship to each other are said to belong to *tīṇṭa kulam* (untouchable clan) *vis-à-vis* each other.

A daughter-in-law shows her respect for her mother-in-law through her every action, but it is not necessary that she should avoid staying in her presence. She does not take the initiative in any work without first seeking the approval of her mother-in-law.

Stresses and Strains between Mother-in-law and Daughters-in-law

The relationship between a mother-in-law and a daughter-in-law is fraught with strains; at times the mother-in-law may abuse her daughter-in-law for any mistake and may carry complaints to her son, and may even persuade him to divorce his wife. An extreme case of such a relationship ended with the daughter-in-law drowning herself in a nearby well, in 1972. At the same time, there are a few instances where the daughters-in-law subdue their mothers-in-laws through their influence over their husbands. But according to informants, subduing the mother-in-law can happen only after the death of the father-in-law.

Relationship between Mullukurumba Households

As mentioned earlier, the male members in a Mullukurumba settlement are agnatic kin because they claim to be descendants of a common ancestor. On the other hand, the married women in the settlement come from different settlements. Till recently, a Mullukurumba looked to his agnatic kin for assistance in all spheres of his life—economic, social and ritual. Even when a Mullukurumba performed certain rituals like the magico-religious diagnosis and treatment of diseases, death rites, etc., he depended on his agnatic kin. On the other hand, the role of the matrikin living in other settlements was limited to the celebration of life-cycle rituals. If, by chance, a matrikin was absent during a ritual occasion, a member of the agnatic kin, belonging to the clan of the person for whom the ritual was being arranged, was made to take up the role of the absent person. Because clan descent is determined through the mother, a clan member is accepted as a substitute for matrikin.

The relationship between two unrelated Mullukurumba households is unstructured and unpredictable, as in many other societies. Any two unrelated households belong to different settlements because all the males of the same settlement are necessarily agnatic kin, belonging to the same maximal lineage. Opportunities for two unrelated Mullukurumbas to become friends are limited, but if such friends visit each other, they enjoy a good time. Drinks and food are served, and a male friend is perhaps taken hunting; if the friend is a female she is taken fishing by the women of the settlement. They attend marriages and funerals in each other's homes.

Marriage

Of the various life-cycle rituals, marriage and funerals are the most important events. A Mullukurumba is permitted to marry any number of times but, at any one point of time, he or she must be monogamous; not a single case of polygyny or polyandry has been reported among them.

Modes of Marriage

Till perhaps forty years ago, marriage by elopement or force was very common among these people. In the former case, the boy and the girl mutually accepted each other and eloped, but in marriage by force, the boy usually kidnapped the girl when she was some distance from the settlement, often when bathing or washing clothes in a stream or river. The former type was locally called *ōṭi kūṭal* ('run and join') and the latter, *atru kaṭavu* ('leaving from the riverside'). There were arranged marriages in the past also, though few, where a boy's father arranged the marriage of his son. This type was referred to as *vīṭu kaṭavu* ('leaving the house').

The *atru kaṭavu* form of marriage seems to have been more popular among Mullukurumbas, for they have introduced a mock fight to symbolize *atru kaṭavu* in their present-day arranged marriages. In this mock fight, the groom's party pulls one hand of the bride, uttering the words, 'Come to our *vīṭu*, we have a big river there for fishing and bathing. Leave this place'. The bride's party resists by holding her other hand, and pulling the bride to their side. This mock fight lasts for a few minutes.

Yet another form of marriage prevalent among the Mullukurumbas is locally known as *mukka vari* (meaning 'three-fourths of the way'); in it the bride, who is always a widow or a divorcée, is met by the groom's party not in her house but somewhere between her house and the groom's. Whichever mode of marriage is adopted, residence after the marriage is patrilocal.

In the first two forms of marriage mentioned, the elder's consent for the marriage is not taken beforehand, but in the other two it is the elders who initiate the negotiations. The Mullukurumbas have a *pittu viruntu* (*dōśa*-bundle carrying) ceremony, to regularize all types of marriages. In the marriages arranged by the elders, this ceremony is performed on the third day after the wedding. For this, the bride's kin hand-pound some rice into flour, mix it with water and buttermilk, and then make 51 or 101 rice cakes (*dōśas*). They tie these in two bundles and carry them to the bridegroom's house along with the bride. In the *ōṭi kūṭal* or *atru kaṭavu* marriages this ceremony is postponed till the bride's people are reconciled. There are instances where people have waited for over a decade to arrange this ceremony, because according to their tradition one cannot marry off one's children unless one has performed this ceremony with one's latest spouse. This ceremony is significant in that it regularizes all marriages at some point in time, whichever the type of marriage chosen.

Marriage Regulations

In all four types of marriages, care is taken not to violate the principles of tribal endogamy, clan exogamy, kin exogamy and settlement (*vīṭu*) exogamy.

If a Mullukurumba married a girl outside his tribe or within his clan, he was excommunicated from his tribe by the tribal council. During the fieldwork in 1965, eight such excommunicated couples were living together in a place called Ambalamūla, in the Wynaad. Members of these eight households took spouses from among themselves or from other tribes in the region who were willing to trespass the norms of their own group.

All the members of a *vīṭu* being agnatic kin, the males and females in it are normally related to each other as mother and son, father and daughter, aunt and niece or nephew, uncle and niece or nephew, brother and sister, grandparent and grandchild, and so on. Beside avoiding their own clan and kin in the selection of a bride, the Mullukurumbas avoid all those settlements into which girls of their native settlement have been married. Even if they negotiate a marriage in such a settlement, they do it only after the death of the woman from their settlement. The reason given does not logically explain this taboo: according to Mullukurumba tradition the groom's party has to pay a traditional brideprice to the bride's party of Rs 5.50 (known as *kannam*). If they take a bride from a settlement to which they had themselves given a bride earlier, then they will be paying *kannam* to the bride's kin of that settlement. They say that this exchange of *kannam* between two such settlements is to be avoided.

Though they avoid cross-cousin marriages now, their kinship terminology presents a different picture and offers clues that marriages among kin might have been prevalent in the past. A Mullukurumba refers to his mother's brother and mother's brother's wife as *māman* or *ammāvan* and *māmi* or *ammāyi* respectively. Father's sister is called *ammāyi* by the children and father's sister's husband is referred to as *cettan* (elder brother) by the mother of the children. As there is an avoidance relationship between mother-in-law and son-in-law among the Mullukurumbas, the daughter-in-law of the house has to attend on the son-in-law of the house, when he comes there alone; and to help a sense of familiarity prevail, this term *cettan* is perhaps used by her while referring to him. He is also expected to refer to his wife's brother's wife as *īttati* (elder sister). Both of them using terms for elder sibling shows that the relationship is based on respect. This fictitious kinship between the two justifies the children of a woman calling their father's sister's husband *ammāvan* and his wife *ammāyi*.

The term *ammāyi* is used by a Mullukurumba to refer to his or her mother-in-law also. Here it is interesting to bring in the various terms of kinship used by the parents of a Mullukurumba to denote the mother-in-law of their child. A woman refers to her son's or daughter's mother-in-law as *natta* (husband's sister) and a man refers to her as *peṅgala* (sister). Both the terms denote the same person, namely sister of the husband. Earlier we have seen that one's father's sister is *ammāyi* to one. Using the same term of reference *ammāyi* for one's mother-in-law, one's father's sister and one's mother's brother's wife may be taken as an indication that at least in the remote past one's father's sister and one's mother's brother's wife were potential mothers-in-law; hence we may take it that cross-cousin marriage was perhaps approved of in the past. Though the actual practice of cross-cousin marriage does not occur now, these kinship terms continue to be in vogue. It is also possible that they never practised cross-cousin marriage but adopted the kinship terminology of the region.

Divorce and Remarriage

Divorce and remarriage are both very common among the Mullukurumbas. Divorce is granted by the *porunnavan* (settlement's headman) and other elders there, once they are convinced that the grounds on which the appeal is made are genuine. If a man does not earn enough or does not feel a responsibility to maintain his wife and children, the wife can divorce him. If a woman does not attend to her household work, to the satisfaction of her husband and in-laws, then she can be divorced. Besides these, indulgence in extra-marital relations which comes to the notice of one's spouse or in-laws becomes a cause for divorce. A woman who has not given birth to children after some years is also divorced. According to reliable reports from some informants, a Mullukurumba girl who does not conceive from her husband within a couple of years of marriage becomes friendly with the classificatory or real younger brothers of her husband and conceives with their help to escape a divorce; unless perhaps she desires a divorce herself. In most male-dominated societies sterility is attributed to females and never to males.

Premarital sex relations are not generally permitted among the Mullukurumbas yet are condoned. If by chance an unmarried Mullukurumba girl becomes pregnant, and if her people are convinced that her paramour is also a Mullukurumba, then it is not very difficult for them to arrange her marriage with some other Mullukurumba. Their tradition has it that a pregnant woman delivers her baby in her husband's home, amidst the kinswomen there. Hence, in the case of an unmarried, pregnant girl also, once married she stays in her husband's home and delivers the baby. At the time of the delivery, the women present compel her to utter loudly her paramour's name thrice. According to them, this is a sufficient humiliation for her, for having indulged in premarital sex relations.

Just as divorces are common among the Mullukurumbas, remarriages are also common. Both men and women are commonly married several times. All the unmarried children of a woman through her earlier husband/husbands move with her to her new husband's house. He is expected to treat them as his own and give a share of his property to them, in case their real father/fathers had left them no property. Because the daughters leave their natal home after their marriages, an old couple depend on their sons and daughters-in-law to look after their property and to do the household work. So if one does not have a son, he usually adopts one from among the sons of his kin. As there is an avoidance relationship between father-in-law and daughter-in-law, even an old and sickly man after becoming a widower wishes to marry and have a wife again. He expects her to prepare hot water and coffee for him at any time he needs them, and to keep some fire near where he sits or sleeps during the cold, rainy season. I came across a case of a Mullukurumba woman of over fifty years, marrying for the fourth time a sick widower of over sixty years, and then becoming a widow within a week. Her two grown-up daughters were already married and settled elsewhere in their husband's villages. The sons of her earlier husbands and the daughters-in-law did not show any affection towards her or interest in her living with them, but at the same time did not drive her out of the home by action or words. So she married this man, obviously not realizing that he would die so soon. In the course of another few weeks, the elders of this village brought one or two offers of marriage for her from some other settlements. This was perhaps the best way for them to send her away, if they did not wish to retain her in their own settlement. There are cases of widows who had been married into a settlement and who had come there with their minor children from an earlier husband/husbands. When such women later lost their husbands again, after having lived there for some years, they were not always sent away from the settlement. On the contrary, if a woman expressed her preference to stay in that settlement, her children through earlier husbands were given a share in the lands left by the deceased man and she lived with them.

Social Control

All these elaborate codes of conduct among the Mullukurumbas were conformed to until quite recently.

Religious beliefs and the authority of a council of elders maintained their well-knit social organization. The authority of elders appears to have been compelling and recognized at almost every level—at the household, minimal and maximal lineage levels, settlement level, regional level and also at the level of the community at large.

It was the traditional practice of the Mullukurumbas to conduct all the rites and ceremonies of their life-cycle collectively in their sacred hall (*tēva pērai*), which is found in every settlement; this helped to promote social solidarity among them. They also believed in survival after death in the form of ghosts, and hence always preferred to keep all the ancestral spirits appeased by collective offerings and worship.

The tribal elders' council of Mullukurumbas, which was locally called *mūmpanmar kūttam* ('meeting of the elder men'), settled all cases of violation of the social norms and so contributed to the maintenance of social solidarity. The Nair Janmi (landlord) was the head of their council, but seems to have been only a nominal figure in the council, because there was not a single occasion when he was present for the council meeting and conducted the cases. Then there was the Mullukurumba chief, the *mūmpan* or *talaical* (elder man), of Appadu settlement. Next to him were two elders from Kottūr settlement and Edūr settlement. Next in the descending hierarchy were the thirteen elders representing the various local regions in which the Mullukurumba population lived. Each of these thirteen elders had not less than five Mullukurumba settlements under his control. Apart from these regional heads who were members of the tribal council, each Mullukurumba settlement which did not have one of the above office-bearers in it had an elder man called *porunnavan* whose words were to be respected by the Mullukurumbas of that particular settlement. All these various offices were hereditary in the male line, and women were not eligible for any of them. The oldest person in the seniormost generation in a settlement succeeded to an office because, it is claimed, all the males in a settlement were agnatic kin.

The duties of a *porunnavan* were to collect and keep cash offerings made to their traditional deity by the Mullukurumbas of his settlement; to be present for all the ceremonies celebrated by his settlement members, in the sacred hall of the settlement; to note whether all in the settlement were following the community's norms; and lastly, to settle divorces and petty conflicts among the members of the settlement.

The duties of the thirteen regional elders were the same as these, but in addition they had to plan and lead the hunting expeditions and also, while making offerings to the deity, give sanction for the marriages of Mullukurumbas living in their region. In addition, they had to report cases of violation of the traditional norms in their respective regions to their chief at Appadu. The three *talaicals* or heads of Appadu, Kottūr and Edūr had the same duties as the thirteen regional elders for their respective settlements and regions, but were also the key persons to decide cases reported to their tribal council. Appadu *talaical*, the chief of them, had the extra privilege of levying fines from guilty persons.

Till four decades ago the council used to meet once a year in the month of Kumbam (February-March) to discuss cases referred to it by the council members. The meetings were held in a thick jungle, a little distance away from Appadu settlement. Women were not allowed to attend: even in those cases where women were the accused ones, their fathers or brothers or husbands were asked to represent them and the women were kept away. When the council met, a diviner (*veliccapati*) was also brought for the meeting. In the course of the case if the person charged with an offence accepted his guilt, he was fined less and the case was thus settled quickly. If he did not accept guilt, they sought the help of the diviner and decided the case according to his words spoken while in trance.

Changes in the Recent Past

For the past twenty-five years this tribal council has been without power, but it continues to meet once a year just to fix a date for the annual hunting festival, *uccala*. Reporting of cases of violation of traditional norms to the tribal council is no longer effective and hence these incidents are just overlooked. Changes in the economic life of the people and also in their attitudes towards traditional religious beliefs have contributed to the loss of power of this tribal council.

The Mullukurumbas used to make offerings of cash in the name of their deity, in moments of crisis. This was done inside the sacred hall which must be there in every Mullukurumba settlement; the *porunnavan* used to keep such offerings in a bundle in the sacred hall itself. Once a year representatives of their tribal council would come and collect the offerings. In this sacred hall there is no idol or picture or object of worship. Though the people went to the nearby temples during certain festive occasions to worship deities installed there, the sacred hall of their own settlement was till recently a place more sacred to them than any other place, including the temple (*ambalam*). This sacred hall was maintained collectively by the Mullukurumbas living in each settlement. Of late, the sanctity of this hall and the significance of all members of the settlement collectively participating in ceremonies conducted by a household in that sacred hall have been reduced considerably.

The traditional practice of unmarried children of either sex above the age of seven or eight staying together and sleeping with their respective age group is also losing ground. When the children who attend residential schools return home for holidays they prefer to stay with their parents. In some houses, children who have received education up to middle or high school level become alienated from their earlier agemates who have not had so much education. In all important matters like marriage, divorce, negotiation of land purchase or sale, shifting of a house from one site to another within or beyond the settlement, etc., Mullukurumbas do not look any more to lineage elders for their counsel or consent. The importance of elders of the maximal lineage has now been lost among the Mullukurumbas. The trend to think and take decisions as the head of the household, regardless of the opinion of one's elder kin living outside one's household, has come to be accepted, and this trend has a definite economic base.

Daily life at present is in contrast to what an old Mullukurumba would have experienced some four or five decades ago. In his days as a youth, life had plenty of adventure, fun, relaxation and mutual concern among the members of the tribe. The Mullukurumba men had hunting and agriculture as their main economic activities till about half a century ago. As the region they lived in was a dense forest, abundant in animals, hunting was quite a gainful occupation. But the presence of wild animals like tiger, wild boar, etc. made hunting a very hazardous activity, particularly if men were alone or in small numbers. Thus apart from the kin affinity that bound them in the past, risky situations in the forest coupled with lack of any powerful modern weapon like the gun made their economic activity a highly co-operative effort; it was strengthened by their deep faith in ancestral spirits and other deities.

Secondly, in the past the whole region was considered as the land of the Mullukurumbas and other native tribes there, but was notionally headed by Nair zamindars, and the issue of having individual titles to land never came up among the Mullukurumbas. Till the appearance of immigrants in the region, there was no dearth of land there for the Mullukurumbas to carry on shifting cultivation and other subsistence activities. But now the situation is different and there can be a tough fight for ownership of even a small plot of land in the region.

Thirdly, in the past the Mullukurumbas procured items like clothes, ornaments, salt, etc., which were not available freely in the forests, by barter from traders who visited them occasionally from the plains. The Ūrali Kuṟumba, an artisan tribe of the region, supplied pottery, tools, baskets, etc., to other tribes in exchange for grain and meat. The Mullukurumbas were thus not exposed to the market or cash economy till three or four decades ago.

Since independence, the region has been cleared, the forests have shrunk and, in the limited area of Reserved Forests, hunting has been banned even for natives. Thus hunting as a regular economic activity had to be abandoned and an important bond amongst Mullukurumbas was severed.

Unlike hunting, which was a group enterprise, agriculture was the business of individual households even in the past, though during certain agricultural operations like transplanting, weeding, and harvesting, the kin readily helped each other. The crops grown on all types of land were mostly meant for eating or for barter purposes; but of late cash crops like coffee, pepper, turmeric and ginger have been grown for sale in the market. This shift to a market-oriented economy on the one hand and the abolition of the Zamindari system with the simultaneous emergence of private ownership of land in the area on the other, have together contributed to the Mullukurumbas's becoming more and more materialistic and individualistic. As many of them have now lost their lands to immigrants, more than 90 per cent of the Mullukurumbas have started doing agricultural labour and earning daily wages for their mere subsistence. Cash transactions have gradually made barter exchange and reciprocal kin obligations meaningless to them.

Kin obligations having thus been eclipsed, the unit that shares the 'we' sentiment among the Mullukurumbas has shrunk from the tribe to the nuclear family. Sharing the gains does still occur among them but only in their occasional hunting expeditions, which they still do jointly. Otherwise, in all other endeavours, each is concerned about his or her immediate nuclear family's gain or loss.

Contact with the Hindu immigrant population has moved people closer to the Hindu religion and Hindu Gods: Mullukurumbas have started worshipping them and celebrating Hindu festivals in the popular ways that the caste populations do; but even now they consciously or subconsciously attribute any type of misfortune, sickness, loss of crops or death to the wrath of their traditional forest deities and spirits.

However, with the exposure to modern education, modern medicine and modern technology, at least some among the younger generation are becoming able to understand the various phenomena behind their difficulties. They are trying to adopt new methods of resolving them, instead of appeasing the traditional deities or summoning a diviner to hear his prophecy.

The new methods they have begun to adopt include getting their children educated and sending them away from home for jobs, taking advantage of the medical units run by either government or private agencies, trying to adopt innovations and modern technology that agricultural science can now offer, and seeking justice at the police stations or magistrates' courts. It is not true that all the Mullukurumbas have turned to modernity or even that those who have done so are fully benefitted by these things—but the fact is that they just cannot avoid the powerful forces now pervading life in South India.

NOTES

1. Although the conventional name of this tribe has always been printed as two words, Muḷḷu Kuṟumba (cf. Kapp and Hockings, above), today it is the practice of the people themselves to pronounce the term as a single word, Mullukurumba, thus playing down somewhat the demeaning etymology that *muḷḷu* (thorn) suggests to them. In my several publications on this tribe I have thus acknowledged their own preference by using the single word.
2. The Mullukurumbas, who are spread widely through the Wynaad, and who recognize themselves as belonging to specific regions there, worship the deities of their respective regions. The male deities of these regions are Atiralanmar, Kariatan, Ārivilli, Karivilli, Venur Talaical, and Vettai Karumahan. There are only two goddesses, Tampiratti and Malaiya Karungali.

Gudalur: A Community at the Crossroads

T. ADAMS

13

Koodalur is a village of Baddagurs, containing between 20 and 30 houses. There are a few Kottaras' houses in the vicinity. Here I was met by the Nambolacota Wáranoor, attended by his dependants...

—T.H. Baber (1830:314)

Gudalur: At the Junction

It is commonly accepted that the community of Guḍalur, located at the very northern- and western-most tip of Tamil Nadu, derives its name from the Tamil words *kūḍal* 'joining, confluence', and *ūr* 'town'. In terms of geography, topography, demography, language, culture and history, Gudalur finds itself at the confluence of multiple and diverse streams of influence.

Geographically, the town is located almost on the converging borders of the three states of Tamil Nadu, Karnataka and Kerala. It stands at the junction of the main roads leading to the Nilgiri plateau from Kozhikode and Sultan's Battery in Kerala to the west, and from Mysore to the north (see Fig. 3.1, p. 80). Long before these roads were constructed, some of the major transportation and communication trails from these areas merged at Gudalur before ascending the higher plateau through Gudalur Pass.

Topographically, Gudalur lies at a point where the tableland of the southeast Wynaad and the slopes of the Nilgiri plateau converge. At 3500 ft. elevation, its climate is somewhat temperate relative to that of the rest of the Wynaad because of its proximity to the Nilgiri plateau. Unlike that plateau, which receives a fairly moderate rainfall through much of the year, spread between both the northeast and southwest Monsoons, Gudalur's rainfall is provided almost exclusively by the southwest Monsoon, with its onset in June, peaking in July and tapering off after August. This pattern is similar to the rest of the Wynaad, except for the fact that the total accumulation is less in Gudalur because of the rain shadow cast by the Kunda Range to the west (cf. von Lengerke and Blasco's paper, above).

Ethnically, the earliest inhabitants on record in the Gudalur area are tribal Paniyas, one small Kota village, and some Kuṟumbas. The former group is distributed across the Wynaad towards the west, extending across the Kerala border, with some members found further to the east as well. The last-mentioned, consisting of several distinct tribes, is related to the Kuṟuba tribals in Mysore and Coorg and to various Kuṟumba groups further to the south (see Chapters 9 and 12). An additional ethnic element is the Wynaad Chettiars, who claim to have migrated to the region from the Coimbatore area of Tamil Nadu several hundred years ago as agricultural cultivators. The 'indigenous' population is thus made up of a number of groups, with extensions in various directions, which coincide in the Gudalur area.

These factors are important for understanding the distinctive and significant aspects of Gudalur's place in the Nilgiris District which encompasses it. The town itself has evolved certain unique features through its status as a community at the intersection of a variety of other economic, political and cultural forces as well. Thus, an examination of the history of the area is essential for an insight into the present character of the town.

Marginal Community

Historical records relating to the area prior to the British imperial period are scanty. However, one fact is quite clear: like the rest of the Nilgiris District, in pre-colonial days Gudalur was only peripheral to the interests of the surrounding political milieu, although it had been a gold-mining region in ancient times (Hockings 1980a: 14). The area was under the nominal control of several princely regimes in Mysore, and alternately or simultaneously claimed by certain ruling families in Malabar. Even after the British East India Company was granted rights to the plateau region in 1799, the legitimate ownership of specific portions of the area, including that around Gudalur, remained in question. Litigation to resolve the issues was only partially successful, and it was not until the capture and execution of the last native Malabar claimant in 1805 that undisputed control was vested in the hands of the Company (Francis 1908: 102-5).

The strategic nature of Gudalur as a point of access to the Nilgiri plateau was not of great significance prior to the development of the plateau as an economically productive and valuable territory (see above, p. 259). As long as the primary traffic channelled through the community was that of local traders carrying on commerce, of relatively minor proportions, between the hill tribes and the Malabar coast and Mysore areas, there was little incentive to attract outsiders into the area. Land was fertile and rainfall plentiful; but only a few Wynaad Chettiars found the returns from the land sufficient compensation for enduring the high risk of malaria endemic in the area, and the region remained the almost exclusive province of the local Paṇiyas and Kuṟumbas.

Subsequent to the penetration of the Nilgiri plateau from the east by the British in the earliest years of the nineteenth century, and the enthusiastic reception given to a climate which offered relief from the sweltering plains, accessibility to this newly discovered haven became an important concern. Many routes were explored, and as early as 1823 funds were made available for improving the path via Gudalur and Gudalur Pass. This track provided an access route for British troops and others stationed in Bangalore or Malabar. However, the threat of malaria in the jungles that surrounded Gudalur kept a damper on the level of traffic funneling through the town.

In 1845, Gudalur proved to be an important point of access to the then-virgin forest area which was about to be transformed into the valuable coffee (and later tea) plantation area of Ouchterlony Valley (O' Valley). This high valley, at the base of the northwestern escarpment of the plateau, lies to the southwest of Gudalur (see Fig. 3.1, p. 80). The population of Gudalur and the neighbouring territories at that time was insufficient to provide labour or foodstuffs for the rapidly developed plantations, and resources had to be sought from Mysore and Kerala. However, being located at the junction of the roads from the supplies source of Mysore and the coffee processing plants of Kozhikode, as well as from the European 'R & R' centre of Ootacamund, brought some additional importance to the community. From that time it began to develop as a centre for administration of the surrounding area, and for provision of services and the distribution of supplies and labour.

Throughout most of the nineteenth century, Gudalur was treated as a sort of step-child to several administrative jurisdictions. For a variety of reasons, it was shuffled from the control of one district to another and back—the shifts at times occurring within the span of a very few years (see Logan 1951: 390-6 for the details). However, throughout these transfers the town of Gudalur remained the only local administrative centre, and lower level officials were stationed there from mid-century onward. Even though the population of O' Valley was almost double that of Gudalur at the beginning of the twentieth century, Gudalur remained the *taluk* headquarters. Its population, though small, was stable when compared with that of the plantation labourers who made up the vast majority of the O' Valley populace, or the tribals who predominated in the southeast Wynaad jungles. Short-lived booms in coffee and cinchona in the Wynaad during the eighteen-sixties and eighteen-seventies and a brief case of gold fever in nearby Devala in 1879-82, re-emphasized the significance of Gudalur as a stable centre for the region.

Throughout this period of Gudalur's history—from the emergence of the Nilgiri plateau as an area of more than local significance, through the opening of the surrounding forest areas to coffee and tea plantations, and even extending well into the twentieth century—the community would appear to have been a prime candidate for expanding its influence in both economic and administrative spheres. However, this did not occur, and the population grew at a slightly slower rate than in other areas of the district until the middle of this century (see Table 13.1).

One of the most significant factors which militated against rapid expansion of Gudalur prior to the 1950s was the continuing extremely high incidence of malaria in the entire Wynaad area. Francis quotes a Badaga man's note to the District Collector in Ootacamund, written in 1894, which makes clear the point that fear of malaria was a powerful disincentive for outsiders to move in and take advantage of the area's potential:

> I have heard that Badagas are to be sent down to Wynaad. I do not know why. Although our throats be cut or we be shot we will not go (with prostrations). I beg and pray. The Badagas are in fear at this rumour. We cannot stand Wynaad. The fever is terrible. We shall die in one day if we go there. We want to die here on the hills (Francis 1908: 366).

The success of O' Valley had served to provide indisputable evidence of the potential of the region for coffee and low-elevation tea cultivation. But it was not until the late 1940s and the mid-1950s that a concerted government effort led to an almost totally successful campaign to wipe out the threat of the 'fever'. With the elimination of this high risk factor—which did not keep all investors away, but certainly inhibited the recruitment of labourers and ensured that most of them would migrate into the area only seasonally—the advantages that Gudalur had to offer far outstripped the costs of exploiting them.

The flood gates opened, and Gudalur saw a population explosion unmatched anywhere else in the district (see Table 13.1). This influx was made

TABLE 13.1

Population Growth in the Nilgiris, 1901-1981

Year	Nilgiris District	Increase	Ootacamund	Increase	Gudalur	Increase
1901	112,882	—	18,596	—	2,558	—
1931	169,330	50.0%	24,616	32.4%	3,268	27.8%
1951	311,729	84.1%	41,370	68.1%	5,207	59.9%
1961	409,308	31.3%	50,140	21.2%	8,328	59.9%
1971	494,015	20.7%	63,310	26.3%	15,558	86.8%
1981	630,169	27.6%	78,277	23.6%	21,983	41.3%

up of immigrants not only from immediately adjacent areas, but from great distances as well. Those with capital to invest found that labourers were now willing to settle permanently in the area, thus providing a steady and available labour force. Merchants, traders and service-oriented businessmen were attracted to the expanding community, and began to establish wholesale and distribution businesses as well as retail trade to provide more immediate access to goods and services than had been available when they were imported from more distant centres.

Caught in the Middle

The period of the 1950s was in many respects Gudalur's debut in regional economic and political spheres. The awakening of ethnic and linguistic regionalism, spurred by the formation of the Linguistic Provinces Commission, prompted multiple claims on this area that had previously been of little concern to anyone outside its immediate boundaries. The Commission's task was to make recommendations on lines demarcating the borders of states in the newly independent nation. Suddenly various diverse groups were voicing concern over the future status of the entire Nilgiris area, and Gudalur was the subject of claims by Kannadigas, Malayalis and Tamils (Linguistic Provinces Commission 1948: 8-9). Each group had a few legitimate arguments, and many spurious ones, to back up their demands.

Since the Gudalur area was the 'homeland' of only a small number of tribal groups, primarily the Paṇiyas and some Kuṟumbas, the majority of the populace at that time were fairly recent immigrants. The question of linguistic predominance, one of the most significant factors being considered by the Commission in proposing boundary demarcations, could be addressed from any of several perspectives. As anyone familiar with attempts to classify Dravidian languages will attest (e.g. Shapiro and Schiffman 1981), the demarcation of boundaries between these languages in a clearcut and conclusive fashion is quite impossible in anything less than an arbitrary fashion. So even if all other factors concerning immigrant populations were laid aside, assertions based purely on the linguistic affinities of the aboriginal (?) inhabitants would be less than definitive. Taking the two above-mentioned tribal groups as examples, we find the following: Diffloth (1968) places the Paṇiya dialect in the Tamil/Malayalam language group; Kuṟumba, on the other hand, is classified by Grierson and Konow (1967) as being a dialect with close ties to Kannada. Neither of these classifications is based on incontrovertible evidence, however; and the strongest claim that seems to be justified is that both of these dialects have developed in close proximity to, and have been significantly influenced by, languages whose speakers have at times been in a dominant position *vis-à-vis* the tribals.

But arguments for the establishment of political boundaries based on linguistic factors did not stop (and at times did not even start) with some formal analysis of language history of the original residents. For several hundred years prior to the arrival of the British in the Nilgiris, Badagas had been migrating from the south of Mysore to the Nilgiris through the Gudalur area. Though few chose to settle on the lower slopes of the plateau, many travelled fairly regularly through the community on their way to and from Mysore for commercial and social reasons. These Badagas spoke a dialect with close affinites to Kannada, and their presence was used as support for the claims of Kannadigas that the region should be made a part of Karnataka (then Mysore State):

> It is quite unnecessary to labour to prove that the languages spoken by the Badagas and the aboriginal tribes are forms of Kannada.... It may be that the Kannada spoken by these tribes has become corrupt by the policy pursued by the Government of severaly [*sic*] excluding the language from the School Curriculum and by lack of intercourse with the fountainhead of Kannada language and literature caused by administrative barriers. But any anthropologist will be able to tell us that the languages now current in the Nilgiris have their roots in Kannada (All Karnataka Unification Sangha 1948: 10).

In addition to the indigenous tribes and Badagas, a number of Wynaad Chettiars (sometimes identified as 'Malabar Chettis') had resided in the Gudalur area since before British times. These individuals often acted as the 'masters' of groups of tribals, supervising their clearing and planting of crops in portions of the Wynaad jungle. They provided food and a few simple amenities for the previously nomadic groups, and this led to the establishment of numerous semi-permanent settlements. Economically, they had a considerable impact on the area. Although they claimed an origin in the Coimbatore area of Tamil Nadu, and followed a number of typically Tamil customs, the fact that they spoke a dialect of Malayalam provided a partial basis for the claim that the Wynaad region should be assigned to the proposed province of Kerala. With the additional weight of argument that the 1931 Census showed a predominance of Malayalam speakers in Gudalur *taluk*—most of whom were migrant estate labourers, therefore not to be considered 'residents' for the Commission's purposes, according to Tamil and Kannadiga claimants—Malayalees asserted the claim that the entire Wynaad, including Gudalur as its administrative centre, should be made a part of Kerala.

Finally, Gudalur had been a *taluk* headquarters in the Madras Presidency since its ping-pong transfers from one administrative division to another finally concluded in 1877. Thus, the predominant political force in the entire Nilgiris District, including Gudalur and southeast Wynaad, had been that of the Tamil majority in the southern reaches of the Presidency for some seventy years prior to the Commission's establishment. Tamils based their claim to the region on the assertion that this long-established administrative region should not be fragmented and/or transferred to another political jurisdiction on questionable linguistic grounds. (There was also a widespread rumour, for which I could find no solid substantiations, that several influential Tamils in Gudalur imported many relatives and friends to the town for a brief period to allow them to assert a claim to linguistic dominance by Tamils and preserve the area under Tamil control.)

A coincidence of historical circumstances had thus brought Gudalur into the middle of disputes between influential forces over rights to its control. The economic florescence of the area coincided with the focus of attention on regional group or linguistic interests. Up to that point, Gudalur had been on the margins of many social and economic interests, but central to none; and now the area was briefly a centre of attention. The location of the community at the 'confluence' continued to be significant.

At the Fringe

Once the decision was reached to include Gudalur in the newly established boundaries of the state of Tamil Nadu, the community reasserted its heterogeneous character. It had been fortunate for the town that most of the potentially divisive polemics concerning group dominance had been voiced by outsiders. For the residents, and the large numbers of newly arriving immigrants, no clearly delineated barriers existed to inhibit social interaction between the multifarious ethnic and linguistic groups which made up the populace.

However, Gudalur remained on the extreme edge of Tamil influence. Migration to the area, and communication with the rest of the state, were inhibited to some extent by the barrier of the Nilgiri plateau. While many individuals did move to Gudalur from other parts of Tamil Nadu, especially from the Coimbatore area on the eastern side of the plateau, their numbers were never sufficient to establish Tamils in a position of numerical dominance.

Accessibility from the Malabar coast of Kerala, on the other hand, was relatively easy; and many Malayalees from that area who were forced by population pressures to look outside for land and/or employment found the economic potential of Gudalur enticing. They might have been less inclined to migrate there if the political climate were laden with a strong sentiment of linguistic regionalism; but Gudalur was looked upon as a sort of half-breed by the Tamil Nadu government, and there were few restraints to Malayalee immigration.

Though they provided the greatest numbers, Tamil Nadu and Kerala were not the only sources of immigrants to Gudalur. The population in 1978 also included almost five per cent each of Telugus, Kannadigas and Sri Lanka Tamil repatriates. (It should be noted that not all the Telugu speakers migrated from Andhra; almost half of them were long-time residents of the Coimbatore area who had retained Telugu as their mother tongue.) Additional small numbers of entrepreneurs came from areas farther to the north. In short, Gudalur attracted representatives of a rather broad range of geographical and linguistic backgrounds; and it attracted them rapidly and in large numbers. During the decade of the nineteen-fifties in which the threat of malaria subsided, Gudalur's population grew at a rate almost double that of the District as a whole, and close to triple that of the largest urban centre in the District, Ootacamund. The town's population almost doubled again in the succeeding decade, when disputes over administrative control had been resolved, and when the town was almost totally free of the dreaded 'fever'. This was a remarkable rate of growth when compared with that found either District-wide or in Ootacamund (see Table 13.1).

Heterogeneity

The population of Gudalur as of 1978 (see Table 13.2) represented a decennial growth rate which had slowed to approximately 54 per cent. This is attributable to several factors, the most significant of which is the fact that employment opportunities had at least temporarily stagnated. However, the earlier explosion, coupled with the present rate of growth, placed a strain on available accommodation. Capital investment in housing during the preceding decade had been minimal. Immigrants seeking employment as labourers, either in town or in the neighbouring plantations, found themselves forced to accept accommodation in tiny cubicles ('doors') in lines of multiplex structures which were sub-divided almost to the limit. (To illustrate: some census house numbers—indicating original single dwelling plots—have been divided by as many as 17 sub-letters, some of these once again divided by as many as 8 sub-numerals, and some even re-divided by further sub-sub-letters; e.g. 'Door No. $26F_{5c}$', a separate residential 'door' in the original dwelling plot No. 26.)

TABLE 13.2

Population of Gudalur Town, 1978 (based on the author's census)

Total population	21,470		
% increase since 1971	38.9%	(≃ 54% /10 years)	
Mother Tongue	**% of total**	**bi/multilingual in Tamil* (% of total)**	**bi/multilingual in Malayalam* (% of total)**
Tamil	39.7%	—	22.3%
Malayalam	44.0%	31.6%	—
Kannada	4.8%	4.4%	2.4%
Telugu	4.2%	3.7%	0.9%
Hindi/Urdu	4.0%	3.9%**	0.8%
Badaga	1.6%	1.6%	1.1%
Other	1.7%	1.1%	1.3%***
Religious Affiliation	**% of total**	**Malayalam speakers (% of total)**	**Tamil/other speakers (% of total)**
Hindu	54.4%	24.6%	29.8%
Muslim	31.1%	15.9%	15.2%
Christian	12.2%	3.4%	8.8%
Other	2.3%	0.1%	2.2%

* no tests were performed to determine degree of fluency

** most of these are Muslims long resident in Tamil Nadu

*** many of these are Paṇiya tribals, whose dialect is closely related to Malayalam

An interesting situation has been generated by this housing scarcity. Newly arriving immigrants are often unable to find vacant accommodation in or near the residences of individuals who have similar ethnic and/or linguistic backgrounds. As families with some accumulated wealth either build new houses or anticipate a move to better accommodations, word of an impending vacancy becomes community-wide knowledge. A general shifting then occurs, with several local families each moving up to slightly better living quarters than they enjoyed before, leaving the poorest housing for new arrivals (and causing untold hardship for any census taker!).

It is important to note that 'better' accommodation, or a more desirable place to live, was defined by most residents as involving (in order of priority): (1) more room; (2) better amenities (*e.g.* availability of electricity, water supply); (3) closer to friends or other 'people like us'. Where these priorities were ordered differently, less than 5 per cent reported that social considerations (no. 3 above) were held to be more important than physical ones. Thus even fairly long-time residents have not established any predominant pattern of communally oriented areas within the town. There were, in fact, only three housing lines containing more than three 'doors' which did not have representatives of at least two different religions or mother tongues living in them. A not unusual example is one line which was occupied by one family each of Telugu Hindus, Malayali Hindus, Malayali Muslims, Kannadiga Hindus and Tamil Christians.

There is a very significant factor, besides housing availability, involved in the lack of an ethnically based spatial subdivision of the community. Since migration to Gudalur over the past fifteen to twenty years included immigrants from many different areas of India, there are only relatively small collections of individuals who have in common a complete set of 'ethnic' attributes. These groups, recognized as distinct both by themselves and others, include the few tribals and a group of low-status sweepers who live in town. Interestingly, it is of benefit to both of these groups to retain a separate identity, since such identity brings government assistance including housing, education, and in some cases land, which are reserved for members of 'Scheduled Castes and Tribes'.

Even when Gudalur's population is classified only in terms of major religious and linguistic features (such as are presented in Table 13.2), it represents a broad and diverse social spectrum, with no significant plurality of any single 'group'. If these gross features are subdivided into some of their components one finds, for example, that the 'Tamil/other, Hindu' category is inclusive of five mother tongues and thirty-nine *jatis*. The largest fairly homogeneous segment is that represented by 'Malayali, Muslim' category, comprising 16 per cent of the population, 86 per cent of whom identify themselves as 'Mappillas'. Significantly, in many respects this is the most cohesive—also the most commonly denigrated—social group in the community. Even so, members of this category are representative of four subdivisions of Mappilas.

Communitas?

In 1978, Gudalur exhibited few clear-cut and regularly applied features of social differentiation in the sense that 'ethnicity' is usually understood. This in itself is not proof that processes of differentiation and/or stratification are not operative in the community. And the lack of a general pattern of territorial localization of identifiable groups—either in terms of self-identification or by way of outside analysis—is not sufficient to justify a claim that Gudalur should be recognized as different in any more than cosmetic detail from other small South Indian communities. However, evidence from other aspects of residents' behaviour helps to strengthen this assertion. One of these is the low esteem with which political groups or leaders are regarded.

The social cohesion exhibited in Gudalur is not based on organized political structure. Long-term residents are well aware of the fact that the community is of marginal concern to outside political forces. In spite of being a *taluk* headquarters, the town has been the recipient of a minimal amount of state government assistance for development. For instance, the 1971 Census shows that the only 'urban' areas in the Nilgiris District which received no outside government assistance in the year 1968-9 were those in the Wynaad area, including Gudalur. In addition, the tax base for the municipality is so small—the same census figures show a per capita revenue of Rs 6.60, compared with Rs 52.12 for Ootacamund—that even locally elected individuals have little in the way of economic resources to use as a basis for political influence.

The apolitical nature of Gudalur is significant in a number of respects. Special interest groups, such as ethnically or linguistically based factions, find no organized basis of support. Politicians, labour organizers, and religious leaders, have from time to time sought to polarize the community in order that support for themselves and their programmes would be backed by a plurality of the population out of fear, disdain, or even patronization of the other residents. Their success has been minimal; and the consensus of the townspeople interviewed is clearly that such people who promote a 'cause' only disrupt the search for harmonious relations within the community. An expression heard with some regularity is 'dustbin elections', a metaphorical expression of the on-going process, as they see it, of sweeping out the accumulation of political refuse.

A clear expression of this feeling of community interests predominating over those of larger political entities or special interests was found in an incident which began on 4 October 1978. On that day, Tamil Nadu police, revenue and forest officers swept through areas of private estate and government land adjacent to Gudalur, evicting squatters who had recently taken up residence and begun to clear the land for homesteads. The vast majority of these encroachers were Malayalis who had moved across the nearby Kerala border in search of unoccupied arable land; and in an attempt to stem the influx, the police had a mandate to remove those who had arrived within the past year. One individual who had cleared and planted fields on government land for almost ten years, and who had constructed a *pakka* dwelling on the land, was evicted and his house and crops destroyed. In a dramatic protest, he immolated himself in front of the Revenue District Officer's office in Gudalur. This act led to a *hartal* and protest march in town, in which all segments of the community participated.

A number of outside partisan organizations latched onto this incident as a vehicle for voicing their particular grievances. Congress (I), DMK, and Muslim League party representatives all condemned this sort of 'repression' by the AIADMK Government in power. Each stressed that such incidents would never have occurred if their party had been in control; and each, in an idiom expressing the orientation of his own party, suggested that Gudalur's residents would be well served to organize themselves against such 'atrocities' in the future. A potentially even more significant protest was filed by the Kerala government, giving the impression that the Tamil Nadu administration was discriminating against Malayalis through its actions.

After a very brief occupation by armed police, the community resumed its characteristically peaceful atmosphere. Rumours of numerous alleged atrocities were shown to be false, and outsiders were unable to sustain a high level of emotional involvement in the broad issues which they argued were the ultimate causes of this incident. The Kerala Government soon voiced satisfaction that the incident was not essentially discriminatory in motivation, and dismissed the issues involved as being internal concerns of Tamil Nadu.

Interviews with residents shortly after this incident pointed to the fact that the local protest, expressed in the *hartal*, was against the destruction of property and crops of a few individuals who were considered by locals to be members of the community. They were glad to see the current wave of squatters evicted, since they felt that if the public lands were to be opened to anyone, they should have first rights of occupation. Also the majority—including Malayalis—felt that the issue was not one of Tamils versus Malayalis; rather, any conflict which existed was between those who participated in Gudalur society, whatever their origins, and those 'outsiders' who might attempt to take advantage of that community's resources.

The force that binds Gudalur's residents together has much in common with what Turner calls 'communitas': 'a relationship between concrete, historical idiosyncratic individuals' (Turner 1969: 131). This type of social relationship is quite different from the organization of a community based on relationships between social positions ('statuses') without regard to the qualities of the individual occupants of those positions at any particular time. The recent and rapid expansion of the Gudalur community, with a four-fold growth in population during the last thirty years or so, means that most social relationships lack historical depth. Most residents must confront and deal with other members of the community on the basis of what they know about them as individuals. In such circumstances, a shopkeeper, for example, must cultivate customers as individuals rather than depend on the existence of a certain clientele who, as members of a group who have always traded with his family, can be depended on to continue the long-established relationship. In other words, it is not possible to say of most social relationships in Gudalur that 'it is so because it has always been so'.

If all the immigrants had come from similar backgrounds, or had been socialized in some common set of beliefs and practices, there might well have been a shared understanding of acceptable (or 'proper') organizational structure, interaction patterns, and even hierarchical ranking. However, understandings of these sorts which are shared throughout the Gudalur community are only very general (i.e., those which are commonly accepted throughout South India). The constraints imposed by these shared general principles of social relationships are much less rigid for Gudalur's residents than those found in most long-established communities, where specific as well as general status relationships, and associated behavioural expectations, are clearly defined and are common knowledge. The dearth of historically justifiable patterns of social interaction, and consequent low level of habitual behaviour based on shared expectations, creates a social environment not unlike that found in some metropolitan areas. This environment not only disallows a regular dependence on local tradition or habit as a guide to behaviour, but it also provides a fertile medium for innovative or entrepreneurial activity. As expressed by one resident, and echoed in substance by many, 'in Gudalur, a man has to make his *own* place'.

Old Languages, New Setting

The demand that Gudalur makes—or the opportunity it provides—for individuals to create social positions for themselves finds clear expression in the use of language in the community. Gudalur represents a microcosm of major Dravidian languages as well as a few Indo-Aryan tongues (see Table 13.2); and the range of social and geographical dialects is, to say the least, extensive. The lack of a common experiential background reduces the degree to which social information is shared; but communication on a general level is not seriously impeded, since virtually the entire population is to some extent conversant in Tamil (86 per cent) and/or Malayalam (73 per cent). There is no need for, nor occurrence of, a 'pidgin' language to bridge the gap between speakers of different languages.

There is, however, one significant sense in which communication in Gudalur has been influenced by historical, linguistic and social factors. An individual, in the process of 'making his own place' in the society, must establish his position relative to others around him. Dravidian languages are like many others in the world in that they incorporate indicators of the relative social status of the communicants into pronouns, verb endings, and other linguistic components. Thus in most common speech events it is necessary to select from among several alternative forms, each of which defines the social position of the speaker *vis-à-vis* the addressee or referent (e.g. low/high, close/distant, young/old, weak/strong).

In a community with clearly defined historical relationships, such specification is fairly regular, accepted and shared as common knowledge. Not so in Gudalur, where most social relationships are still flexible and relative rank is often ill-defined. For example, there are a number of individuals who, in terms of ascriptive ('caste') rank, would find themselves very low in the social hierarchy. However, some of these individuals own land in Gudalur which they acquired at a time when that land was of little value to most people because of the threat of malaria. As the town grew, these lands were turned into residential plots and became a source of considerable income. Two sets of ranking criteria were thus in conflict, since these individuals were ascriptively low but economically strong.

Perhaps in a small traditional community this group of people might have attempted to modify their ascriptive rank through Sanskritization: 'the process by which a 'low' Hindu caste...changes its customs, ritual, ideology, and way of life in the direction of a high...caste' (Srinivas 1966: 6). Such changes in behaviour might then provide a basis for a claim to a higher social rank than would have been acknowledged by others in the community prior to their implementation. However, in Gudalur the social hierarchy was not yet clearly formulated, and rather than assert a high traditional rank which might or might not be conceded by the rest of the community, these individuals chose to downplay *any* system of ranking.

Being in an ambivalent social position, this group of people now employs a strategy of manipulating their speech so that they are rarely required to specify a social differential (either higher *or* lower) between themselves and others. A clear illustration of this behaviour is the deletion of most pronouns and personal verb endings. For example, phrases like (a) and (b), which specify relative status, are often transformed into ambiguous phrases like (c):

(a) *(ni:nga) enge po:ni:nga?*
'Where did you (polite) go?'

(b) *(ni:) enge po:ne:?*
'Where did you (impolite) go?'

(c) *enge po:ya:ccu?*
'Where did going happen?'

In this way they avoid any implications of relative ranking and side-step the problem of either asserting a high rank based on economic power or accepting a low one by virtue of their ascriptive status.

There are other members of the community who also find it advantageous to avoid specifying relative social rank. As noted above, Gudalur is very much a free market as far as shopkeepers are concerned. They must establish relationships with customers rather than depend on long-established associations. One way of attracting clients is to reduce the social distance between merchant and customer by using linguistic forms that do not imply a difference in social rank between the two. This allows both seller and buyer to maintain private images of their own social positions without making them public. Any public assertion of higher rank by the merchant might either alienate the customer or lead to a challenge of that rank; and admission of lower rank could bring about a lessening of respect or loss of face. Eliminating language forms which specify rank may not be entirely satisfactory to the participants, but in the Gudalur context it serves a useful purpose. (This type of language manipulation is also found in some cities, where individuals are either trying to avoid specifying relative rank or where it is not clear to the conversants what their relative ranks actually are.)

On the Verge

Much of what has been described above makes Gudalur appear to be a small-scale version of an urban centre. It is 'cosmopolitan' in make-up, dependent on exchange with outsiders for its economic stability, and relatively open to anyone who contributes to the economic well-being of the community. It also provides an environment conducive to social 'passing'—allowing a person to assert a particular social status based on his own achievements.

But Gudalur is still a small town! The above facts must be viewed with the recognition that the individuals who make up this community interact on a regular basis, and know a great deal about each other on the personal level. One person cannot, for instance, expect to take advantage of another for short-term gain, and then move on to another unsuspecting victim. The commingling of individuals with different ethnic, linguistic and social status origins which began during the rapid expansion of the community continues, although some criteria for the structuring of relationships are emerging. While the ultimate product of this process is by no means certain, some trends are apparent and others less clear.

First, the ideology expressed by many that an individual in Gudalur is 'self-made' is becoming more social fiction than fact. There is still the potential for exceptional persons to step out of whatever ascribed identity they were tied to in their community of origin and here become respected both socially and economically for what they make of themselves. But the establishment of roots in the community has led to the assignment of social rank (and, to a limited extent, the passage of power) to the younger generations based on the performance of the preceding ones. This longitudinal allocation of social position is quite pervasive in Indian philosophy—legitimated by tradition, and institutionalized in 'caste'—and is unlikely to be discarded entirely by a single small community surrounded by a society which continues to hold to it. As yet, this stratification is not generally expressed in the idiom of caste ranking; but the possibility remains that it may eventually be reinterpreted in that way.

For some time, the nature of social relationships in Gudalur had been determined primarily by the benefits each individual sought to derive from a particular context involving specific individuals. There was little social interaction based on habits, behaviour patterns developed through long and intimate associations, or generalized expectations about what other individuals or groups 'usually do'. As the population stabilizes, such habitual behaviour is becoming more common. Children are often the most prolific generalizers, and in Gudalur this is clearly true.

First- or second-generation adults continue to emphasize the importance of evaluating and dealing with others on a personal and individual basis, and in most cases continue to deny the existence of definite 'groups' and/or social levels in the town. Economic criteria are important to them, but they commonly identify these as variable and based on individual abilities. For second- and third-generation children, group boundaries are fairly clear and significant. Stereotypes of adult group identities are common, and these are often accompanied by value judgments concerning the group as a whole. This phenomenon may be explained in part by the fact that there are separate primary schools using Tamil, Malayalam and English mediums, as well as private Christian schools and Muslim classes teaching Urdu. In addition, there are some poor families who cannot afford any school, and whose children are thus separated from the others both physically and to a considerable extent socially. Thus the boundaries between segments of the community, which are often indistinct or irrelevant to adults, are made artificially explicit for the children.

While school-aged children may be formulating notions of significant differences between groups in the town, women find themselves in a position of being much more free to interact with others of different backgrounds. The norm for women in most of India is to remain at home much of the time, and to co-operate with other women of their household in dealing with domestic chores. They may participate in larger-scale activities on the occasion of some festival, ritual or other group gathering; but rarely do they commingle on a regular basis with women unrelated to them either by kinship or ethnic identity. In Gudalur, the randomization of residence has led to circumstances in which co-operation between women living in adjacent 'doors' often entails sharing of chores between members of traditionally non-associating groups. Women's communication networks are greatly expanded as ideas, feelings and gossip are transmitted between adjacent residences and then along the more distant lines between members of related kin or ethnic groups. Thus while children conceptualize the community as being made up of bounded groups, women find it much less so.

Leading the Way?

Up until 1978, it had not proved to be in the best interest of any significant segment of the population to attach primary importance to either factual or fictionalized 'origins' of the members of the Gudalur community. Opportunities existed for practically all highly motivated persons to expand their spheres of influence. Those who came to the town with images of self which constrained them to doing manual labour found that they could survive well in that role; and those who wanted to establish new identities found the opportunity, with few questions asked. There was little in the way of a socially defined 'we' versus 'they' conflict of interest.

The potential for unlimited expansion in Gudalur is proving to be as much a myth as it has been shown to be in the American West. Gudalur is 'modern' in many respects, and may be illustrative of the direction in which small-town India is heading. Many of Gudalur's residents admit freely that they moved there as much to escape from constricting demands and expectations from family and traditional associations in their home communities as to improve their economic and social status. Throughout the rest of India, transportation and communication are becoming more accessible; and as the security of known, but limited, opportunities at home becomes less attractive than the challenge and potential rewards of 'making one's own place' by moving to another locale, more and more communities may find themselves in a situation similar to that of Gudalur.

The results of such increased mobility would probably not prove identical in every community to those occurring in Gudalur, since unlike Gudalur, many (if not most) centres which would attract such immigrants already have an established power base which would have to be dealt with—either challenged or accepted—by the new arrivals. The relative anonymity with which such a mobile individual would first confront the new community, however, and the challenge of formulating a self-image and an image of self in others, would be comparable to that found in Gudalur. The social flexibility gained in such a process is surely not without potential hazards, as is clearly exemplified in the poverty and despair found among some immigrants to India's cities. And deeply ingrained and pervasive traditions such as fairly specific rules of endogamy will continue to set limits on the individual's freedom to formulate his own social identity.

Much of the Nilgiri plateau has been deeply influenced by the presence of Europeans. Towns like Ootacamund, Wellington and Coonoor (which were all founded by them) exhibit many trappings of western culture; but these attributes are in most cases only slightly filtered borrowings. After all, the area was for more than a century developed for the almost exclusive benefit of the colonialists. Gudalur, on the other hand, was influenced by some economic and other interests of foreigners; but the process by which it developed, and the people involved, were predominantly Indian. It is for this reason that Gudalur may well be more indicative of the future direction of rural India than the other modernizing areas in the Nilgiris District.

British Society in the Company, Crown and Congress Eras

PAUL HOCKINGS

14

...we are terribly observed, and of course I doubt not pulled to pieces, but thank God we are still quite English, and domestic, taking our walk together every evening, our tea and our bath afterwards...

—Letter from Lady West to Mrs Lane (6 March, 1824)

The series of cultural changes which affected the Nilgiri Hills in the eighteen-twenties was so rapid, so far-reaching and so permanent as to make one pause and wonder whether the term 'cultural revolution' can usefully be applied to such a small region. But even if it can, the term 'revolutionary' certainly could not characterize the men and women who came to the Nilgiris from Britain—with one possible exception, John Sullivan.

The Founder

Born in London on 15 June, 1788, and baptized on July 2, in the very fashionable Parish Church of St. George, south of Hannover Square, John Sullivan was the Englishman destined to have a greater cultural impact on the Nilgiri Hills than any other single person, Indian or European, in their entire history. His parents, John and Ann, had him privately educated in arithmetic and merchant accounting with a certain Mr C. Mathias, of George Street just north of Portman Square (which at that time was on the very outskirt of London), and then in August 1803 the 15-year old youth was nominated for a Writership—the lowest position—in the Madras Establishment of the East India Company. This nomination came from the influential Jacob Bosanquet, then Chairman of the Board of Commissioners, on the recommendation of the boy's own father Stephen John Sullivan, who had previously been in the Madras establishment himself. Before his son's birth, he had been commissioned by Lord George Macartney, the Governor of Madras, to conclude a treaty with the Maharani Lakshmi Ammanni, so as to restore the ancient Royal House of Mysore (Hayavadana Rao 1930:3161). The only other facts known about

Stephen J. Sullivan are that he joined the Madras Establishment of the East India Company in 1778, became a Persian translator and secretary in 1780, and was resident at Tanjore in 1782, where he concluded the treaty with Tirumala Rao on behalf of the Maharani on October 28, 1782 (Hayavadana Rao 1930:2556-7, 2560, 3146). After this date we hear nothing further of S. J. Sullivan, and apparently he retired later to London's West End. His father, Laurence Sullivan, had been an influential politician in the Company too. That Stephen Sullivan had some of his son's progressive characteristics is attested by Hough's comment on his 'enlightened zeal in the cause of native education' (Hough 1860: 410, n.2; the same writer confirms that the Coimbatore Collector, John Sullivan, was indeed his son.)

A summary of the official records (Prinsep 1885) gives this outline of John Sullivan's career:

1804	Writer, in Madras.
1805	Assistant to the Secretary, Revenue and Judicial Department.
1806	Court Registrar, Chittaput, South Arcot District.
1807	Assistant to the Chief Secretary in the Secret, Political and Foreign Department.
1809	Acting Assistant to the British Resident at Mysore.
1811-14	In England.
1814	Collector at Chingleput.
1815	Special Revenue Commissioner in Coimbatore.
1815-30	Permanent Collector of Coimbatore (including the Nilgiris)
1818	On furlough at Cape Colony, South Africa.
1819	First visit to the Nilgiris.
1819-21	Administrative work in Madras.
1820	Member, Board of Revenue; married Henrietta. Harington on February 2 (she immediately became pregnant).
1821	First visit to Ootacamund, February 22; young son died in Coimbatore in July.
1822	Started building at Ootacamund.
1823-7	Mostly in Ootacamund with family.
1828	Ootacamund made a military cantonment, and thus taken from Sullivan's control and placed under a commandant.
1830-5	On absentee allowance in England.
1835-6	Resigned his appointment as Member of Council and was made Judge of the Faujdari Adalat and also Senior Member of the Board of Revenue. Being permitted to reside where he liked, he chose Ootacamund.
1836-41	President of the Revenue, Marine, and College boards.
1838	His wife and eldest daughter Harriet died in Ootacamund.
1841	Retired in May from the Madras Civil Service, with an annuity from the Company's fund; had seven children to bring up.
1855	Died on January 16, in England.

Sullivan was by no means the first European to visit the Nilgiri Plateau. Jacome Finicio, an Italian priest, had been there in 1603, and nearly a dozen

Englishmen had made brief trips to the area in the decade prior to Sullivan's arrival. It was in fact two subordinates in the Collector's office at Coimbatore, John C. Whish and Nathaniel W. Kindersley, who so appreciated the possibilities of the cool plateau climate, when they climbed up the eastern slopes late in 1818, that they persuaded Sullivan to go and see for himself.

This he did in January of 1819, and again in May, when he visited the eastern part of the plateau for three weeks accompanied by a noted French naturalist with the imposing name of Jean-Baptiste-Louis-Claude-Théodore Leschenault de la Tour. The Frenchman had been sick in the plains, but soon after reaching the higher elevation he became more vigorous than he had been in months, and energetically set to collecting samples of over 200 species of plants (many of them then unknown). The healthiness of the plateau climate was to be a constant theme in Mr Sullivan's subsequent recommendations to the East India Company in Madras. 'C'était le paradis terrestre: zéphyrs embaumés et fraicheur l'année entière' (Blavatsky 1925: 36). A century later the beauty and salubrity of the Nilgiris were still being extolled by a few expatriate poets (see Appendix).

The story of Sullivan's first house near Kotagiri (started in 1819, the same year that the first house was being built in Simla), and the subsequent development of another hill-station for Europeans at Ootacamund, have been fully recorded elsewhere and need no reiteration here (Francis 1908; Grigg 1880; 'Panter-Downes' 1967; Price 1908). But the role of John Sullivan himself in changing the face of the entire district has never been properly evaluated.

It was probably on February 22, 1821 that Mr Sullivan first visited 'Wotokymund' with a Badaga guide. Emeneau has however recorded a Toda tale which cannot be taken too literally, and which makes a Toda named Te:ty responsible for introducing Sullivan to the valuable site.

> He had to go to the plains for rice, salt, sugar, etc..... As he and his companion were going down, they met Sullivan coming up. They asked with gestures, since they spoke no Tamil, where he was going, and said that they would carry him up the mountains. They made him a stretcher with cloaks and brought him up. When they reached what is now Stonehouse Hill, he asked shelter in the Toda mund that was situated there. They gave it in a small tent... . He asked all the Todas to be called and made with them an agreement for so much ground as could be covered by a sheepskin. Then he cut the skin round and round into a long strip as thin as hair. It was ten miles long. He measured ground with it and said that all this ground was his, and the Todas kept their agreement. Afterwards he started to build (Emeneau 1963:189).

The ancient tale of the sheepskin (as used by Dido to acquire Carthage and the Portuguese to acquire Bassein) is probably a fiction here, particularly as there were no sheep on the Nilgiris at that time; and it is doubtful if Sullivan could express such a complicated agreement in any language the Todas knew; we know that in fact he paid them roughly a rupee an acre for land he bought from them. It is true however that Stonehouse Hill was a funerary mund (and thus not inhabited by Todas) at the time he acquired it and started building there. During the four months April-July 1822 he made considerable progress on his stone house. In this work he was helped by his Scottish gardener, Mr Johnston, and doubtless by some local Badaga or Kota villagers.

Other buildings appeared in quick succession. The settlement's first church was a small Roman Catholic chapel, built in 1823 just after 'Stonehouse' was completed. An Anglican chapel was also erected between 1826 and 1828, but before that a dozen or more houses had sprung up. Sullivan and Major William Kelso, the commandant, had a heated argument about the location of bazaars. A 'native village' of immigrants from the plains grew up at Kandal Bazaar in 1829-34, and a separate 'Brahmin village' also appeared then.

From 1820 to 1827 Sullivan made frequent pleas to the Directors of the East India Company to develop the Nilgiri Hills as a sanitarium for sick European troops in India. Hardly was 'Stonehouse' completed when Sullivan had to bury Cornet H. Harrington of the 7th Light Cavalry; he had died in April 1823 at the age of 18 years, thus prompting Sullivan to form an unconsecrated cemetery near the new house which is still in existence. Quickly appreciating the possibilities of Ootacamund, the government followed a policy of developing the roads leading to it, renting houses for convalescent soldiers in the growing town, and appointing a resident medical officer and a cantonment commandant. John Sullivan encouraged this development, and indeed rented out 'Stonehouse' and other places he had subsequently built to the government for the use of sick soldiers. We must remember, however, that it was only in 1827 that 'Lord William Amherst, then governor general, set the precedent for the upper strata of British Indian society...by vacationing in the hills for recreational purposes only, perhaps on grounds of preventive medical therapy...', and in Darjeeling, not Ootacamund (Dickason 1975: 115).

By 1829 the town's population had reached 500; and Sullivan then held five times as much Nilgiri land as did all other European inhabitants put together. At the end of 1822 he had enclosed three square miles of land, all of it unoccupied, for agricultural experiments. In seeking the Company's permission for this, he had soothingly argued that 'the experiments may eventually prove useful to the public, and the expense of making them will be my own.' Until 1830 he was still officially Collector of Coimbatore, to which district the Nilgiris belonged until 1868.

The massacre of fifty-eight Kurumbas suspected of sorcery in 1835 prompted Sullivan and the Governor of Madras, Sir Frederick Adam, to give the commanding officer of Ootacamund police authority over the hills, and appoint him magistrate; thus affording protection 'to the lives and property of all classes of the inhabitants' (Grigg 1880: 229).

From its earliest days Ootacamund has been blessed with a lake some two miles long. This, despite its natural setting amidst trees in the bottom of a valley, is not a natural lake but another creation of Mr Sullivan's. He dammed the stream early in 1825 to make a tank for irrigation of fields in the plains below.

This scheme never came to pass; yet Sullivan did create a most attractive serpentine stretch of water which is today one of the chief tourist attractions in the town. The government, for reasons of expense, was never to develop this into the headwater of an irrigation system. It regrettably became the sewer, bath, and drinking-water source for the townspeople (Price 1908), and half of it was subsequently filled in.

The developments we have already attributed to John Sullivan by no means exhaust his influence upon the Nilgiri community. He was directly responsible for agricultural change that revolutionized the local economy, for the three main Nilgiri export crops of tea, potatoes, and cabbages were first encouraged by him, and he introduced many other European fruits, grains, and vegetables besides (Robertson 1875).

When Sullivan settled on the hills the only other residents there were, according to an 1821 census, 222 Todas, 317 Kotas, and 3778 Badagas. The latter two communities were then millet-farmers, whose rising fortunes were soon to be linked with the needs of the expanding European town population and plantation economy. John Sullivan was obviously a keen horticulturist, for in 1821 he had Mr Johnston sent out from England to help him and to bring British seeds for experimental planting on the hills. The list of species which Sullivan introduced over the next few years is especially remarkable because all of these items are still grown in the district. They include new varieties of oats, wheat and barley; the market crops of beetroot, turnip, radish, cabbage, potato, strawberry, peach and apple; ornamental flowers, including laburnum, sparaxis (actually from South Africa), rose, heliotrope, violet, mignonette; and several other important species, including oak, hemp, flax, vetch, lucern and geranium. As the noted French anarchist, Dr Elisée Reclus, was later to observe of the Nilgiris:

> ...the European appearance of the landscape was greatly enhanced in the parks and gardens by the introduction of most of the scents of England; in various spots the illusion could be complete: the Englishman finds himself once more in the hills of Malvern or Devon; his house is covered with the same climbers; his garden contains the same flowers, and trees of the same species are planted around his compound; European birds let loose in the wood have bred there, and the little lakes of the plateau, where there used to be just one species of fish, have received colonies of carp, tench and trout (Reclus 1883:516, trans.).

Sullivan farmed 200 acres in Ootacamund himself, and also distributed seeds of various cereals and vegetables to interested Badaga farmers. He gave them European varieties of wheat and barley, both of which degenerated by cross-fertilization with the local varieties; but long afterwards the barley was called *Sullivan gañji* by Badagas. In 1839 he sent to the Madras Agri-Horticultural Society excellent samples of hemp and flax that he had grown, expressing the hope that these could prove of commercial value. At about the same time he sent many specimens of insect pests in timber to Madras, for identification. During his last years in the district (1835-40) he repeatedly offered the opinion that the plateau was well suited to tea cultivation, and even sent good samples of cured tea to Madras from 1836; yet it was only after 1865 that tea became commercially important on the Nilgiris.

Sullivan and others, perhaps thinking of the Industrial Revolution during their boyhoods, believed that the Nilgiris might become another Yorkshire: hence '...the inexhaustible supply of water-power afforded by the streams upon them would lead to the establishment there of mills and factories of every kind' (Francis 1908: 223). Yet this laudable hope bore no fruit; industrialization in the nineteenth century was impossible because of the high cost of unskilled labour and the lack of railway transportation in the region. Far-flung markets were not yet accessible, and the economy of India was based altogether on farming and plantations.

The building of the early roads up into the hills was done by army sappers, though it was much encouraged by Mr Sullivan's advice. The very first improved track originated with his request of March 1819: this was the Kotagiri Ghat. In 1823 he obtained government financing to improve a pass leading from the plateau westwards to the Wynaad; and in 1824 he began improving the routes across the Wynaad to Malabar and to Mysore. In 1826 he improved another pass up the southern side of the hills, which was later known as Sullivan's Ghat.

When the Wynaad plateau was transferred to Malabar (an area with which it was not intimately connected) in 1830, it was John Sullivan who initiated the move to re-transfer it to his own Coimbatore District (which was effected in 1843).

His interests were thus very broadly concerned with the district's development, and not solely with the growth of the one town. He wanted the Nilgiris to become an area of British colonization (perhaps by pensioned soldiers), and to be integrated into the economy of Madras Presidency. His reports and letters constantly point out the financial advantages to the Hon'ble Company of developing the resources on the plateau.

Although Sullivan repeatedly feuded with other government officials, most notably with the commandant at Ootacamund, he was benevolent towards the tribal and peasant peoples with whom he came into contact. While it would not be unfair to say that he exploited his own government to develop his new house and line his own pockets, he certainly did not exploit the native peoples whom he found on the Hills, but on the contrary was long remembered as their benefactor.

The Todas still sing of Te:θy as the man who held the reins of John Sullivan's horse; but it was another man, Te:ty, whom Sullivan appointed as Toda headman in 1822. The tribe had no indigenous headmen; yet for the past 160 years they have mediated with the government through this official, called the *monegar*. This headmanship descended in the male line from Te:ty to his fourth son (Mutefïn), to *his* eldest son (Ïfyefïn) and then another son, Piɫya:ṛ

(Emeneau 1971a: 331, 369), and then to Te:θyxe:n, next to Piḷyxuḍ, and finally to Te:θyxe:n's son Sinwïṭṇ, the *lasmonegar* (Hockings and Walker, 1983). After it was instituted by Sullivan the position was first in Kï:wïṛ clan, but as Piḷya:r̲ was childless it was transferred with Te:θyxe:n to To:r̲o:ṛ, a clan of the opposite Toda moiety. This change is perhaps not as radical as might appear, since Te:θyxe:n's mother (Pu:pi:), though 'formally married' to a To:r̲o:ṛ man, was the daughter of Kwï:s̲yaḷf, the extremely influential brothers' son of Mutefin and a man whom Rivers (1906:556) described as 'the chief representative of his family on the *naim*' (or tribal council). It was apparently this person who effected the transfer of the headmanship to his daughter's son Te:θyxe:n. The latter was followed however by Piḷyxuḍ, another brother's son to Piḷya:r̲ and hence a member of Kïwïṛ, clan. Both in Kwï:s̲yaḷf's time and, nearly a century later, the *monegar* was not an influential person in Toda affairs. His duty was to collect taxes for grazing and, supposedly, to keep a record of Toda births and deaths. The position was recently abolished.

Sullivan also attempted in 1820, without success, to change Toda custom in another direction; for he was the first of many Europeans to decry the practice of female infanticide, which did not finally stop till the middle of this century.

The most far-reaching of his dealings with the Todas was on the matter of land tenure. The land on which Ootacamund and Coonoor were built was originally devoted to Toda grazing, hamlets and temples. Sullivan had the foresight to realize that without some sort of recompense for taking such lands, the British might later have legal problems with Toda claimants. He strongly endorsed the absolute proprietary rights of the Toda tribe to the entire Nilgiri plateau, on the presumption that they were the earliest settlers there. The Governor of Madras, the Rt. Hon. Stephen Lushington, led a rival school of thought with the argument that throughout India proprietary right in land belonged to the government. From the purchase of Stonehouse Hill until 1828 private persons like John Sullivan had paid the Todas cash for plots bought from them, and the government tacitly recognized such land-titles. In 1828 it ordered that Todas must be paid for land at the rate of sixteen times the annual revenue assessment for the grazing of Toda buffalo. Yet by 1831 this arrangement had been conveniently forgotten, and the government got into the habit of granting 'waste land' to British settlers without any compensation going to the Todas. It was John Sullivan who, as a Member of Council in 1835, opened the question again on behalf of the Todas. Ill-informed though his arguments were, he carried the Court of Directors with him, and it was ordered that a sum of Rs 150 be paid to the Todas annually by the government as compensation for the land in Ootacamund—as it still was until recently.

Mr Sullivan's relations with the Kota tribe were of a different kind. These people as well as being farmers were the ironsmiths and craftsmen for the other communities; they had been obliged to buy scrap iron in the plains to fulfil their customary obligations. John Sullivan showed some Kotas how to extract iron from ores occurring on the Nilgiri Plateau; yet the idea did not catch on, apparently because of a religious conservatism that prevented the Kotas from innovating in their technology.

He had much more impact on the Badagas. These people claim it was they who invited Sullivan to the plateau to escape from the lowland heat, and to re-survey their lands, for which they believed they were paying the Company too much in revenue. It is typical of Mr Sullivan that he seems to have taken the side of the cultivators against his own government's interests, since he arranged for a revenue survey of Badaga lands in 1820.

We have already seen how Sullivan distributed new seeds and vegetables to Badaga farmers, in the hope that they could improve their agriculture. He also encouraged improvement in another novel direction; for in 1820 or 1821 he established the first Nilgiri school, in the Badaga village of De:na:ḍu. Ten years later he repeated the experiment in another Badaga village. These two schools taught many Badaga youths literacy in Tamil and Kannada, no doubt so that they might become clerks (Harkness 1832: 69).

Although most of his concerns were administrative and agricultural, John Sullivan showed intellectual predilections as well. Thus during his first four months in Ootacamund he faithfully recorded meteorological observations five times each day. It was partly on the basis of these that he built his case for the healthfulness of the hills, at a time when all Anglo-Indians believed the closest healthy spot was in South Africa!

Another of his intellectual interests, understandable in view of the great lack of entertainment in the new town, was to initiate a subscription for a public reading room. This got started in 1829; it was a lending library as well as a public meeting-place.

The very early photograph of Sullivan published by Sir Frederick Price (1908:24) shows a short, middle-aged man in a frock coat and flannel trousers. There is a look of tired disillusionment around his eyes and mouth. It seems as if he has long been sitting in that half-relaxed position, staring placidly from the picture (indeed, early photographic exposures demanded as much); yet there is something rather obstinate and querulous about his compressed lips which contradicts his soft, aimless hands and the droop of the shoulders. His clothes have the crumpled appearance of one who spends many sleepless nights at the office where he dozes fitfully at his desk. Yet he was a good family man and was repeatedly accused of neglecting his official duties in Coimbatore.

Unlike some Englishmen serving in India at that time, Sullivan was able to have his family with him in Ootacamund. Mrs Sullivan was carried up from the plains in 1820, his then 17-year old bride and the first European lady to visit the Nilgiris; in 1822 Harriet Ann, her first child, was born; and in 1823 she started housekeeping in Ootacamund. Henrietta Cecilia Sullivan was in fact destined to die there, in 1838. There is a memorial to Mrs Sullivan and her two children in St. Stephen's Church. Their son, born in Coimbatore but baptized as 'Stonehouse' in May of 1823, was the first recorded baptism in Ootacamund.

Mr Sullivan retired to England in 1841, a tragic man who had buried his young wife and two children in Ootacamund, but still had seven more children to bring up. One of them was later to become a Collector of Coimbatore and to play host to the infamous Madame Blavatsky. Sullivan's house remains as a part of the Government Arts College, an affiliate of Madras University. But he also left behind a flourishing new town, India's first hill-station. It is indeed rare for us to know so much about the man who started a town from the bare earth.

Tea and Scandal

John Sullivan was somewhat unusual for his time in the extent of innovation that he was able to combine with his administrative work. His public roles, as we have seen, were numerous: magistrate, administrator, meteorologist, entrepreneur, town planner, engineer, capitalist, farmer, churchman, and protector of the poor. This is a curious contrast with what we know of all the Europeans who were to follow Sullivan to the Hills; for there was an almost caste-like discreteness, a separate lifestyle and a minimum of intercommunication between the administrative officials, the army officers, the planters, the tradesmen, the teachers, the Protestant missionaries, the Catholic priests, the retired and the tourists; altogether a dozen hermetically sealed units only briefly to be united when and if they sat in a church together. Their interests, after all, were utterly different from one group to the next: some hunted jackals and shot woodcock, others sought souls; some pursued profits, other young ladies; some poured out their feelings in memoirs and poetry (see Appendix), while others waited only for pensions, peace and better health. All stripes of men have visited the Nilgiris, from the founders of modern India, like Macaulay, Besant, Nehru and Gandhi, to soldiers of fortune like Sir Richard Burton and the Duke of Wellington, to eccentrics like Madame Blavatsky and Edward Lear; not to mention vacationers like Maria Montessori, the educator, Margaret Cousins, the Irish suffragette and freedom-fighter, and a steady stream of scholars investigating everything under the sun (Hockings 1978). Generally these people were Western Europeans, and mostly English, Scottish and Irish; but not all were of the same social background nor the same Church. One hardly need point out that the impact these several status-groups had on the Nilgiri District was highly variable. Even in a single group one would find occasional innovators but numerous conformers. Sir Richard Burton catalogued English society at Ootacamund in the nineteenth century in these words:

> Among the ladies, we have elderlies who enjoy tea and delight in scandal: grass widows...and spinsters of every kind, from the little girl in bib and tucker, to the full blown Anglo-Indian young lady, who discourses of her papa the Colonel, and disdains to look at anything below the rank of a

> field-officer. The gentlemen supply us with many an *originale*. There are *ci-devant* young men that pride themselves upon giving ostentatious feeds which youthful gastronomes make a point of eating, misanthropes and hermits who inhabit out-of-the-way abodes, civilians on the shelf, authors, linguists, oriental students, amateur divines who periodically convert their drawing-rooms into chapels of ease rather than go to church, sportsmen, worshippers of Bacchus in numbers, juniors whose glory it is to escort fair dames during evening rides, and seniors who would rather face his Satanic Majesty himself than stand in the dread presence of a "woman". We have clergymen, priests, missionaries, tavern-keepers, school-masters, and scholars, with *précieux* and *précieuses ridicules* of all descriptions (Burton 1851: 295-6).

Ninety years later Margaret Cousins left us a similarly tongue-in-cheek account of society in Kotagiri (Cousins and Cousins 1950: 723-45).

The class structure of these Europeans in pre-independence days seems to have embraced a hierarchy of four social classes (five, if we include the Eurasians or Anglo-Indians). The names which have been applied to these five in Table 14.1 have no fixed meaning in sociology, and are merely convenient as labels for discussion purposes. Nonetheless the class divisions themselves were real, and were regularly exhibited through marriage preferences for spouses from one's own class. They were equally well to be seen in cities like Madras or Bangalore, and were evident too in the other British hill-stations throughout India.

Here a slight digression is in order. Throughout the British Isles social identity has long been founded on the three obvious and highly important bases of religion (Protestant, Catholic, Jew), language (English, Gaelic, Welsh), and national character ('There was an Englishman, a Scotsman, and an Irishman…'). But in England proper, the Anglo-Saxon realm, the criteria of identity have been primarily religion, region and social class. Often cited as 'typically' English in popular books and magazines, for example, are the Cockneys, actually a tiny minority of the nation. In fact, far from being your typical Englishman, a Cockney is rather peculiar: he was born at the centre of one of the biggest cities on earth, he is probably always a Protestant, he belongs to the working class, and he speaks an easily recognizable dialect that is admittedly picturesque in vocabulary though short on dentals and aspirates. One cannot think of a Cockney speaking Yiddish, taking his holidays in Monte Carlo, or even buying land in Cornwall. His strong social identity finds support only in the company of other Cockneys.

What happened to the foundations of social identity when British people settled in the Nilgiris? A fair proportion of these people were from Scotland or Ireland, but were not peasants of those realms and did not (according to the Census of India) have Gaelic, Welsh or similar languages as mother-tongues. English was the *lingua franca*, even for the handful of French and German missionaries. Nor was regional identification at home important to people who would spend most of their adult lives in India. One might joke at the

Ooty Club about the rural origins of so-and-so, or mention that his wife was a Geordie, but in the business of day-to-day living in the Nilgiris District social class was the only aspect of social identity that still meant anything. Virtually everybody who 'counted' was conscious of being a subject of the Queen and an Anglican too.

A generalization heard early in this century had a certain amount of truth behind it till at least the 1920s. It was that Kotagiri was the preserve of missionaries, Coonoor of boxwallahs (traders), and Ootacamund of I.C.S. and retired military personnel. It was never strictly true, of course, particularly in the case of the large district headquarters (1921 population, 19,467). But patterns of ownership and domestic architectural styles show some territorial variability that lends credence to the assertion. And even in Ootacamund one found that some out-of-the-way corners were almost reserved for particular social categories. Thus Lovedale and Fernhill on the south were where the retired Anglo-Indian officers of the Forest Department, Public Works Department and Post and Telegraphs were to be found. Homes built close to Government House, on the east, belonged to a much higher category of people. Even the sloping graveyards of St. Thomas' and St. Stephen's Churches reflected social stratification.

While much gossip and some more substantial published evidence support the validity of our classification of social units, two queries perhaps arise from the Table 14.1; namely, how we distinguish upper from lower officials, and why we differentiate between Protestant and Catholic missionaries. The upper administrators were in general the heads of all units of local government; i.e. the Collector (of Revenues), the Tahsildar, the Police Inspector, the Sub-divisional Magistrate, the Chief Forest Officer, the District Medical Officer, and other I.C.S. people. Most of these upper males, like some of the upper middle class, had university educations. The lower officials were basically all the Europeans working under these people. Below them again came the Protestant missionaries who (as we shall see below in the case of Henry Gulliford) tended to have working-class origins in Britain, as did the traders. Catholic missionaries were definitely considered a step lower than this because, first, they were usually 'foreign', i.e. Irish, French or Belgian; secondly, most had village origins and poor education; and thirdly, theirs was an unpopular, minority religion for the British. All missionaries found themselves ranked below the government's chaplains. One long-time resident of the Nilgiris and neighbouring districts, after praising the selfless and rough existence of missionaries, added: 'The Government chaplains, on the contrary, were supposed to make a pretty good thing of it; in no sense was their work a 'mission', nor did it lead them into the by-ways of hardship and risk. Socially they considered themselves many cuts above their missionary brethren, who devoted themselves entirely to their calling, hardly taking any part in the social life or amusements of a station' (Handley 1911: 84). Admittedly some of our other categorizations are questionable; the retired, for example, often

TABLE 14.1
The Class System among Nilgiri Europeans, 1850-1950

Class	Social Units
'Upper'	Upper Administrators (I.C.S.); Generals; Lawyers; Bishops and Archdeacons, if Anglican
'Upper Middle'	Lower Officials; Planters; the Retired; Chaplains; Army Officers
'Lower Middle'	Traders ('Boxwallahs'); Protestant Missionaries; Teachers; N.C.O.s
'Upper Lower'	British Other Ranks (Army Privates); Catholic ('Foreign') Missionaries
'Lower Lower'	Anglo-Indians (i.e. Eurasians; most urban occupations)

found themselves through relative poverty in the lower middle rather than the upper middle class.

During the century 1850-1950 society in England was stratified into seven classes, as both sociologists and English respondents seem generally to agree. The most usual labels for these were aristocracy, upper middle class, middle class, lower middle class, skilled working class, unskilled working class, lower working class (Gorer 1975:160). We cannot assume an identity, however, between any one of these classes and a similarly labelled class among the Nilgiri Europeans. For them the upper class of powerful and well-educated administrators was still a far cry from the aristocratic upper class at home (who were never more than one per cent of the population there); and at the other end of the spectrum the lower class of Anglo-Indian workers in the railways, schools, post offices, etc. was in most respects recognized as higher in qualifications and achievements than the vagrants, petty criminals, day labourers and physically incapacitated who made up the lowest ranks of British society and who, for obvious reasons, could never migrate to India.

Social identity in Britain during this period was founded on the three pillars of faith, class, and region of origin. The English living in the Nilgiris felt a much more precise and unitary identity, on the other hand, and for several reasons. Virtually all were Protestants of one sect or another (although Irish and French Catholics were scattered through the European community). Jews were not encountered. Further, one's region of origin became of lessened significance as one's years in India lengthened and one's dialect perhaps changed. Primary adherence as a consequence was to a particular class identification, with the other two factors of faith and region of origin usually holding little relevance. Interestingly, social identity in Britain itself has moved in much the same direction since about 1950.

Class was largely determined by birth, education and profession. In Britain the class system was never thought to extend to foreigners: French or Italians, even if settled in England, were not thought of as 'fitting into' the English classes. Similarly in India nobody considered giving the Indian natives a ranking in British-Indian society. Even the occasional prince, of whom several were to be found around Ootacamund, was a social oddity rather than a member of the upper class. It has long been true of the English middle classes, whether at home or abroad, that it was the married women who usually initiated invitations for hospitality in their homes. They would normally invite neither their Indian acquaintances (who would never dream of bringing their wives or daughters) nor the members of other British social classes, since one tried to invite people one would feel at ease with in conversation and outlook. There was perhaps a little more latitude among the men when in their clubs or masonic lodges; but not much more. In formal situations the English will normally choose for companions members of their own social class (Gorer 1975:163).

It is interesting to see how the hierarchy was quite transformed by Indian independence (1947): essentially none but the former middle classes remained. Now the planters (sometimes called the plantocracy) sprang into the position of an extremely wealthy and worldly upper class, while the retired, the schoolteachers and the missionaries (both Protestant and Catholic) became the new middle class with the total disappearance of British administrators, traders and army personnel. In fact, most of the British population left around 1947 for Cyprus and the Channel Islands, which were virtually tax-free, or Britain and Australia, which were not. A few were to return later.

We should not however over-emphasize the effect of this structural change. What the Nilgiri community was faced with after independence was less a simplification of the European classes and a re-allocation of the social categories, than a great depopulation of Europeans of all classes. At the time of writing, some forty years after independence, it seems that social class is hardly relevant to the few remaining Europeans (many of them now in their eighties). Instead, what we find in the Nilgiri towns is a cast of memorable characters, whose idiosyncracies were so well portrayed by Mollie Panter-Downes in her 1967 book, *Ooty Preserved: A Victorian Hill-Station in South India* . For these people, elderly or widowed as they all are, no weddings await them in the future and hence one prime arena for class discriminations no longer enters into their lives.

Whiskey and Weeds

The Reverend Henry Gulliford's diary reveals that the social life of the British just before the Second World War was scarcely any different than it had been in Indian hill-stations sixty years earlier. Edmund C. P. Hull, who had been a coffee-planter in this area in the mid-nineteenth century, recorded in his rather quaint handbook, *The European in India*..., that there was plenty to keep one occupied on the hills—if one was English, that is:

> Within easy distance will be found different points commanding exquisite and extensive views of the country below. Such points are made the resort of frequent and hilarious picnic parties, which form one of the chief delights of hill-life. In addition to such enjoyments, dancing and croquet parties, concerts, flowershows, and even sometimes "exhibitions," break the monotony. And by keeping always on the move in the cool and bracing air,—now gun on shoulder, on the look out for hares or partridges or larger game, or, again, to escort some of the gentler sex to admire a waterfall or some newly discovered prospect,....the visitor may well contrive to enjoy himself thoroughly (Hull 1878: 62).

Gulliford seems to have amused himself with much the same pursuits (except for the hunting). Further details that correspond precisely with middle class British life in Ootacamund may be found in a chapter of Hull's book devoted to 'Social Custom', which covers in outline:

> Calling—Whom to call on. "Griffins". [These are people who call on everybody; thus p. 176: 'Hunting in couple, and having first taken down the names of the people living in every house, from the Directory, off they set in a "gharry" one morning, and send in their cards from house to house with the usual degree of success', and so on.] Then the chapter continues with Clubs—Government House—Governor's levees, breakfasts.—Receptions.—'At homes'.—Dinner parties.—Borrowing system.—Precedence.—Garden parties.—Ladies' lunches.—Bandstands.—'Social mourning'.—Anglo-Indian hospitality fifty years ago and now.—Parvenu.—Advice to Ladies.

Altogether a thoroughly worthwhile guide!

Its author shows his hand all too clearly here when discussing the parvenus, English people of lower class origins who are trying to pass as respectable member of 'Anglo-Indian' society by every ruse they can think of. Not approving of such social mobility, Hull gives his readers all the clues necessary for identifying such presumptuous folk. But he, like most writers on the English in India, really tells us nothing about the lower-class Britons, especially common soldiers, who were always present in some abundance.

Nor for the matter does the noted traveller, Sir Richard Burton, in his account of Ooty society, written just a couple of years prior to his famed visit to Mecca disguised as an Arab:

> A brief account of the Neilgherry day will answer your inquiry about the existence of amusement. We premise that there are two formulas, one for the sanitarian, the other for the pleasure-hunter.
>
> And first, of Il Penseroso, or the invalid. He rises with the sun, clothes himself according to Dr. Baikie, and either mounts his pony, or more probably starts stick in hand for a four mile walk. He returns in time to avoid the

sun's effects upon an empty stomach, bathes, breakfasts, and hurries once more into the open air. Possibly, between the hours of twelve and four, his dinner-time, he may allow himself to rest awhile in the library, to play a game at billiards, or to call upon a friend, but upon principle he avoids tainted atmospheres as much as possible. At 5 P.M., he recommences walking or riding, persevering laudably in the exercise selected, till the falling dew drives him home. A cup of tea, and a book or newspaper, finish the day. This even tenor of his existence is occasionally varied by some such excitement as a pic-nic, or a shooting-party, but late dinners, balls, and parties, know him not.

Secondly, of L'Allegro, as the man who obtains two months' leave of 'absence on urgent private affairs' to the Neilgherries, and the Penseroso become a robust convalescent, may classically and accurately be termed. L'Allegro, dresses at midday, he has spent the forenoon either in bed or *en deshabille*, in dozing, tea-drinking, and smoking, or, if of a literary turn of mind, in perusing the pages of the 'The Devoted' or, 'Demented One.' He dilates breakfast to spite old Time, and asks himself the frequent question What shall I do to-day; The ladies are generally at home between twelve and two, but L'Allegro, considering the occupation rather a 'slow' one, votes it a 'bore'. But there is the club, and a couple of hours may be spent profitably enough over the newspapers, or pleasantly enough with the assistance of billiards and whist. At three o'clock our Joyful returns home, or accompanies a party of friends to a hot and substantial meal, termed tiffin, followed by many gigantic Trichinopoly cigars, and glasses of pale ale in proportion.

A walk or a ride round the lake, is now deemed necessary to recruit exhausted Appetite, who is expected to be ready at seven for another hot and substantial meal, called dinner. And now, the labours of the day being happily over, L'Allegro concludes it with prodigious facility by means of cards or billiards, with whiskey and weeds.

The routine of life is broken only by such interruptions, as a shooting-party, an excursion, a pic-nic, a grand dinner, *soirée*, or a ball.... A dinner where ladies are admitted is, by L'Allegro, considered an unmitigated pest; and those who dislike formality and restraint, scant potations, and the impossibility of smoking, will readily enter into his feelings.

The Ootacamund *soirée* happens about once every two months to the man of pleasure, who exerts all the powers of his mind to ward off the blow of an invitation. When he can no longer escape the misfortune, he resigns himself to his fate, dresses and repairs to the scene of unfestivity...

There are about half-a-dozen balls a year on the Neilgherries, the cause of their infrequency being the expense, and the unpopularity of the amusement amongst all manner and description of men, save and except the 'squire of dames' only...(Burton 1851: 297-301)

At about the time when Burton wrote this unenthusiastic account, a music hall was opened in Ootacamund. It seems to have existed as a place for light entertainment and balls throughout the latter half of the nineteenth century, and is still standing in dilapidation today. It undoubtedly catered to the lower middle and upper lower classes, particularly as these people were not acceptable at the Ooty Club. Founded a few years earlier, in 1842, that Club's initial code of rules, drawn up by Dr Baikie and some military officers, states quite clearly: 'All members of H M and the Hon'ble Company's Civil, Military and Naval Services, Gentlemen of the Mercantile or other professions, moving in the ordinary Circle of Indian Society, are eligible as Members....' (Price 1908: 246).

Horse-racing was conducted at Wellington from about 1875 to 1905, at Pandalur during the 'gold rush' of 1879-82, and at Ootacamund from 1882 until the present. An unhealthy section of John Sullivan's Lake was filled in to create the race-course now known as Hobart Park. Amateur theatricals were also very popular, especially during the period 1880-1940. 'The Season' in particular has always been an annual attraction in the Nilgiris: 'In May and June there is a constant round of balls, theatricals, dinners, races, gymkhanas, polo, tennis, cricket, etc., but people may live as quietly as they please, giving themselves up to an outdoor life of hunting, shooting, and fishing' (Savory 1900: 335-6).[2] Such 'sportsmen' were constantly depleting the wildlife on the hills. (One remembers the ungrammatical critic who recalled, 'We know from Orme that some of the men shot tigers in wide-brimmed top hats, sometimes with a *pugri* draped round them'; Stanford 1962: 81.) The famed Ooty Hunt was established too for the upper and upper middle classes (and the occasional prince). It hunted jackals with English hounds in the best Leicestershire style. Late in the nineteenth century a novel way of killing time came to the Nilgiris: Sir Neville Chamberlain invented snooker in Jabbalpur, and it was first played publicly in 1877 or 1879 by Sir Neville and his friends in the Ooty Club, whence it was to spread around the world. In all of these activities class segregation was an important consideration. Religious worship and the several masonic lodges were perhaps the only arenas where it was unemphasized.

The origins of the Anglo-Indian community (*sensu stricto*) were frankly discussed by Major Henry Bevan, an early nineteenth-century resident of the Wynaad, in the following language:

> I knew only of one instance of a marriage between a European and a native woman, though I have known of several between Europeans and the Anglo-Indians, or half-castes. The isolated marriage to which I refer, was contracted between a gentleman and a Mussulman woman whom he had previously kept as a mistress....
>
> The paucity of English ladies in India, more especially at the remote stations, has led to the formation of unmatrimonial connexions between European officers and native women. Partly from the necessity of the case and partly from the difference of national customs, these *liaisons* are not deemed so immoral in India as they are in Europe. The mistresses are obtained both from the Hindu and Mussulman races, and they are often sold to their masters by their needy relatives. Offspring is anxiously desired by the mothers, as it establishes a kind of claim to continued protection, but this often proves a source of great anxiety to the fathers in after-life....

> The progeny from the illicit intercourse of Europeans and natives is very numerous, especially about Madras. Many of these Anglo-Indians have attained wealth, rank, and respectability; the stigma which government affixed to them as a class by the regulations respecting the military fund, has been recently removed, but it will be long before the prejudices of the white aristocracy at Madras can be wholly effaced. There are male and female asylums for orphans, most, if not all, of whom are Anglo-Indians; these institutions are admirably managed; the children are well educated, and, at a proper age, apprenticed to different trades (Bevan 1839, I:18-21).

These comments, and the existence by the end of the century of a large Eurasian population, with English as its mother-tongue, show very clearly that there were other diversions for the Englishmen in India than those that Hull and Burton saw fit to suggest. The fact is underlined by the Reverend Henry Gulliford, who in 1917 sometimes preached to British and Commonwealth troops at an isolation hospital near Doddabetta Peak: 'In this camp the men suffering from venereal disease are isolated. There are five or six hundred men here, and 150 came up last week. Most of these men get the disease when they are on furlough' (i.e. on leave in one of the South Indian towns). Further, according to the Bishop of Madras (5 August, 1918), 'they say no arrangements are made for the soldiers to visit women who have been examined [medically] in Bangalore, while it is well known that this is done'. No doubt the presence of large numbers of troops contributed to an increase in Anglo-Indian offspring. Again during the Second World War some 700 British troops were stationed in Ootacamund, and one of their favourite pastimes was to take Indian girls to the large pine forest on One Cairn Hill, two miles from town, where they found a natural bed of moss and pine needles. As one wit observed long ago, 'Necessity is the mother of invention and the father of the Eurasian'. What the extent of such military activities in the Nilgiris really was is impossible to say; history is silent; but there were certainly prostitutes—then and now—in Ootacamund and Coonoor. Wellington Cantonment once even had a regulated brothel. And there are low-status Anglo-Indians everywhere, many of them born during the Second World War. They were in an ambiguous situation, for while they liked to emphasize their English speech and customs, the British definitely did not think of them as fellow-countrymen. Indeed the British, as the creators of this community, felt guilt towards them; while the very Englishness of their behaviour and the steadfastness of their service brought in question the British belief that only those of 'pure race' were superior enough to be in positions of authority in India (Greenberger 1969:54). In early days British disdain for the Eurasians doubtless was fed by the knowledge that many were descended from slaves purchased by European residents of India, especially of Goa, during the eighteenth and early nineteenth centuries.

Here we can usefully turn to the mute statistics in Table 14.2 which demonstrates all too clearly that in the Nilgiris District (1) whenever a census shows a big increase (at least 9 per cent) in European males over the previous

TABLE 14.2

Change in the European and Eurasian Population

famine ↓ immigration

	1871	1881	% of change	1891	% of change	1901	% of change
EUROPEANS							
Males	818	1,127	+38%	922	−18%	2,200	+139%
Females	521	571	+10%	873	−53%	1,564	+ 79%
TOTAL	1,339	1,698	+27%	1,795	6%	3,764	+110%
EURASIANS							
Males							
Females	523	527	+ 1%	748	+42%	678	− 9%
TOTAL	273	485	+78%	489	+ 1%	539	+ 10%
	796	1,012	+27%	1,237	+22%	1,217	− 2%
DISTRICT TOTAL	49,501	91,034	+84%	101,138	+11%	112,882	+ 12%
S.E. WYNAAD	25,440		—		—		—

	1911	% of change	1921	% of change	1931	% of change	1941	% of change
EUROPEANS								
Males	2,553	+16%	1,649	−35%	1,836	+12%	1,436	−22%
Females	2,074	+33%	1,876	−10%	1,410	−25%	1,825	+29%
TOTAL	4,627	+23%	3,525	−24%	3,246	− 8%	3,261	0%
EURASIANS								
Males								
Females	707	+ 4%	649	− 8%	644	− 1%	942	+46%
TOTAL	626	+16%	636	− 2%	984	−55%	867	−12%
	1,333	+10%	1,285	− 4%	1,628	+27%	1,809	+11%
DISTRICT TOTAL	118,618	+ 5%	126,519	+ 7%	169,330	+34%	209,709	+24%

↑ World War I

↑ World War II

(Source: *Census of India*)

census figures, it also shows a big increase (9 per cent) in Eurasian females—1881, 1901, 1911, 1931; (2) whenever a census shows a big decrease in European males since the previous census—1891, 1921, 1941—it also shows a big decrease in Eurasian females—1941—or else the tiniest increase of 1 per cent or 2 per cent in Eurasian females—1891, 1921—far lower than the natural decennial rate of increase; (3) there is scarcely a comparable correlation between the statistics for European females and Eurasian males: a big increase was noted among European females in 1881, 1891, 1901, 1911, and 1941, whereas big increases occurred among the Eurasian males only in 1891 and 1941. Similarly, a big decrease was recorded for European females in 1921 and 1931, while a very slight decrease was recorded for Eurasian males in 1901, 1921, and 1931. It should further be noted that the entire District population increased between 1871 and 1881 by 84 per cent, partly by the transfer of the southeastern Wynaad from Malabar to the Nilgiris, partly by the increase of births over deaths, and partly because of heavy immigration occasioned by a famine on the South Indian plains in 1877. Finally (4) we should note that at no census did the trend for Eurasian females closely follow that for Eurasian males; 1921 is the only case of a vaguely similar trend in the two sexes of this group, and the males then showed a decrease of -8 per cent since the previous census while the females showed an increase of +2 per cent. On all other occasions Eurasian trends for the two sexes were distinctly further apart than this.[3] (The big increase in Eurasian male workers in 1941 arose from immigration to replace European males who had gone elsewhere for war duties.)

Plantations were one setting in which Eurasians were born and grew up, although they were more commonly an urban population. Their appearance in plantations has to be understood in the light of the isolation that European planters experienced. These men were as rigidly separated from other European society as were the missionaries, though for quite different reasons. It was relative poverty combined with moral scruples about drinking and several other European amusements that served to keep the missionaries almost completely segregated from other Englishmen. The planters found themselves living and working in quite isolated tracts often 20 to 50 miles away from the main towns of the Nilgiris. These people did not suffer from poverty, of course; one planter is said to have boasted that he spent more on booze in a month than his missionary neighbour earned in a year. Some of the planters 'went native' and formed families of Anglo-Kuṟumbas, for example, or indulged themselves in similar liaisons of a more temporary nature with selected tea-pluckers. Quite recently, when a British planter announced to his plucking force that he was finally going to marry an English lady, the chorus that greeted the news can only be translated as, 'What about us.'

While lonely planters and British Other Ranks might easily slip into such liaisons, the other European groups found it much more inconvenient to do so, and in general did not want close relations of any kind with the natives. The social and racial segregation of the English seems to have begun in infancy:

> Left too much among [Indian nurses], an English child will only too soon learn the native language, and, as a necessary consequence, lose much of the innocence of childhood. The acquaintance of English children with the vernacular tongues, cannot, in my opinion, be too strongly deprecated... (Hull 1878: 141).

One wonders whether respectable citizens like Edmund Hull were even aware of the loss of childhood innocence by young Indian girls living at the various British army cantonments.

Nothing illustrates the racial bias of many British settlers more clearly than the story of Badaga Brown. That wasn't his real name, of course. He was a ne'er-do-well Etonian who had fiddled his way on a ship to Australia at the beginning of the century, but in later life found himself in Ootacamund. There he took up the unheard-of-profession—for an Englishman—of being a usurer to the Badagas. It is said that he also contrived to sire a number of 'Badaga' children. When quite an old man he was reclaimed by an English spinster, whom he married, and who made him a Roman Catholic. After a while he started disposing of his property and was preparing to set up his final home in Bangalore (a city blessed with segregated cemeteries), when somebody asked him why he could not see out his remaining days in the peace and beauty of the Nilgiri Hills. 'What,' he retorted, 'and be buried among all these black buggers?'

One of the more persisting social problems of Ootacamund was that it was not on the route of the 'fishing fleet', an annual invasion that brought out boat-loads of nubile English and Australian girls to spend the winter months in such promising spots as Bombay, Colombo and Calcutta. As a result, it seems there was something like a permanent shortage of marriageable English ladies in the Nilgiris. And those who might be considered young enough to be 'available' were sometimes of such an independent bent that marriage with them would be no easy accomplishment. This of course runs counter to the more general image we have from literature of long-suffering, meek, faithful Victorian womanhood in India.

Thus there was in the Nilgiris, over a century ago, a very wealthy British planter with a very nubile daughter. The neighbourhood of their estate seems to have been infested with 'creepers', i.e. English apprentices and other young hangers-on. One day it became apparent to the *pater familias* that his daughter was in the family way, although not yet married. He therefore called a meeting of all the creepers within reach and asked whoever it was, as an English gentleman, to own up. Immediately one rather astute young fellow stepped forward, and quickly found himself married to the young heiress. It was only after she gave birth to twins, both 'very black' (as the locals say), that people realized how astute he really was. The two boys died mysteriously within a day of their birth and were buried in the back garden; a curious instance of the principle of death before dishonour which inspired the imperial British in the nineteenth century. *Sic transit gloria Ootacamundi*![4]

A more orthodox marital arrangement on the plantations is illustrated by the story of the way in which the Morris family and their British managers opened up six coffee plantations on the nearby Biligiri Rangan Hills (Morris 1953). One sees there a pattern that must have been repeated many times over in the early history of the Nilgiri and Coorg plantations too. A pioneer would work there, first with his brother; then one or two English 'creepers' would join them; next the plantation owner would marry a 'creeper's' sister; and some of their children would remain on their plantations to repeat the process.

Men of the Cloth

The first Protestant missionaries, those of the Church Missionary Society in Ootacamund, the American Arcot Mission in Coonoor and the Basel Mission in Ketti, all came to the Hills in the 1830s and 1840s because of indifferent health (Francis 1908: 127), and the Basel people were clearly attracted by the great similarity between the scenery of the Ketti Valley and that of Appenzell in north-east Switzerland. The missionaries of the nineteenth century seem to have carried into their work an arrogant self-righteousness which can only have made it more difficult for them to win converts. The Reverend J. F. Metz, for example, was the first scholar to learn Badaga properly, and was also the author of a still useful ethnography based on his own observations of the Badagas and the hill tribes. His practice was to tramp from village to village—a different one almost every day—preaching as he went. While this brought him into intimate contact with the Badagas and some other groups, his contempt for them bordered on paranoia. Once, when challenged to walk on fire, Metz 'replied that a much greater wonder than that, was, that I should not cease to love such squalied fellows as they were; and that too, in spite of all the abuse they daily heaped upon me' (1864: 55). His great lack of empathy was exemplified by his own account of how, 'one day, after four Badagas had been burned in one house...people were eager to hear the Word and ask questions about it, so that it was quite refreshing to me' (Basel Mission 1866: 78-9). He used to tell his friends that the fact that he 'was congenitally deprived of the senses of taste and smell' was what allowed him to work for so long with the Badagas. The Badagas nicknamed him 'three-quarters God'. Not all of the missionaries were taken seriously even by their European congregations. A certain Reverend Dr Sayers complained from a pulpit in Madras around 1870: 'What I want in the bag is rupees, not a scrap of paper saying "How are you, Sayers, old cock?"'

By the early part of the present century, many of the missionaries had withdrawn into the circle of their own acquaintances, and had minimal contact with Indians except when duty called them before a congregation of South Indian Christians. Take Henry Gulliford, for example, a Methodist who spent about twenty years in the Nilgiris, and whose 48-volume personal diary (1917-36) is an untapped mine of social trivia. He begins it in the style of a conventional eighteenth century novel:

> I, Henry Gulliford, was born at Wheddon Cross, in the parish of Cutcombe, near Dunster, Somerset, on the 15th of November, 1852. My father is called Robert, and my mother Jane. He is a watchmaker by trade. I was christened at the Wesleyan Chapel, Wheddon Cross, but was sent to the Church school, as there was no Wesleyan school there. I felt early religious impressions, but they wore off. I lived at home till I was about eleven years old, when I went to Luxborough with my sister. Before I went I was again christened, at Cutcombe church, by the request of the Revd. M. King, and answered for myself, though I know but very little about it. I stayed there

nearly twelve months, when I returned to Cutcombe. I remained there till the year 1866, when I went to school at Cullompton, for a year. On leaving there I was apprenticed to the printing at Mr Cox's, *Free Press* Office, Williton, where I am now. I shall have reason to thank God, through all eternity, I trust, for ever coming here; for here it was I became converted. On coming to Williton first, I was, what was considered, a good lad, but there was no religion in me....

Within a few pages he is alternately preaching at local prayer meetings and castigating himself for his want of love of the Lord. This goes on for three volumes, and then begins 'My Voyage to India.—Friday, September 21, 1877—About nine o'clock a party of ten of us assembled at the Mission-House, and soon after 9.30 we started for the Docks....' and so it went on. The later volumes, dealing with Gulliford's years in the Nilgiris, document an unending round of tea-parties, picnics, committee meetings, and preaching engagements. I have analysed two six-month periods in his diaries, one at the beginning of his Nilgiri stay and another towards the end. During early 1917 the writer recorded 223 separate encounters with Europeans, 4 with Tamils, one with Badagas, one with a Kanarese, and two with a North Indian. During the same months of 1934, he had 346 encounters with Europeans, 6 with Tamils, one with a Kanarese and two with North Indians. The Todas and other tribesmen are never mentioned. During the entire period examined Gullford never once ate a meal with Indians, either in their home or his. (I have not here counted as 'encounters' the mention of activities with his wife, nor the congregations of unspecified membership to which he preached.) Gulliford may have been effective as a church administrator and public speaker, but as a spiritual ambassador to the local people he was a dismal failure. And there were plenty more like him; some are still around (Cousins and Cousins 1950: 725-6).

In defence of the missionaries, however, it must be pointed out that the Basel Mission in numerous villages, and the Roman Catholic orders in all of the towns, made a valiant and rather effective attempt to make education available to a previously non-literate people. There are a few Badaga villages which have had schools in them now for a century-and-a-half. This was no mean achievement, whatever motivations may have lain behind it. There were also sporadic attempts to run mission clinics and hospitals for the rural population.

In summary, we can see that the impact of the European settlers was an extremely varied one. As missionaries they made an ecclesiastical, if not a spiritual change, in a small minority of the native population; as soldiers and planters they introduced new genetic material to the local people; as planters and administrators they developed and greatly expanded the cash economy; as military and retired people they brought in new standards of etiquette, and new types of public entertainment; as officers and administrators they built towns and roads. And in all of these activities, except for the religious and genetic changes, John Sullivan, the founder and first settler, played a key role.

LONGWOOD SHOLA

Kotagiri, Nilgiris

Dawn, that calls the soul from sleep,
Brings the hungry bulbul's *yeep*...
Yeep...and that most final sound,
Peaches plumping on the ground
As his reckless slashing sabre
Ends a season's cosmic labour,
Laying low high fruitage ripened
For the early riser's stipend.
  And, in truth, why should there be
Less than prodigality
When the wakening woods are choric,
And the firmament plethoric,
With the promise of abundance
After all things step to one dance
When the domineering rain
Drowns all else in its own strain?
  Meanwhile, morning's invitation
Calls, through garden and plantation,
To the shades of Longwood *shola*,
Where the feet weigh scarce a *tola*
Poised on pathways thickly strown
With the leaves of seasons gone,
Stirring from deciduous death
Nature's vitalising breath.
  Overhead slim branches swirl
As the bright-brown barking squir'l
Plays at gentleman-and-lady
To and fro, and shy cicadae
Bandy wiry-shrill persuading
(*Twing twing TWANG*, such serenading!)
Universal invitation
Myriad-masked throughout creation.
  Here and there fawn-flowered spirea
Stands as an inspired idea
In the brain of earth, a relic
Of the ministry angelic
From whose touch all beauty springs
Into joy of leaves and wings.
  Even such rapture raises me,
All becomes a Mystery.

Water among pebbles tinkling
Needs no ceremonial sprinkling
Here to consecrate an altar,
Holy scripture, holy psalter.
Yea, past all dogmatic fission,
Here is ritual provision:
Multi-coloured cloths and bands,
Holy water for the hands
Flowing neither cold nor torrid,
Sacred ashes for the forehead
Gathered where the flame of day
Burns a glory into clay.
Yea, when hearts have learned the craft
That can break love's casual shaft,
And can rise in quiet woods
Into Love's immortal moods,
Such exalted tenderness
Bends the dreaming brow to bless
That the dragons of desire
Vanish in creative fire,
And futilities of thought
Scatter dustily to nought,
While the soul, in deep repose
Lifted into vision, knows
Spirit-freedom, loosed from sense,
Joy that needs no penitence.
Ah! such moments yet must fade
Till the soul all debts has paid
Unto darkness, and can look
On the earth as on a book
Shining, throbbing with the hymn
Of its heavenly paradigm.
Still, a reminiscent beat
Times a poet's homing feet
Where a mountain-forest river,
Swift on sand, round granite sliver,
Chants high deeds for panegyrics,
Lilts alluring themes for lyrics,
Under boughs that richly shed
Nourishment for heart and head.
Fancy's fruitage roundly ripened
As the early riser's stipend.

—JAMES COUSINS[5]
(from Cousins 1940: 432-5)

NILGIRI SUNSHINE

Outside the Clubhouse
The old fellows sit,
Feeling ready for tiffin
And wonderful fit;
The morning round over—
A decent round too—
In the Nilgiri sunshine
Sheer gold out of blue.

They're old and they're past it?
By no means at all!
At the fourteenth Sir James
Drove a deuce of a ball,
And the Colonel's approaching
Was great, on my oath—
But the Nilgiri sunshine
Had a big hand in both.

The Nilgiri sunshine!
Sheer gold out of blue,
Heart-winning, heart-warming,
Heart-waking anew;
The flowers for its fairies,
The birds for its friends,
And its message—'Life's excellent;
Life never ends.'

The sunshine of England
It shows in its way
That winter's not summer,
That night isn't day;
But it can't call the colours,

It hasn't the art
Of the Nilgiri sunshine
That kindles the heart.

Outside the Clubhouse
The sun blazes down
On the jolly old stagers
Rose-ruddy and brown;
Happy as sandboys
They sit there and thrive
In the Nilgiri sunshine—
That keeps 'em alive.

There's a breath in the wattle,
A stir in the gums;
Down his bright stairway
The-sun-god comes—
Down to his Nilgiris,
Azure and gold,
Where nobody ever
Need fear growing old

—HILTON BROWN
(from Brown 1936: 28-9)

NOTES

1. These facts about Sullivan's childhood have only recently come to light, and I am indebted to the India Office Library and H. Lincoln Townsend, the Hon. Secretary of the Nilgiri Library, for them. The rest of my discussion of Sullivan has already been presented elsewhere (Hockings 1973).
2. For a fuller account of the May activities, 'the Season' in Ooty, see Stone (1925: 85-105).
3. Rank order correlations do not tell us much, because they do not properly take into account the size of demographic variations. In Table 14.3, showing values for rho, however, it will be noticed that Eurasian female fluctuations have a higher correlation with European male fluctuations than with Eurasian male fluctuations; but not as high as the correlation between European male and female fluctuations. It should be noted that in the nineteenth century 'Anglo-Indian' referred to British People born in India, whereas after 1900 it meant 'Eurasian'.

TABLE 14.3
Value of rho

	European male	Eurasian male
European Female	+0.79	+0.60
Eurasian Female	+0.60	+0.43

4. This pun must be quite ancient (Fletcher 1911: 19).
5. From Cousins 1940: 432-5. *Bulbul,* not the Persian nightingale, but a favourite crested bird in India. *Shola,* ancient forests on the Nilgiri Mountains, South India. *Tola,* Indian unit of weight, three-eighths of an ounce.

The Cultural Ecology of the Nilgiris District

PAUL HOCKINGS

15

There are strong grounds for supposing that the Kurumbas once occupied and cultivated the plateau of the Hills, and were driven thence by the Todas into the unhealthy localities which they now inhabit, on the pretext of their being a race of sorcerers.... Several spots near the Badaga villages bear the name of 'Motta' to this day, and traces of houses are still visible; and in one place a stone enclosure for buffaloes is to be seen, which, as I gather from an old piece of Badaga poetry, formerly belonged to a rich Kurumba, who was murdered by the Todas, at the instigation of the Badagas.

—J.F. Metz (1864: 122-3)

The study of the relationship between populations and their environments has always been a concern of anthropology. Nearly a century ago it was common for many ethnologists and human geographers to explain the pattern of a culture as being environmentally determined. Such views are no longer current, and for a variety of reasons. The interest in man's environment still persists, nonetheless, and in recent years has been expressed in the theory of cultural ecology. In this approach, certain environmental features are seen to interact with aspects of a particular culture, but in no sense are they considered to have caused those cultural aspects: 'The immediate causes of cultural phenomena are other cultural phenomena' (Kroeber 1939:1). The approach of cultural ecology allows one to explore how culture is influenced by environment, and how an environment is modified by a culture. It analyses equilibrium processes in an ecological system which tend to keep it from fluctuating too much, but which may lead into irreversible processes of change amounting to cultural evolution on a local scale. The processes of equilibrium and change can be seen operating in the Nilgiris.[1]

Few districts in the world approximate the ideal natural laboratory for behavioural scientists more closely than does this one. Although a very small area, it has long been famous in the annals of anthropology, botany and travellers' journals for the variety and interest of the conditions to be found there. The paradoxes of the region have even been noted in a curious expatriate

English literature that grew up during the mid-nineteenth century: here we have a place with a temperate climate that is squarely in the tropics; here a plateau that is comfortable and homely in a very English sense, yet surrounded by forests long the home of tigers and elephants; here a territory that the British came to regard as their special preserve—for had they not built its towns, public institutions and roads?—yet it was a place they shared comfortably enough with over a dozen of the most varied tribal cultures.

From a more scientific point of view, the Nilgiris present other, unique attractions. This mountainous district, located in a corner of Tamil Nadu State at the juncture of the Eastern and the Western Ghats, ranges in elevation from 300 to 2,638 m (1,000 to 8,640 ft). Its flora concomitantly ranges from tropical through semi-tropical to what is called tropical montane, and is indeed very like what can be found by botanists in the lower parts of the Himalayas some 2,500 kms to the north. The zoologist may see a rich collection of the larger mammals here, including the tahr or 'ibex', the bear, the wolf (probably now extinct), the gaur or jungle buffalo, the tiger, the leopard and the elephant, not to mention a variety of deer, canines, and wild cats.

It is with the primates, however, that this book is concerned; and here too as rich a variety is encountered as might be hoped for anywhere. The list is impressive: tree shrew,[2] slender loris,[2] lion-tailed macaque, bonnet macaque, black langur, common grey langur, and humans (some eighteen tribes, plus immigrants).

All of these occur within an area of about 40 by 65 km; scarcely larger than the Lake District, and slightly smaller than Rhode Island State. On all sides of the Nilgiris the presence of mountain slopes or of sparsely inhabited forest has the effect of setting the Nilgiri population off from that of neighbouring districts, so that we are dealing with a clearly bounded, relatively isolated population unit.

In the vast literature on the Nilgiri Hills that is now available, there is some discussion of the demographic features of this area, but only in the valuable reports connected with the Census of India (cf. Hockings 1978). The District, as well as being the most exhaustively studied of any comparably-sized locale in Asia east of the Holy Land, can boast one of the longest census runs in any region of the non-western world, with relatively reliable population counts every few years from 1812 (Keys) to 1981 (not yet published). While this is not the place to enter into an analysis of this statistical goldmine, some observations about the demographic character of the District may be illuminating.

Were we to make a thorough comparison between this and any other mountainous zone of the Old World, we would soon be struck by the exceptional character of the Nilgiris. For in many of the other mountainous terrains some of the local inhabitants, whether they be Basques, Gurkhas, Norwegians, Scottish Highlanders, Armenians or others, have found it necessary to emigrate from their homeland, either permanently or seasonally, to relieve the population pressure on their farmlands. No such pattern is evidenced among

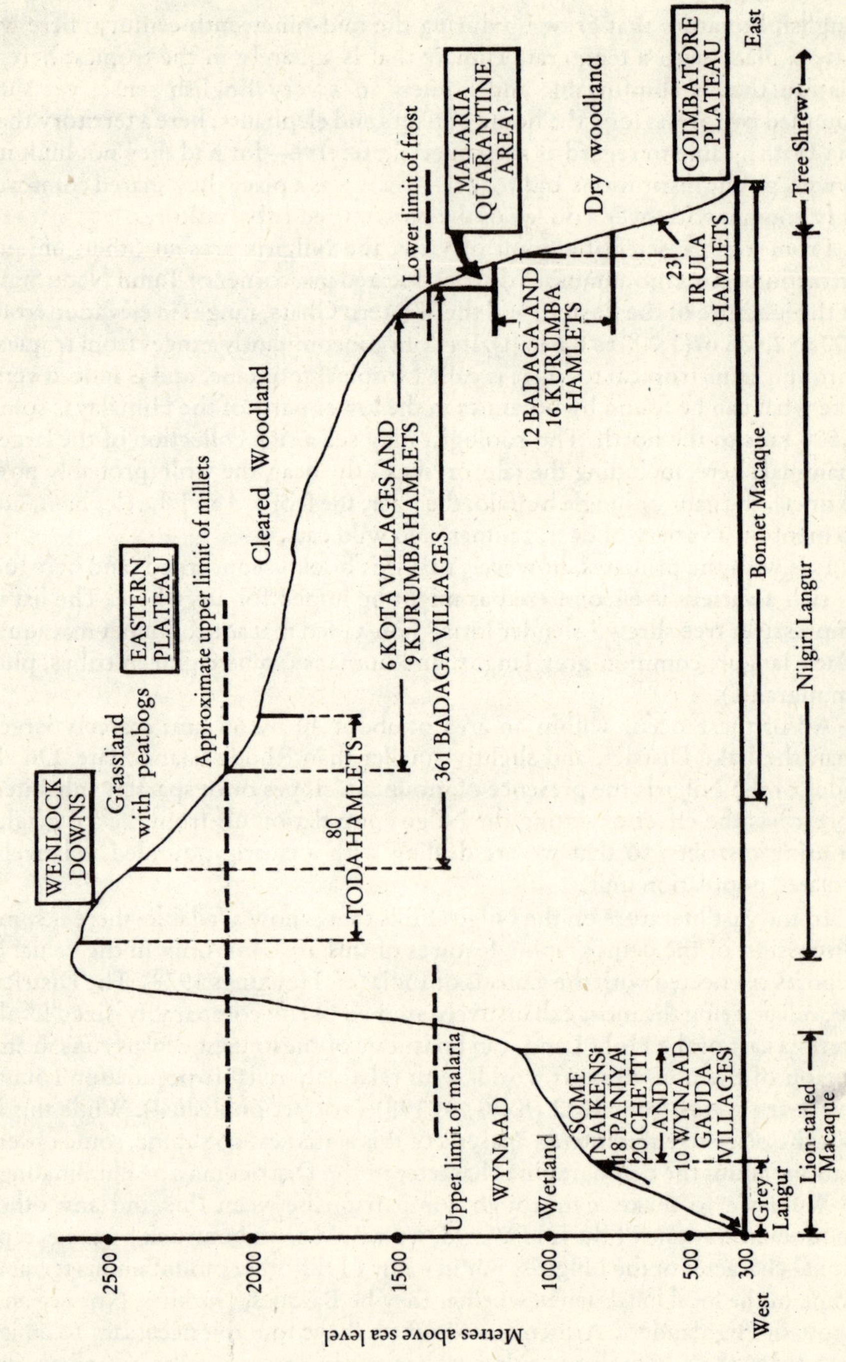

FIG. 15.1. Schematic Cross-Section of the Nilgiris.

the Nilgiri communities. A few Todas do engage in a seasonal migration from drier parts of the massif to wetter during the summer months, but this does not take them away from the mountains. The other communities discussed in this book never moved away from their homes at all. Admittedly the Kotas and Badagas once engaged in annual trading expeditions that took them down to the plains for a few days; and until recently some did also migrate their cattle during the hot season; and perhaps the Kuṟumbas and the Iruḷas engaged in hunting treks through the foothill forests. But in no case was there either a permanent or a seasonal movement of population away from the Nilgiri Hills. In fact, quite the opposite: in 1821,100 per cent of the population consisted of the indigenous tribe-like communities we have been discussing in the preceding chapters; by 1961, in contrast, only 25 per cent of the population were *not* immigrants and descendants of immigrants (Hockings 1980a:247). What made the difference in *this* mountainous zone was primarily the existence of plantations and several factories on the massif which required substantial amounts of labour: these factors reversed the more usual migration pattern, and kept the better educated or more ambitious local people at home while attracting immigrants from the plains.

Another common phenomenon of mountainous areas is the profusion of monasteries, seminaries, nunneries or lamaseries.[3] Auvergne, Tibet and the Ceylon highlands are just a few well-known examples where such institutions almost literally have locked up a segment of the adult population to improve the religious standing of the community while reducing Malthusian pressures. True, the half-dozen Catholic convents and sanatoria in the Nilgiris are devoted to teaching and recuperation and house a few dozen nuns and priests. But virtually all of the inhabitants are immigrants from beyond the Hills, and they make no demographic difference to a District population now surpassing half a million. (Buddhist, Jain and Lingayat monasteries or nunneries are non-existent here.)

There are however two population-regulating institutions that some Nilgiris communities have shared in common with other mountainous zones of the world. The most widely known of these is polyandry, the simultaneous marriage of one woman to several men, usually brothers. It is a well-known (although nearly defunct) feature of Toda society, and fraternal polycoity was reportedly also practised by Badagas and Kuṟumbas in earlier times (Hockings 1980a: 20-1, 41). Elsewhere in the world the Tiyans and several other castes of Kerala, the Marquesans (a non-fraternal case), the Aleuts, the Warau of Guyana, some Eskimos, and certain Orinoco tribes seem to be the only lowland exceptions to a general rule that polyandry is a highland phenomenon.[4] Whatever may have been its social origins, its effects in most polyandrous societies are to reduce the number of children born into the population, to provide a resident male householder for each woman even when some of her husbands may be away working in the plains, to provide better care of the mother and her newborn, and to encourage the practice of female infanticide.

This latter institution, necessary if a permanent imbalance in the male/female ratio is to be maintained, was another well-known (and highly criticized) feature of Toda society until the middle of this century. The cessation of Toda infanticide and polyandry a generation ago, coupled with notable improvements in their health care, have had the effect of producing a distinct advance in Toda population totals. Consider the figures in Table 15.1.

TABLE 15.1
Toda Population over the Past Fifty Years

	Actual Total	Toda Males	Toda Females	Females/ 1,000 Males
1931	597	340	257	756
1941	630	342	288	842
1950	689	373	316	847
1960	762	409	353	863
1971	945	503	442	879

(Nambiar 1965: 28-9, etc.)

The gradual change in the imbalance of the sexes reflects the slow abandonment of female infanticide; while the steady increase in the population of the tribe allows us to assert that there have never been more Todas alive at any time in the past two centuries than there are today (including Christian Todas). It is consequently amazing to find so many contemporary Indian journalists, officials and even anthropologists speaking of the Todas as 'the vanishing tribe' (Raju 1972), 'a dying race' (Singh 1973), with 'only 879 left' (Viets 1970).

The unconscious concern which seemed to underlie traditional institutions of polycoity, polyandry and female infanticide was that in the near future there would not be sufficient farmland (in the Badaga and Kota cases), sufficient grazing (in the case of the Todas), or sufficient game and jungle produce (in the case of the Irulas and the Kuṟumbas) to support the next generations if the population became too numerous. (These concerns, if indeed recognized, seem now to have been lost sight of in the rapid economic change and population growth of recent years.) One other solution to this demographic conundrum, namely raiding, although widely encountered in mountain provinces, has played no part in the Nilgiri economy. Whereas cattle theft and brigandage were a constant way of life in the Hindu Kush, the Sub-Himalayas, the Caucasus, Scotland and many other mountainous zones of the Old World, the Nilgiri tribes lived in relative amity, possessed no militia or weapons of war, and used to operate a complicated system of intertribal economic and ritual exchange that was based firmly on trust (Hockings 1980a: 99-133). Since the Nilgiri food-supply was never too depleted, marauding was never a necessity here.

Mandelbaum, in a well-known paper (1941), concentrated on the inter-ethnic relations that linked four Nilgiri communities in a kind of economic and ritual symbiosis. His account of this system drew in part on the earlier observations of Harkness (1832), Breeks (1873), Rivers (1906) and others, but he was able to bring them up to date and offer a more refined view of the social and economic conditions in which they operated and—more recently—have failed to operate. Hockings (1980a) also draws on these earlier accounts, but points out that the four communities discussed by Mandelbaum (1941) represent only a part, indeed a simplification, of a distinctly broader network. Thus, while Mandelbaum was concerned with ties between the Todas, Kotas, Badagas and Kuṟumbas, Hockings mentions the important economic input of the Chettis living on the Coimbatore Plain, and also discusses (1980a: 126-33) the role of Irulas, Kasuvas, Ūrālis, Wynaad Gaudas and Ha:sanu:ru Badagas. A different but no less interesting tribal interrelationship has been discussed by R. Misra (1972) in the far west of the Wynaad Plateau. There, in Erumad village, she encountered Mullu Kuṟumbas, Ūrāli Kuṟumbas, Kadu Nayakas, Paniyas and Wynaad Chettis (whom she simply calls 'Chetty'), all forming a 'tribal system'. More questionably, Fürer-Haimendorf (1954) identified a community on the Kerala border where he suspected that a formal relationship existed between Todas and Mudugas. One of the 1961 Census of India 'Village Monographs' also discusses intertribal relations in another particularly remote village in the lowland forest, without implying wider systemic relations between the tribes (Nambiar and Bharathi 1965).

The ecological dimensions of this cultural complexity are fascinating. Although a relatively small region, the Nilgiri Massif and the associated Wynaad Plateau show quite considerable variability in land-forms, soils, flora, fauna, microclimates, primates, and human occupance patterns. None of the human communities discussed in this book subsists exclusively by one economic activity; all have additional sources for some food or income (Table 15.3). These sources were fewer in number before the advent of the British, greater after the latter established towns, railways, markets, small industries, a local bureaucracy, and the important tea and coffee plantations. Modernization for the Badagas and (to a lesser extent) other tribal groups, as for the British residents themselves, has meant increasing involvement in the capitalist economy which these newly introduced factors represented. Some tribal communities have resisted the impact of modernization until very recently though, and have been able to maintain something close to their indigenous subsistence patterns. This has been possible because those particular patterns were the cultural adaptation of specific communities to their

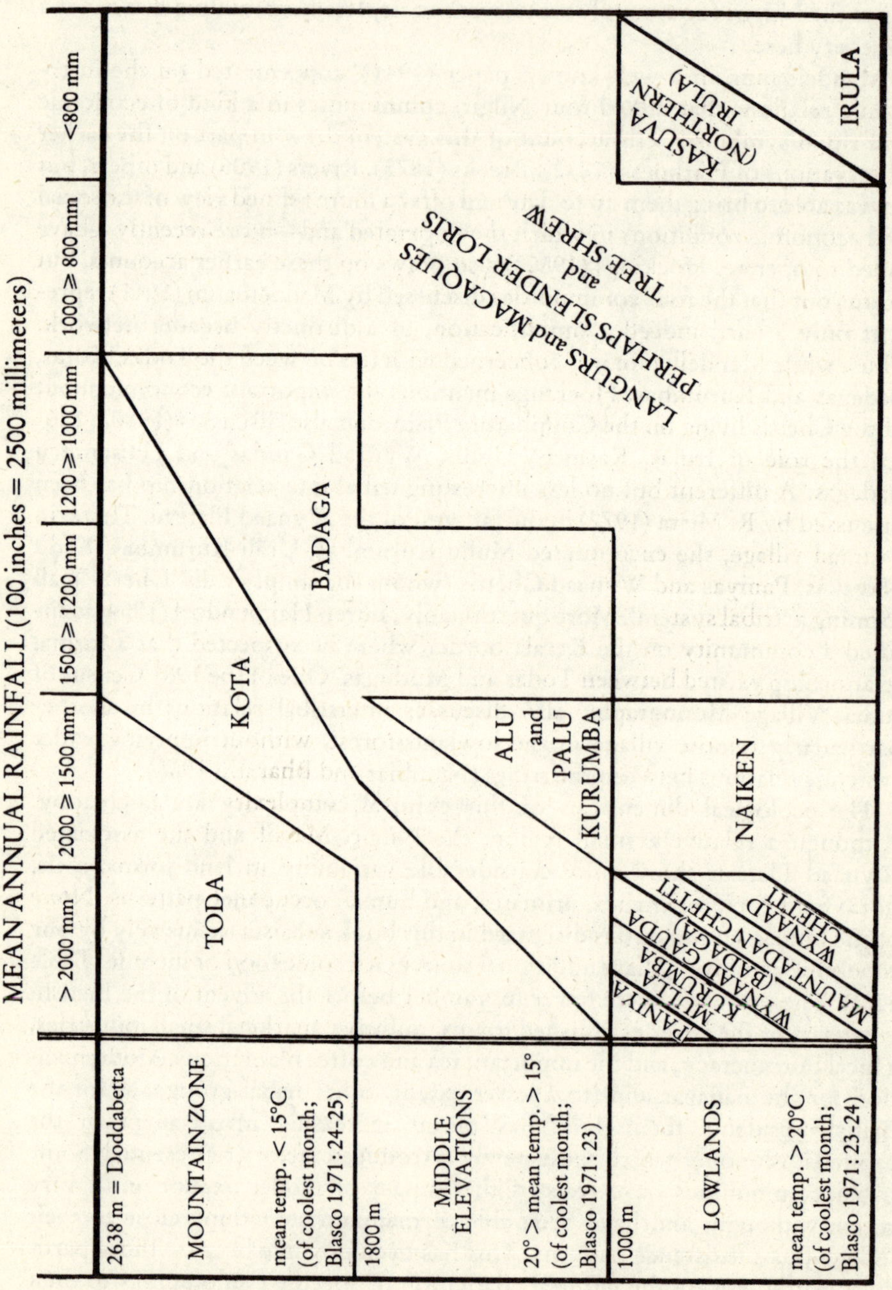

Fig. 15.2. Nilgiri Ecological Niches

particular ecological niches; and because there was no more powerful claimant to the resources available in those niches.

TABLE 15.2
Nilgiri Memorial Ceremonies

Tribe	Original Means of Disposal	Memorial Ceremony	Its Means of Performance	Reference
Badaga	cremation unless Lingayat	many years later; very elaborate	held when one entire generation of a village has died out	Thurston & Rangachari 1909, I: 121-23.
Iruḷa	burial	annual food offering	for all who died during the year	Thurston & Rangachari 1909, II: 380.
Naiken	burial	annually	done individually for every person who died during the year	(Bird-David)
Kota	cremation	annually	for all who died during the year	Thurston & Rangachari 1909, IV: 28.
'Kuṟumbas'	burial or cremation	many years later		Thurston & Rangachari 1909, VI: 169.
Paṇiya	burial if young; adults cremated	every 3-4 years	for all who died in the interim	Thurston & Rangachari 1909, VI: 66.
Toda	cremation	some months later	only held for a prominent adult	Thurston & Rangachari 1909, VII: 147-51.
Ūrāli Kuṟumba	burial	2-3 years later	bones exhumed, displayed, and reburied	Thurston & Rangachari 1909, VII: 256

Traditionally there were at least seven distinct ecological niches on the Nilgiri and Wynaad Plateaux, as well as several others occupied by the non-human primates. Fig. 15.2 illustrates the distribution of the communities in these niches, which have been defined simply in terms of mean temperature of the coolest month (or elevation), and mean monthly rainfall.

In studies of Swat in Pakistan (Barth 1959) and other parts of the world where the ecological aspect of tribal interactions has been emphasized, it has thus far been usual to talk about the tribes which make up the local 'system' as being populations which occupy the same biome, or which are found in close and very similar ecological niches. Here the Nilgiri and the Wynaad Plateaux constitute two biomes, or biotic areas, 'major regions in which distinctive plant and animal groups usually live in harmony with each other, and are also well adapted to the external conditions of environment' (Watts 1971:186). These biomes are made up of a number of separable ecosystems. Fig. 15.2 probably represents the first attempt in this part of Asia to specify in precisely what ways the neighbouring ecological niches differ from each other. With the recent detailed climatological mapping, primarily by von Lengerke and Blasco, it has become possible to show exactly how these niches are differentiated. It will be seen that a difference of as little as 200 mm (8 ins) in the mean annual rainfall can spell the difference between a niche being exploited by monkeys or by mankind! The diagonal lines in many segments of Fig. 15.2 show that, even defined by these very specific criteria of mean annual rainfall and mean temperature, there are some niches which were traditionally shared by more than one tribal group, especially in the Wynaad.

In such situations there are several kinds of relationships which may develop between the competing groups: feuding or warfare; spatial separation; a complex ritual and economic symbiosis; slave/master relationships; enforced separation through the threat of sorcery (or even of cannibalism); parasitism, as in begging, or predator/prey relationships.

In the Nilgiris region each of these possibilities has been realized to some extent. Thus we can find isolated cases during the nineteenth century in which Badagas, aided by Todas, burnt down the huts of Kuṟumba families (Hockings 1980a: 300-1). Feuding with Christian converts and with opposed factions has also arisen from time to time within the Badaga community. Mauntadan Chettis and Wynaad Chettis occupy a niche characterized by its being in low elevations and with a mean annual rainfall of over 2000 mm; but these people are found in different parts of the Wynaad, and hence different parts of the niche. The complex ritual and economic symbiosis of Todas, Kotas, and Badagas, particularly where rainfall is in the range of 2000 ⩾ 1500 mm. in the mountain zone, has been extensively discussed by Mandelbaum (1941), Hockings (1980a) and others. In the Wynaad lowlands we have found Paṇiyas traditionally acting as slaves, or agrestic serfs, on land owned by the Wynaad Chetti farmers. It is the wettest and most fertile land in the District, which goes some way towards explaining why and how four other groups, the Mullu Kuṟumbas, Mauntadan Chettis, Naikens and Wynaad Gaudas (an offshoot of the Badagas) are also found in the same niche. We may also note the special mechanism of sorcery which characterized everybody's view of the Kuṟumba and Irula tribes, and which has had the effect of keeping competing groups out of their ecological niches. Parasitism is a widespread adaptation in Asia, where it usually takes the form of privileged begging. In the Nilgiris it scarcely occurs, however, except as predator/prey relations which mark the behaviour of those non-human primates (usually bonnet macaques) who move in too close to human settlements.

One of the more fascinating aspects of the distribution of Nilgiri Plateau tribes into their separate niches is the discovery that, to judge by the location of 80 Toda, 373 Badaga, 7 Kota, 23 Irula and 25 Kuṟumba settlements marked on the one-inch survey maps, there are *no* tribal settlements anywhere in the District between 1200 and 1300 m, and only four small Badaga ones between 1300 and 1400 m elevation. The lowest level of 1400 m or thereabouts at which winter frosts occur, is also the highest level at which the malaria-causing Plasmodium parasites can survive, that being the limit for their Anopheles mosquito vector (Watts 1971: 327-8). We seem to have here evidence of a *cordon sanitaire*, an uninhabited zone which has had the effect, at least in the past, of reducing the chances of tropical diseases, and primarily of malaria, reaching the plateau communities. (With such an uninhabited zone, the Anopheles would be unlikely to ingest the Plasmodium parasite from the blood of infected people around the Nilgiri escarpment.) At the same time it would seem that the Kuṟumba sorcerers needed to keep their own settlements well away from Badaga ones, either out of fear—fear at least that familiarity might breed contempt—or for hunting purposes. This points up in dramatic fashion the island-like character of the Nilgiri Massif, already discussed in another context by von Lengerke and Blasco (pp. 62-3).

Just as there are a number of plant species that are unique to the Nilgiri Hills, so are there some cultural features that are either peculiar to this small region or are found there in much greater concentration than elsewhere in southern India. It is not possible to present here a complete list of such traits—indeed, trait-lists are quite unfashionable in anthropology today—but we can point out that considerable evidence has been amassed by Emeneau and Zvelebil for the existence of distinctively Nilgiri linguistic forms shared by several of the local languages but not found more widely in the Dravidian family. Of social institutions the most distinctive in the past was the polyandry of Todas and a very similar institution of fraternal polycoity found among Badagas, and possibly the Kuṟumbas, Irulas, and Kotas too (Peter 1963:122, 126). Polyandry did occur elsewhere in South India, but very rarely: ethnographic sources located it only among the Nayars, Tiyans and Kammālans of Kerala and some Kandyans of Ceylon, as well as 'some Tamils and Telugus' (Peter 1963:122). In none of these communities could polyandry be considered the dominant or normal marriage practice, even if it were in these cases a legitimate set of unions. Its rarity in the entire world is underlined by the fact that in 'a representative sample of 186 of the world's societies' (Murdock and Wilson 1980:75), Todas are the only group besides Marquesans to appear as cases where 'The predominant form of family organization is an independent polyandrous family or a stem family with polyandry' (ibid.:82)—and that with some exaggeration too, using Rivers' data from 1902!

In exactly the same sample the Todas and the Chenchus are the only two Asian groups treated which have 'A deliberative body representative of several or all of the state's [i.e. tribe's] major class or ethnic components' (i.e. Toda *no:ym*; Tuden and Marshall 1980:121). There are, of course, some other known cases, among them the Badagas.

A further feature of the Nilgiri cultures that sets them apart from Hindu India in general is the elaboration of memorial ceremonies. Table 15.2 summarizes the circumstances of these important events. It is true that some of the lower-elevation tribes do not elaborate the memorial much: a simple annual food offering at the graveyard suffices for the Irulas and Paniyas, for example. In this respect they are similar to some lowland Hindus, who are expected to perform the ceremony of *śraddhâ*, an offering of food and prayers to the spirits of departed parents, on fixed occasions (Thurston and Rangachari 1909, I: 304).

But it should be noted that in some of the major agricultural castes occupying the lowland countryside near the Nilgiris one does not even encounter the performance of *śraddhâ*; for example, among the numerous Gangadikara Okkaligas from whom many of the Badaga clans are descended (Anantha-krishna Iyer 1930, III: 183), and similarly among the Koṅgu Vellalas (Thurston and Rangachari 1909, III: 421). It is certainly the case that elaborate events like the 'dry funerals' of the Kotas and Todas, and the now discontinued *manevale* of the Badagas, find no parallel among Hindu communities elsewhere.

Whether there was a *jajmani*-like relationship amongst the plateau communities is a matter that has never been fully resolved, although the positive evidence is strong. There were fixed patterns of economic, social and ritual exchange between the Todas, Kotas, Badagas, Irulas and Kuṟumbas, though these have become moribund since about 1930. They were based on a generally accepted ranking which placed the Todas at the top of a hierarchy and found the Kotas at the opposite extreme. One has the feeling that the Badagas were a later, intrusive element in this schema, for while they acknowledged the ritual superiority of the wholly vegetarian Toda tribe, they found themselves accorded high status by all of the tribes, including the Todas, on grounds of their economic power as the dominant farmers, and on grounds of their politico-juridical power, which was expressed in the fact that the Badaga paramount chief and his council of headmen constituted the court of last resort for disputes involving Todas, Kotas and Kuṟumbas as well as Badagas.

While this chapter and indeed many others in the book have tended to concentrate on what is distinctive about the culture of the Nilgiri communities, one cannot overlook the fact that they are still distinctively Indian. In writing about 'The Nilgiris as a Region', Mandelbaum has already made this point. It cannot be overlooked that all of the communities speak Dravidian languages and have a Dravidian kinship system; that cross-cousin marriage is very widespread; that Saivite Hinduism is a dominant religious persuasion in many of the communities.

Modernization of the Nilgiri economy has repeatedly caused chains of ecological reaction that have drastically, and most often irreversibly, changed the life of man, other primates, and the flora and other fauna which occupied the relevant ecological niches. Thus British hunters in the nineteenth century decimated the megafauna, probably making the wolf extinct in this area and nearly bringing tigers and leopards to the same fate. Despite protective legislation, local Indian hunters have had a similar effect on all of the smaller primates, grey langurs and bonnet macaques being the only species to have survived in considerable numbers. Canadian and Indian government planners have now built so many large hydro-electric dams in the District that significant amounts of grazing and forest cover have gone under water, microclimates have altered and the habitats of innumerable animals have been destroyed, thus reducing traditional food sources for the foraging tribes.

Many members of those tribes have been obliged to join the labour pool on plantations or elsewhere. The establishment of the plantations in the last century had an effect similar to that of the dams on the indigenous flora and its dependent fauna. Thus food chains have been altered quite radically, and food procurement strategies of the human residents have of necessity changed with time. The chapters in this book which deal with Todas and Irulas substantiate this most poignantly, as does the economic and social history of the Badagas (Hockings 1980a).

Table 15.3 makes it clear that the various Nilgiri communities are distinguished as much by their mix of food procurement strategies as they are by language, dress or other customs. It is noteworthy that the high-status Todas and British were dependent on a minimum of local food species for their economic support (the British being a special case in that theirs was not a subsistence economy); the Badagas were dependent on just a few domesticates; while the low-status jungle-dwelling tribes, the Irulas and Kuṟumbas, were dependent on a very wide range of animal and plant species, and consequently on a greater number of diverse procurement strategies than any of the other communities. From these circumstances we draw the general conclusion that food production strategies have been evolving in the direction of a more favourable ratio of per capita energy expenditure to food or cash gained. Thus the higher-status communities in the Nilgiris have been serving for some decades as emulation models, not only in questions of lifestyle but also for subsistence patterns. We now find the Irulas, Naikens and Kuṟumbas in the process of giving up hunting altogether, reducing their foraging activities, and seeking daily-wage employment on the plantations. We find some Todas venturing into agriculture, or at least into the profit-making of land rentals. We find Badagas following in the footsteps of the British, first by shifting from subsistence millet cultivation to the cultivation of European vegetables for the market; and then when pests became insurmountable problems, switching from potato terraces to tea plantations. The Kotas have done likewise, though more slowly. A few Kotas and many Badagas have now moved right out of

TABLE 15.3

Foraging, Hunting, Pastoral and Cultivation Strategies

Procurement Strategy	Communities Involved	Manufacture and Maintenance of Tools and Facilities		Direct Procurement	Processing	Storage
1. Plantations (two species)	British settlers Badagas	*Tools*-sprays, vehicles, etc. *Facilities*-sheds, tea-drying factories, chemicals, insurance	Move to and from area of activity on foot.	Plucking by paid labourers	Drying; packing; shipping	Shipped to the auctions and world market.
2. Pastoralism (one or two species)	Todas Badagas Kotas	*Tools*-dairy vessels. *Facilities*-cattle-sheds, calf byres, stone pens, sacred dairies.		Milking cows & buffaloes (no slaughter)	Churning; heating; clarifying butter	Metal and bamboo vessels
3. Agriculture (a few domesticates)	Irulas Kurumba tribes Badagas Kotas	*Tools*-hopes, dibbles, winnows, baskets, forks, (ploughs now rarely used), cooking implements. *Facilities*-some terracing, irrigation ditches (rare), hearths.		Clearing jungle cultivation (i.e. soil preparation, rarely irrigation, fertilizing, planting, wedding); harvesting.	Cleaning; processing; cooking.	Granary basket; pots.

TABLE 15.3 (*Contd.*)

Foraging, Hunting, Pastoral and Cultivation Strategies

Procurement Strategy	Communities Involved	Manufacture and Maintenance of Tools and Facilities		Direct Procurement	Processing	Storage
4. Hunting (a few wild species)	Irulas Kurumba tribes Naikens	*Tools*-nets, spears, traps, snares, knives. *Facilities*-pits for trapping, natural strychnine for fish position.	Move to and from area of activity on foot.	Search and capture (locating, stalking, trapping, killing prey).	Butchering; cooking; drying.	In rafters of hut.
5. Foraging (any edible species)	Irulas Kurumba tribes Naikens Non-human primates	*Tools*-nets, digging-sticks, baskets. *Facilities*-knowledge of local habitat (paws and teeth)		Search and collection —	Shelling, grinding; leaching; cooking. —	Pots

the farm economy by joining the urban middle-class in such respectable professions as medicine, law, teaching and administration. Here they are following a westernizing model.

TABLE 15.4
Average Daily Food Consumption

Tribe	Protein (gms)	Total Protein (gms)	Animal Cereals (gms)	Milk & Products (gms)	Food Energy (calories)
Toda ($N = 119$)	28.5	75.1	540	664	3100
Kota ($N = 104$)	2.0	88.9	574	344	3060
Paṇiya ($N = 128$)	3.1	52.7	448	8	1975
Iruḷa ($N = 110$)	9.0	50.3	440	154	1860

(Roy and Biswas 1964: 128; Sen Gupta 1980: 158-9)

Some recent nutritional figures for samples from four of the Nilgiri tribes (Table 15.4) reflect the superiority of the Toda and Kota diets over those of the Paniyas and Irulas. The clear dichotomy between high, adequate caloric levels (Todas and Kotas) and the low, less adequate levels of Paniyas and Irulas should not perhaps be construed as indicating that the traditional Toda and Kota strategy of exploiting relatively few food sources is more successful than the strategy of exploiting many species through diverse techniques. Rather does the dichotomy reflect the current economic situation which gives Todas and Kotas sufficient cash income to purchase adequate quantities of rice and millets; whereas the other two tribes are mainly poorly paid plantation labourers at the present time, and thus find themselves deprived of basic requirements.

In broad ecological terms, the Nilgiris District has undergone a drastic and quite irreversible transformation since the advent of the British more than a century-and-a-half ago. Back in those times substantial portions of the territory were still the preserve of nature, while today every part is either under human occupance or is conceived to be within the habitat range of one community or another. Even two extensive tracts which acquired an inviolable character for most earlier British administrators and armed, self-styled 'sportsmen', namely the Wenlock Downs and Mudumalai Forest, are now being partially occupied by the land-hungry. In Mudumalai this process has involved the recent extension of farming and lumbering. Elsewhere too on Wenlock Downs and in the Wynaad squatting has been a recent problem (cf. Adams' chapter). On Wenlock Downs different State Government departments have severally taken their toll, while exercizing their powers of development and control. We have already seen how the Electricity Board has created numerous large hydro-electric dams. An environmentally degrading

Protein Products Factory has been located amidst various Toda residential and ceremonial sites. Yet more territory has been taken over for the somewhat cleaner but just as uncongenial Raw Film Factory. Elsewhere, many square kilometers have been planted with acacia and eucalyptus by the Forest Department, thus instantly destroying long-established buffalo-grazing tracts that had been used by Todas, Badagas and Kotas for centuries past, and producing a quick-growing timber that is virtually useless for any purposes but dye and fuel. Resource management which does not consider the total systemic impact of development projects, taken both individually and together, may regrettably lead to the destruction of numerous rare and fascinating forms of life throughout the Nilgiris District. The world will thereby be a poorer place. It is clear that both the benefits and the costs to the entire local population must be calculated before such development projects can be legitimated.

The history of the various strategies of resource exploitation (summarized in Table 15.3) illustrates an inexorable cultural development away from a neat interdigitation of occupance patterns and resource exploitation. Stated simply, it is the law of cultural dominance at work: 'That cultural system which more effectively exploits the energy resources of a given environment [a] niche] will tend to spread in that environment at the expense of less effective systems' (Sahlins and Service 1960: 75). Different Nilgiri cultures have constituted different sociocultural environments. The changes that have occurred through the centuries, and especially those introduced by the British, have brought about an important shift in the management and use of both natural and manpower resources, and are consequently requiring drastic adjustments nowadays—by humans as well as other life forms—to the new Nilgiri environment that has been created.

In summary, it can be shown that the plantation people (no. 1 in Table 15.3) took over some land from the agriculturists (2) and the pastoralists (3), just as in an earlier period the agriculturists (2) had taken over some land from the pastoralists (3) and the hunters and foragers (4, 5). It is also arguable that, in prehistoric times, the Proto-Todas or some other pastoral group (3) created vast grasslands by first burning off the brush, and then annually burning the grass itself, thereby removing it from the primary area exploited by early hunters and foragers (4, 5). Cultural development here has clearly been a unidirectional process, the substitution of strategies that exploit a few food species for older strategies that exploit many. At each substitution, kilocalory expenditure has increased dramatically.

In recent decades, the newer strategies have introduced and then managed intensively a few food species, for sale as well as subsistence. In the process environments necessary for self-sustaining symbiotic diversity have been reduced in an area or destroyed, while the propagation of patterns of economic dependence and concentration of control over resources have been in the hands of relatively few people. Thus, with a dominance hierarchy replacing tribal interdependence, the Nilgiris in the twentieth century are undergoing a very substantial but not altogether satisfying redistribution of access to and hence benefits from what are of course fairly limited resources.

NOTES AND REFERENCES

1. This paper has benefited greatly from the criticism of Ted Adams, Dieter B. Kapp, Hans J. von Lengerke, William A. Noble, and Anthony R. Walker. Ray Brod gave valuable help by preparing the diagrams.
2. The status of these two species, *Tupaia ellioti* and *Loris tardigradus*, respectively, as prosimians is currently under review.
3. There are of course plenty of such institutions spread through lowland areas too.
4. While the Todas and Tibetans are certainly the most often cited mountain societies having polyandry, the institution has also been reported in hilly areas among Kashmiris, certain Australian Aborigines, the Maoris, various central Himalayan tribes, the Koryaks of N.E. Siberia, the Lele of the southern Congo, some Bantu tribes in Angola, the Nfachara of Buji in northern Nigeria, the Hima of Rwanda, and some Baganda people in the Ankole region of Uganda.

Bibliography

(A) NATURAL SCIENCES

Note: This section of the Bibliography refers only to Chapters 2 and 3. Fuller references for most items may be found in Hockings 1978.

BAGNOULS, F., AND HENRI GAUSSEN

1957 Les climats biologiques et leur classification. *Annales de Géographie* 66: 193–220.

BAIKIE, ROBERT

1857 *The Neilgherries: Including an Account of Their Topography, Climate, Soil, and Production; and of the Effects of the Climate on the European Constitution.* With maps... extracts from other writers incorporated; and statistics to the present time—collected by the editor on a late visit. 2nd edition. Calcutta: J. Thomas.

BECKMANN, WALTER

1972 Soil Formations in Southern Indian Mountains. [Third International Working-Meeting on Soil Micromorphology, Wrocław 1969.] *Zeszyty Problemowe Postępów Nauk Rolniczych* 123: 399–406.

BEDDOME, RICHARD HENRY

1876 The Forests and Flora of the Nilgiris. *Indian Forester* 2: 17–28.

BELLAN, M.F.

1985 [Map of the Main Vegetation Types from Satellite Imagery: Nilgiri Hills.] Toulouse: Institut de la Carte Internationale de la Végétation.

BLANFORD, WILLIAM THOMAS

1891 Mammalia. In *The Fauna of British India, including Burma anc Ceylon.* Vol. II. London: Taylor and Francis; Calcutta: Thacker Spink & Co.; Bombay: Thacker & Co., Ltd.; Berlin: R. Friedlände & Sohn.

BLASCO, FRANÇOIS

1971 Montagnes du Sud de l'Inde: forêts, savanes, écologie. *Institu Français de Pondichéry, Travaux de la Section Scientifique e Technique*, X (1). Pondichéry: Institut Français.

, AND G. THANIKAIMONI

1974 Late Quaternary Vegetational History of Southern Region. I *Aspects and Appraisal of Indian Palaeobotany* (Krishna Rajarai Surange, Rajendra Nath Lakhanpal, and D.C. Bharadwaj, eds. 632–43. Lucknow: Birbal Sahni Institute.

BLYTH, EDWARD

1843 Report of the Zoological Curator (Museum of the Asiatic Socie of Bengal). *Journal of the Asiatic Society of Bengal* 12: 766–82.

BOR, NORMAN LOFTUS

1938 The Vegetation of the Nilgiris. *Indian Forester* 64: 600–9.

BROCKWAY, LUCILE H.

1979 Kew and Cinchona. In *Science and Colonial Expansion: The R of the British Royal Botanic Gardens*, 103–39. New York and L don: Academic Press, Inc.

CHAMPION, HARRY GEORGE, AND SHIAM KISHORE SETH
1968 *A Revised Survey of the Forest Types of India.* Delhi: Manager of Publications.

DANIEL, J.C., AND P. KANNAN
1967 The Status of the Nilgiri Langur [*Presbytis johnii* (Fischer)] and Lion-Tailed Macaque [*Macaca silenus* (Linnaeus)] in South India. *Bombay Natural History Society Report 1967*: 1–9. Bombay: Bombay Natural History Society.

DEMANGEOT, JEAN
1973 Une montagne tropicale, Les Nilghiri (Inde du Sud). *Finistera: Revista Portuguesa de Geografia* 8: 292–309.
1975 Recherches géomorphologiques en Inde du Sud. *Zeitschrift für Geomorphologie* (N.F.) 19: 229–72.

DEMAREST, WILLIAM J.
1977 Incest Avoidance among Human and Nonhuman Primates. In *Primate Bio-Social Development: Biological, Social, and Ecological Determinants* (Suzanne Chevalier-Skolnikoff and Frank Eugene Poirier, eds.): 323–42. New York and London: Garland Publishing Co.

FRANCIS, WALTER
1908 *Madras District Gazetteers, The Nilgiris.* Madras: Superintendent, Government Press.

FYSON, PHILIP FURLEY
1915-20 *The flora of the Nilgiri and Pulney Hill-Tops (above 6,500 feet) Being the Wild and Commoner Introduced Flowering Plants around the Hill-Stations of Ootacamund, Kotagiri and Kodaikanal.* Madras: Superintendent, Government Press; 3 vols.

GAMBLE, JAMES SYKES
1915-25 *Flora of the Presidency of Madras.* Calcutta, London: Secretary of State for India in Council; 7 pts.

GAUSSEN, HENRI, PIERRE LEGRIS, AND M. VIART
1962 *Notes on the Sheet Cape Comorin.* New Delhi: Indian Council of Agricultural Research (An unabridged translation of French explanatory notes on Sheet 2 of the *International Map of the Vegetation and of Environmental Conditions*, NC 43/44, Cape Comorin, ICAR ed., Madras 1961).

GOVINDARAJAN. S.V., AND NIHAR RANJAN DATTA BISWAS
1964 Topo-Sequence of the Soils of Nilgiris and Potentialities of their Use [*Proceedings of the Symposium on Fertility of Indian Soils* (Madras 1962)]. *Bulletin of the National Institute of Sciences of India* 26: 117–25.

GREEN, STEVEN, AND KAREN MINKOWSKI
1977 The Lion-tailed Monkey and its South Indian Rain-forest Habitat. In *Primate Conservation* (Prince Rainier III, of Monaco, and Geoffrey Howard Bourne, eds.): 289-337. New York: Academic Press.

GUPTA, R.K.
1960 Ecological Notes on the Vegetation of Kodaikanal. *Journal of the Indian Botanical Society*, 39:601–7

HOCKINGS, PAUL EDWARD
1974 Place-names and Cultural Ecology. *Journal of Asian and African Studies* (Tokyo), 9: 193–211.
1978 *A Bibliography for the Nilgiri Hills of Southern India* (Revised edition). New Haven: HRAF Press.

1980b *Sex and Disease in a Mountain Community.* New Delhi: Vikas Publishing House (Pvt.) Ltd.

HOLLAND, THOMAS HENRY
1900 The Charnockite Series, a Group of Archaean Hypersthenic Rocks in Peninsular India. *Memoirs of the Geological Survey of India* 28: 119–249.

HORNADAY, WILLIAM TEMPLE
1885 *Two Years in the Jungle: The Experiences of a Hunter and Naturalist in India, Ceylon, the Malay Peninsula and Borneo.* New York: Charles Scribner's Sons.

HORWICH, ROBERT H.
1972 Home Range and Food Habits of the Nilgiri Langur, *Presbytis johnii. Journal of the Bombay Natural History Society* 69: 225-67.

HRDY, SARAH BLAFFER
1974 Male-Male Competition and Infanticide among the Langurs (*Presbytis entellus*) of Abu, Rajasthan. *Folia Primatologica* 22: 19–58.
1977 *The Langurs of Abu: Female and Male Strategies of Reproduction.* Cambridge: Harvard University Press.

HUTTON, ANGUS F.
1949 Notes on the Snakes and the Mammals of the High Wavy Mountains, Madura District, South India. Part II—Mammals. *Journal of the Bombay Natural History Society* 48: 681–94.

ILSE, D.
1955 Olfactory Marking of Territory in Two Young Male Loris, *Loris tardigradus lydekkerianus*, Kept in Captivity in Poona. *British Journal of Animal Behaviour* 3: 118–20.

INDIA. METEOROLOGICAL DEPARTMENT
1962 Monthly and Annual Normals of Rainy Days. *Memoirs of the Indian Meteorological Department* 31.
1967 Climatological Tables of Observatories in India (1931–60). New Delhi: Meteorological Department.

ITANI, JUNICHIRO
1972 A Preliminary Essay on the Relationship between Social Organization and Incest Avoidance in Nonhuman Primates. In *Primate Socialization* (Frank Eugene Poirier, ed.); 165-71. New York: Random House, Inc.

JAY, PHYLLIS CAROL JOCELYN (afterwards DOLHINOW)
1965a Field Studies. In *Behavior of Nonhuman Primates: Modern Research Trends* (Allan M. Schrier, Harry F. Harlow, and Fred Stollnitz, eds.); 2: 525–91. New York and London: Academic Press.
1965b The Common Langur of North India. In *Primate Behavior, Field Studies of Monkeys and Apes* (Irven DeVore, ed.); 197–249. New York; Holt, Rinehart and Winston.
1968 *Primates: Studies in Adaptation and Variability.* New York: Holt, Rinehart and Winston.

JEYADEV, T.
1954 *Working Plan for the Nilgiris Forest Division.* Madras: Madras Forest Department.
1957 *Working Plan for the Nilgiris Forest Division. 1st April 1954 to 31st March 1964.* Madras: Superintendent, Government Press.

KARR, JAMES RICHARD
1973 Ecological and Behavioural Notes on the Liontailed Macaque (*Macaca silenus*) in South India. *Journal of the Bombay Natural History Society* 70: 191–3.

KRISHNAMURTHI, SUNDARAM
1953 *Horticultural and Economic Plants of the Nilgiris.* Coimbatore: Coimbatore Co-operative Printing Works, Limited.

KRISHNAN, MADHAVIAH
1971 An Ecological Survey of the Larger Mammals of Peninsular India. (Part I). *Journal of the Bombay Natural History Society* 68: 503–55.

KURUP, G.U.
1975 Distribution, Habitat and Present Status of the Rain Forest Primates of Western Ghats, India. [Abstracts of the Winter School on the Use of Nonhuman Primates.] *Biomedical Research*: 40–1. New Delhi: Indian National Science Academy.

LEGRIS, PIERRE
1963 La végétation de l'Inde: Écologie et Flore. *Institut Français de Pondichéry, Travaux de la Section Scientifique et Technique*, 6.

, AND FRANÇOIS BLASCO
1969 Variabilité des facteurs du climat: Cas des Montagnes du Sud de l'Inde et de Ceylan. *Institut Francais de Pondichéry, Travaux de la Section Scientifique et Technique* 8 no. 1.

LENGERKE, HANS JÜRGEN VON
1977 *The Nilgiris. Weather and Climate of a Mountain Area in South India (Beiträge zur Südasienfo schung*, 32.) Wiesbaden: Franz Steiner Verlag.
1978a On the Short-term Predictability of Frost and Frost Protection—a Case Study on Dunsandle Tea Estate in the Nilgiris (South India). *Agricultural Meteorology* 19: 1–10.
1978b Frost in den Nilgiris. Klimatologische und ökologische Beobachtungen in den kalten Tropen Südindiens. *Erdkunde* 32: 10–28.

MACAULAY, THOMAS BABINGTON
1877 *The Life and Letters of Lord Macaulay* (George Otto Trevelyan, ed.). Detroit: Belford Brothers; 2 vols.

MCCANN, CHARLES
1933 Observations on Some of the Indian Langurs. *Journal of the Bombay Natural History Society* 36: 618–28.

MAHALINGAM, P.K. AND D. JOHN DURAIRAJ
1968 A Study of the Physical and Chemical Properties of High Level Soils of the Nilgiris. *Madras Agricultural Journal* 55: 314–19.

MATTHEW, K.M.
1969 The Exotic Flora of Kodaikanal, Palni Hills. *Records of the Botanical Survey of India*, XX (1), 241.

MEHER-HOMJI, VISPI MINOCHER
1965 Ecological Status of the Montane Grasslands of the South Indian Hills: A Phytogeographic Reassessment. *Indian Forester* 91:210–15.

MOHNOT, S.M.
1968 Interactions and Social Changes in Troops of the Langur, *Presbytis entellus*, in India. *Abstracts of Papers, Symposium on Natural Resources of Rajasthan*: 620–1.
1971 Some Aspects of Social Changes and Infant-Killing in the Hanuman Langur, *Presbytis entellus* (Primates: Cercopithecidae) in Western India. *Mammalia* 35: 175–98.

Noble, William Allister
1977 Settlement Patterns and Migrations among Nilgiri Herders, South India. *Journal of Tropical Geography* 44: 57–70.

Odum, Eugene Pleasants
1971 *Fundamentals of Ecology.* 3rd edition. Philadelphia: W.B. Saunders Co.

Pardhasaradhi, Y.J. and R. Vaidyanadhan
1974 Evolution of Landforms over the Nilgiri, South India. *Journal of the Geological Society of India* 15: 182–8.

Parthasarathy, Mandayam D.
1972 Some Comparative Aspects of the Socioecology and Behaviour of the Hanuman Langur (*Presbytis entellus*) and the Bonnet Macaque (*Macaca radiata*). In *Abstract Book, Fourth International Congress of Primatology, Portland*, 56.
1975 Some Observations on Ecophysical and Behavioral Factors in the Hanuman Langur, *Presbytis entellus*, and the Bonnet Monkey, *Macaca radiata*. In *Abstracts of the Winter School on the Use of Nonhuman Primates in Biomedical Research*, 39. New Délhi: Indian National Science Academy.

Pascal, J.P.
1984 Dense Evergreen Forests of Western Ghats. Pondichery: Institut Français. *Travaux de la Section Scientifique et Technique*, no. XX.

Poirier, Frank Eugene
1967 The Ecology and Social Behavior of the Nilgiri Langur (*Presbytis johnii*) of South India. Eugene: University of Oregon: Ph.D. Dissertation in Anthropology.
1968a Analysis of a Nilgiri Langur (*Presbytis johnii*) Home Range Change. *Primates* 9:29–43.
1968b The Nilgiri Langur (*Presbytis johnii*) Mother-infant Dyad. *Primates* 9: 45–68.
1968c Nilgiri Langur (*Presbytis johnii*) Territorial Behavior. *Primates* 9: 351–64.
1969a The Nilgiri Langur (*Presbytis johnii*) Troop: Its Composition, Structure, Function and Change. *Folia primatologica* 10:20-47.
1969b Nilgiri Langur (*Presbytis johnii*). Territorial Behaviour. In *Proceedings of the Second International Congress of Primatology, Atlanta, Ga. 1968. Vol. 1, Behavior* (Clarence Ray Carpenter, ed.); 1:31–5. Basel and New York: S. Karger.
1969c Behavioral Flexibility and Intertroop Variation among Nilgiri Langurs (*Presbytis johnii*) of South India. *Folia primatologica* 11: 119–33.
1970a Characteristics of the Nilgiri Langur (*Presbytis johnii*) Dominance Structure. *Folia primatologica* 12: 161–87.
1970b The Nilgiri Langur (*Presbytis johnii*) of South India. In *Primate Behavior: Developments in Field and Laboratory Research* (Leonard Allen Rosenblum, ed.); 1: 251–383. New York and London: Academic Press.
1970c The Nilgiri Langur Communication Matrix. *Folia Primatologica* 13: 92–137.
1971 The Nilgiri Langur—a Threatened Species. *Zoonooz* 44, no. 7: 10–16.
1972a Introduction. In *Primate Socialization* (Frank Eugene Poirier, ed.); 3–28. New York: Random House.

1972b Nilgiri Langur Behavior and Social Organization. In *For the Chief: Essays in Honor of Luther S. Cressman by some of his Students* (Frederick William Voget and Robert Lloyd Stephenson, eds.); 119–34. Eugene: University of Oregon Press.

1973 Socialization and Learning among Nonhuman Primates. In *Learning and Culture* (Solon Toothaker Kimball and Jacquetta Hill Burnett, eds.); 3–41. Seattle: University of Washington Press.

1974 Colobine Aggression: A Review. In *Primate Aggression, Territoriality, and Xenophobia: A Comparative Perspective* (Ralph Leslie Holloway, ed.); 123–57. New York and London: Academic Press.

1975a The Human Influence on Subspeciation and Behavioral Differentiation among Nilgiri Langurs and St. Kitts Green Monkeys. *American Journal of Physical Anthropology* 42: 323-4.

1975b Socialization of Nonhuman Primate Females. In *Being Female, Reproduction, Power and Change* (Dana Raphael, ed.); 13–19. The Hague, Paris: Mouton Publishers.

1977 Introduction. In *Primate Bio-social Development: Biological, Social, and Ecological Determinants* (Suzanne Chevalier-Skolnikoff and Frank Eugene Poirier, eds.); 1–39. New York: Garland Publishing.

PRABHAKARA RAO, P.

1969a Progress Report for the Field Season 1967–8. Geology of Parts of Nilgiris District, Madras State. Madras: Geological Survey of India (typescript).

1969b Progress Report for the Field Season 1968–9. Geology of Parts of the Nilgiris and Coimbatore Districts, *Tamil Nadu*. Madras: Geological Survey of India (typescript).

PRATER, STANLEY HENRY

1965 *The Book of Indian Animals*. 2nd edition. Bombay: Bombay Natural History Society.

RANGANATHAN, CONDRAMANICKAM RAGHUNATHA

1938 Studies in the Ecology of the Shola Grassland Vegetation of the Nilgiri Plateau. *Indian Forester* 64: 523–41.

RAO, C., AND R. NARAYANA

1927 Observation on the Habits of the Slow Loris *Loris lydekkerianus*. *Journal of the Bombay Natural History Society* 32: 206–8.

1932 On the Occurrence of Glycogen and Fat in Liquor Folliculi and Uterine Secretion in *Loris lydekkerianus* (Cabr.). *Half-yearly Journal of the Mysore University* 6: 140–70.

ROONWAL, MITHAN LAL, AND S.M. MOHNOT

1977 *Primates of South Asia: Ecology, Sociobiology, and Behavior*. Cambridge and London: Harvard University Press.

SHANKARNARAYAN, KANIMANGALAM ANANTHANARAYAN

1958 The Vegetation of the Nilgiris: The Shola and Grasslands. *Journal of Biological Sciences* 1: 90–8.

SIMONDS, PAUL EMERY

1965 The Bonnet Macaque of South India. In *Primate Behavior, Field Studies of Monkeys and Apes* (Irven DeVore, ed.); 175–96. New York: Holt, Rinehart and Winston.

1972 Outcast Males and Social Structure among Bonnet Macaques. In *Abstract Book, Fourth International Congress of Primatology, Portland*; 69.

SOUTHWICK, CHARLES HENRY, AND M.F. SIDDIQUI
1970 Primate Population Trends in Asia, with Specific Reference to the Rhesus Macaques of India. In *Problems of Threatened Species* (Colin W. Holloway, ed.); 135–47. Morges: International Union for the Conservation of Nature and Natural Resources (IUCN Publications, n.s., no. 18; Eleventh Technical Meeting, Papers and Proceedings, Vol. 2).

SPATE, OSKAR HERMANN KHRISTIAN, ANDREW THOMAS AMON LEARMONTH, AGNES MOFFAT LEARMONTH, AND BERTRAM HUGHES FARMER
1967 *India and Pakistan, A General and Regional Geography*. 3rd edition. London: Methuen and Co. Ltd.

SUBRAMANIAN, K.S., AND M.V.N. MURTHY
1976 Bauxite and Hematite Cappings in the Nilgiris, Tamil Nadu—Study from Geomorphic Angle. *Journal of the Geological Society of India* 17: 353–8.

SUGIYAMA, YUKIMARU
1964 Group Composition, Population Density and some Sociological Observations of Hanuman Langurs (*Presbytis entellus*). *Primates* 5: nos. 3–4: 7–37.
1965a Behavioral Development and Social Structure in Two Troops of Hanuman Langurs (*Presbytis entellus*). *Primates* 6: 213–47.
1965b On the Social Change of Hanuman Langurs (*Presbytis entellus*) in their Natural Condition. *Primates* 6: 381–418.
1967 Social Organization of Hanuman Langurs. In *Social Communication among Primates* (Stuart A. Altmann, ed.); 221–36. Chicago: University of Chicago Press.
1968 The Ecology of the Lion-Tailed Macaque [*Macaca silenus* (Linnaeus)]: A Pilot Study. *Journal of the Bombay Natural History Society* 65: 283–92.
1971 Characteristics of the Social Life of Bonnet Macaques (*Macaca radiata*). *Primates* 12: 247–66.
———, KENJI YOSHIBA, AND M.D. PARTHASARATHY
1965 Home Range, Mating Season, Male Group and Inter-Troop Relations in Hanuman Langurs (*Presbytis Entellus*). *Primates* 6: 73–106.

TANAKA, JIRO
1965 Social Structure of Nilgiri Langurs. *Primates* 6: 107–22.

TROY, J.P.
1979 *Pédogénèse sur les Roches charnockitiques en Région de Montagne du Sud de l'Inde*. Paris: ENGREF.

VERMA, KUSUM
1965 Notes on the Biology and Anatomy of the Indian Tree-Shrew, *Anathana Wroughtoni*. *Mammalia* 29: 289–330.

VISHNU-MITTRE
1974 Late Quarternary Paleobotany and Palynology in India: An Appraisement. *Birbal Sahni Institute of Paleobotany, Special Publications* 5: 16–51.

WEBB-PEPLOE, C. GODFREY
1947 Field Notes on the Mammals of South Tinnevely, South India. *Journal of the Bombay Natural History Society* 46: 629–44.

WIGHT, ROBERT
1845-51 *Spicilegium Neilgherrense; or a Selection of Neilgherry Plants.* Madras: Franck & Co.; American Mission Press; Calcutta: Ostell, Lapage & Co.

YOSHIBA, KENJI
1968 Local and Intertroop Variability in Ecology and Social Behavior of Common Indian Langurs. In *Primates: Studies in Adaptation and Variability* (Phyllis Carol Jocelyn Jay, ed.); 217–42. New York: Holt, Rinehart and Winston.

(B) HUMAN SCIENCES

Note: This section of the Bibliography covers all parts of the book except Chapters 2 and 3. For further details on most items, see Hockings 1978.

AGASTHIALINGAM, S., AND S. SAKTHIVEL
1977 Ethnographic Study of the Todas of Nilgiri Hills. *Bulletin of the Institute of Traditional Cultures, Madras*, Jan-June: 420–35.

AIYAPPAN, AYINIPALLI
1948 *Report on the Socio-economic Conditions of the Aboriginal Tribes of the Province of Madras.* Madras: Government Press.

ALL KARNATAKA UNIFICATION SANGHA
1948 *Replies & Memorandum Submitted to the Linguistic Provinces Commission of the Constituent Assembly on Behalf of the People of the Nilgiris.* Mangalore: The All Karnataka Unification Sangha.

ANONYMOUS
1819 To the Editor of the Government Gazette (Probably from John Sullivan). Copy of a letter dated January 30, published in *Madras Courier*, February 23, 1819; reprinted in GRIGG 1880: lii-lv.

ANANTHAKRISHNA IYER, L. KRISHNA
1930 *The Mysore Tribes and Castes.* Mysore: The Mysore University 5 vols.

ATHREY, N.B.
1953 Coffee Industry in Nilgiris. In *Horticultural and Economic Plants of the Nilgiris* (Sundaram Krishnamurthi, ed.); 11–18. Coimbatore: Coimbatore Co-operative Printing Works, Limited.

BABER, THOMAS HERVEY
1830 Journal of a Route to the Neelghurries from Calicut. *Asiatic Journal* (n.s.) 3:310–16.

BAHADUR, KRISHNA PRAKASH
1978 *Caste, Tribes and Culture of India, Vol. IV: Karnataca, Kerala and Tamil Nadu.* New Delhi: Ess Ess Publications.

BARTH, FREDRIK
1959 *Political Leadership among the Swat Pathans.* London: London School of Economics Monographs on Social Anthropology, No. 19.

BASEL MISSION
1866 *Report of the Basel German Evangelical Missionary Society... Report of the Basel German Evangelical Mission in Southwestern India*, No. 27. Mangalore: Basel Mission Press.

BECK, BRENDA E.F.
1972 *Peasant Society in Konku.* Vancouver: University of British Columbia Press.
1979 *Perspectives on a Regional Culture: Essays about the Coimbatore Area of South India* (Brenda E.F. Beck, ed.). New Delhi: Vikas Publishing House.

BELLI GOWDER, M.K.
1923–41 A Historical Research on the Hill Tribes of the Nilgiris. Ketti (manuscript).

BENBOW, JESSICA
1930 The Badagas—Beliefs and Customs. Bangalore: United Theological College (manuscript).

BEVAN, HENRY
1839 *Thirty Years in India: or, a Soldier's Reminiscences of Native and European Life in the Presidencies, From 1808 to 1838.* London: Pelham Richardson; 2 vols.

BHOWMIK, KANAI LAL, *et al.*
1971 Toda. In *Tribal India: A Profile of Indian Ethnology*; 181–93. Calcutta: The World Press Private Ltd.

BIRCH, DEBURGH
1838 Topographical Report on the Neilgherries. *Madras Journal of Literature and Science* 8: 86–127.

BIRD, NURIT
1983 Wage-Gathering: Socio-Economic Changes and the Case of the Naiken of South India. In *Rural South Asia: Linkages, Changes and Development* (Peter Robb, ed.); London: School of Oriental and African Studies. (Collected Papers on South Asia Vol. 5.)

BIRD-DAVID, NURIT HAYA
1987 The Kurumbas of the Nilgiris: An Ethnographic Myth? *Modern Asian Studies* 21: 173-89.

BLAVATSKY, H.P. [ELENA PETROVNA BLAVATSKAIA]
1925 'Letter No. XXII' to 'Letter No. XXVI'. In *The Letters of H.P. Blavatsky to A.P. Sinnett and other Miscellaneous Letters Transcribed, Compiled, and with an Introduction by A.T. Barker* (Alfred Trevor Barker, ed.); 43–55. London: T. Fisher Unwin.

BREEKS, JAMES WILKINSON
1873 *An Account of the Primitive Tribes and Monuments of the Nilagiris.* (Susan Maria Breeks, ed.); London: India Museum.

BROWN, CHARLES HILTON
1936 *The Gold & the Grey: Some More Collected Verses 1930–1935.* Oxford: Basil Blackwell.

BUCHANAN, FRANCIS (afterwards BUCHANAN-HAMILTON)
1870 *A Journey from Madras through the Countries of Mysore, Canara, and Malabar*...2nd edition in 2 vols. (originally 1807). Madras: Higginbotham and Co.

BURL, AUBREY
1980 *Rings of Stone: The Prehistoric Stone Circles of Britain and Ireland.* New York: Ticknor & Fields.

BURROW, THOMAS AND MURRAY BARNSON EMENEAU
1984 *A Dravidian Etymological Dictionary* (Revised edition). London: Oxford University Press [DEDR].

BURTON, RICHARD FRANCIS
1851 *Goa, and the Blue Mountains; or, Six Months of Sick Leave.* London: Richard Bentley.

CALDWELL, ROBERT
1856 *A Comparative Grammar of the Dravidian or South-Indian Family of Languages.* Ist edition. London: Trübner & Co.
1875 [2nd edition of Caldwell 1856.]
1913 [3rd edition of Caldwell 1856.]

CAMMIADE, L.A.
1930 Urn-burials in the Wynaad, Southern India. *Man* 30: no. 135: 183–6.

CHAKRAVARTI, N.P.
1936 Section III: Epigraphy. *Archaeological Survey of India Annual Report 1935–6*: 88–116.

CHILDE, VERE GORDON
1948 Megaliths. *Ancient India* 4:4–13.

CHOCKALINGAM, K.
1978 *Special Tables on Scheduled Castes and Scheduled Tribes. Census of India, 1971, Series 19, Tamil Nadu, Part V-A.* New Delhi: Controller of Publications.

CLEGHORN, HUGH FRANCIS CLARKE
1861 *The Forests and Gardens of South India.* London: W.H. Allen & Co.

CONGREVE, HARRY
1847 The Antiquities of the Neilgherry Hills, including an Inquiry into the Descent of the Thautawars or Todars. *Madras Journal of Literature and Science* 14, pt. i:77–146.
1878 On Druidical and other Antiquities between Mettapoliam in Coimbatore and Karnul on the Tungabhadra. *Madras Journal of Literature and Science* 1878: 150–68.

COON, CARLETON STEVENS
1958 An Anthropological Excursion around the World. *Human Biology* 30: 29–42.

CORNELIUS, J.T.
1963 A New Probe into the Origin of the Todas. In *Anthropology on the March: Recent Studies of Indian Beliefs, Attitudes and Social Institutions* (L. Krishna Bala Ratnam, ed.); 372–81. Madras: The Book Centre.

COUSINS, JAMES HENRY SPROULL
1940 *Collected Poems (1894–1940).* Madras: Kalâkshetra.
, AND MARGARET E. COUSINS
1950 *We Two Together.* Madras: Ganesh & Co. (Madras) Ltd.

DANIEL, GLYN EDMUND
1980 Megalithic Monuments. *Scientific American* 243: 78–90.

DAS, GOPI NATH
1957 The Funerary Monuments of the Nilgiris. *Deccan College Postgraduate Institute, Bulletin* 18: ix-x, 140–58.

DHARMALINGAM, ANDI
1971 The Hethaimman Festival, *Hindu*: January 29.

DICKASON, DAVID G.
1975 The Indian Hill Station. *Geographical Review* 65: 115–17.

DIFFLOTH, GÉRARD FÉLIX

1968 The Irula Language: A Close Relative of Tamil. Los Angeles: University of California; unpublished Ph.D. dissertation in Linguistics.

1975 The South Dravidian Obstruent System in Irula. In *Dravidian Phonological Systems* (Harold F. Schiffman and Carol Mary Eastman, eds.); 47–56. Seattle: University of Washington Press.

DUBOIS, JEAN-ANTOINE

1906 *Hindu Manners, Customs, and Ceremonies.* 3rd edition (Henry K. Beauchamp, trans.) Oxford: Clarendon Press.

ELMORE, WILBER THEODORE

1915 Dravidian Gods in Modern Hinduism: A Study of the Local and Village Deities of Southern India. *University of Nebraska Studies* 15: 1–149.

EMENEAU, MURRAY BARNSON

1938 Toda Culture Thirty-Five Years After: An Acculturation Study. *Annals of the Bhandarkar Oriental Research Institute* 19: 101–21. (Reprinted in Emeneau 1967a: 303–17.)

1941 Language and Social Forms: A Study of Toda Kinship Terms and Dual Descent. In *Language, Culture, and Personality: Essays in Memory of Edward Sapir* (Leslie Spier, Alfred Irving Hallowell, and Stanley Stewart Newman, eds.); 158–79. Menasha, Wis.: Sapir Memorial Publication Fund (Reprinted in Emeneau 1967a: 233–57.)

1944 Kota Texts. Part One. *University of California Publications in Linguistics* 2: i–vi, 1–191.

1946a Kota Texts. Part II. *University of California Publications in Linguistics* 2: 193–390.

1946b Kota Texts. Part Three. *University of California Publications in Linguistics* 3: 1–190.

1946c Kota Texts. Part Four. *University of California Publications in Linguistics* 3: 191–335.

1953 Proto-Dravidian *c-: Toda t-. *Bulletin of the School of Oriental and African Studies* 15: 98–112 (Reprinted in Emeneau 1967a: 46–60.)

1956 India as a Linguistic Area. *Language* 32: 3–16.

1958 Toda, a Dravidian Language. *Transactions of the Philological Society* 1957: 15–66 (Reprinted in Emeneau 1967a: 1–36.)

1963 Ootacamund in the Nilgiris: Some Notes. *Journal of the American Oriental Society* 83: 188–93.

1965 India and Historical Grammar. (Annamalai University, Department of Linguistics Publication 5.)

1966 Some South Dravidian Noun Formatives. *Indian Linguistics* 27: 21–30.

1967a *Dravidian Linguistics, Ethnology and Folktales: Collected Papers.* Annamalainagar: The Annamalai University.

1967b The South Dravidian languages. *Journal of the American Oriental Society* 87: 365–413.

1971a Dravidian and Indo-Aryan: the Indian linguistic area. *Symposium on Dravidian Civilization* (Andrée F. Sjoberg ed.); 33–68. Austin, New York: Jenkins Publishing Co.

1971b *Toda Songs.* Oxford: The Clarendon Press.

1974a The Indian Linguistic Area Revisited. *International Journal of Dravidian Linguistics* 3: 92–134.
1974b Ritual Structure and Language Structure of the Todas. *Transactions of the American Philosophical Society* (n.s.) 64, no. 6.
1979a Toda Vowels in Non-initial Syllables. *Bulletin of the School of Oriental and African Studies* 42: 225–34.
1979b Linguistic Archaisms in Toda Songs. *South Asian Languages Analysis* 1: 31–45.
1980 *Language and Linguistic Area: Essays*....selected and introduced by Anwar S. Dil. Stanford: Stanford University Press.
1984 *Toda Grammar and Texts.* Philadelphia: American Philosophical Society (*Memoirs* 155).
, AND THOMAS BURROW
1962 Dravidian Borrowings from Indo-Aryan. *University of California Publications in Linguistics* 26: i–x, 1–121 [DBIA].

EVANS, IVOR H.N.
1937 *The Negritos of Malaya.* Cambridge: Cambridge University Press.

EVANS-PRITCHARD, EDWARD EVANS
1937 *Witchcraft, Oracles, and Magic among the Azande.* London: Oxford University Press.

FERGUSSON, JAMES
1872 *Rude Stone Monuments in all Countries: Their Age and Uses.* London: John Murray.

FERROLI, DOMENICO
1939 Padre Fenicio's Expedition to Todaland (1603) and to the Kingdom of the Salt (1610). In *The Jesuits in Malabar* 1: 472–80. Bangalore: The Bangalore Press.

FINICIO, JACOME
1603 Two Mss. on the Mission of Todamalâ (Numerous versions, listed fully in Hockings 1978: 59–60.) English trans. by A. de Alberti (cited here) in Rivers 1906: 719–30.

FIRTH, RAYMOND WILLIAM
1951 *Elements of Social Organization.* London: Watt & Co.

FLETCHER, FRANCIS W.F.
1911 *Sport on the Nilgiris and in Wynaad.* London: Macmillan and Co. Ltd.

FOLKE, STEEN
1966 Evolution of Plantations, Migration, and Population Growth in Nilgiris and Coorg (South India). *Geografisk Tidsskrift* 65: 198–230.
1967 Central Place Systems and Spatial Interaction in Nilgiris and Coorg (South India). *Geografisk Tidsskrift* 66: 161–78.

FOOTE, ROBERT BRUCE
1901 *Catalogue of the Prehistoric Antiquities in the Government Museum, Madras.* Madras: Superintendent, Government Press.

FRANCIS, WALTER
1908 *Madras District Gazetteers. The Nilgiris.* Madras: Superintendent, Government Press.

FUCHS, STEPHEN
1973 *The Aboriginal Tribes of India.* Delhi: Macmillan India.

FÜRER-HAIMENDORF, CHRISTOPH VON
1943 *The Chenchus: Jungle Folk of the Deccan.* London: Macmillan & Co.

1952 Ethnographic Notes on some Communities of the Wynad. *Eastern Anthropologist* 6: 18–36.

1954 Hereditary Friendship and Inter-tribal Sex Relations between Todas and Mudugas. *Man* 54: no. 24: 28–9.

1962 *The Apa Tanis and their Neighbours.* London: Routledge and Kegan Paul.

1969 *The Konyak Nagas.* New York: Holt, Rinehart, and Winston.

, AND ELIZABETH VON FÜRER-HAIMENDORF

1945 *The Aboriginal Tribes of Hyderabad. Volume II. The Reddis of the Bison Hills.* London: Macmillan & Co., Ltd.

1948 *The Aboriginal Tribes of Hyderabad, Volume III. The Raj Gonds of Adilabad. Book 1, Myth and Ritual.* London: Macmillan & Co., Ltd.

GARDNER, PETER M.

1966 Symmetric Respect and Memorate Knowledge: the Structure and Ecology of Individualistic Culture. *Southwestern Journal of Anthropology* 22: 389–415.

1976 India's Changing Tribes: Identity and Interaction in Crisis. In *Main Currents in Indian Sociology 3* (Giri Raj Gupta, ed.); 289–318. New Delhi: Vikas Publishing House.

1982 Ascribed Austerity: a Tribal Path to Purity. In *Man* 17: 462-9.

GAUSSEN, HENRI, PIERRE LEGRIS, AND M. VIART

1962 *Notes on the Sheet Cape Comorin.* New Delhi: Indian Council of Agricultural Research [An unabridged translation of French explanatory notes on Sheet 2 of the *International Map of the Vegetation and of Environmental Conditions*, NC43/44, Cape Comorin, ICAR ed., Madras 1961.]

GHURYE, GOVIND SADASHIV

1950 *Caste and Class in India.* Bombay: Popular Book Depot.

GOPALAN NAIR, C.

1911 *Wynad: Its Peoples and Traditions.* Madras: Higginbotham & Co.

GORER, GEOFFREY

1975 English Identity over Time and Empire. In *Ethnic Identity: Cultural Continuities and Change* (George De Vos and Lola Romanucci-Ross, eds.); 156–72. Palo Alto: Mayfield Publishers.

GOUGH, E. KATHLEEN

1955 The Social Structure of a Tanjore Village. In *India's Villages* (Mysore Narasimhachar Srinivas, ed.); 82–92. Calcutta: West Bengal Government Press.

GREENBERGER, ALLEN JAY

1969 *The British Image of India: A Study in the Literature of Imperialism 1880–1960.* London: Oxford University Press.

GRIERSON, GEORGE ABRAHAM, AND STEN KONOW

1967 *Linguistic Survey of India. Volume IV: Munda and Dravidian Languages.* Delhi: Motilal Banarsidass (Reprint of 1906 edition).

GRIGG, HENRY BIDEWELL

1880 *A Manual of the Nílagiri District in the Madras Presidency. Compiled and Edited by H.B. Grigg.* Madras: E. Keys, Government Press.

GULLIFORD, HENRY

1917–36 (Personal Diaries; numbered VPC G39–G48.) Bangalore: United Theological College; 10 notebooks.

GURURAJA RAO, BAIRATHNAHALLI KRISHNAMURTHY RAO
1972 *Megalithic Culture in South India*. Mysore: University of Mysore.
HANDLEY, M.A. (MRS)
1911 *Roughing It in Southern India*. London: Edward Arnold.
HARKNESS, HENRY
1832 *A Description of a Singular Aboriginal Race Inhabiting the Summit of the Neilgherry Hills, or Blue Mountains of Coimbatoor, in the Southern Peninsula of India*. London: Smith, Elder and Co.
HAYAVADANA RAO, CONJEEVERAM
1930 *Mysore Gazetteer Compiled for Government. Volume II, Historical*. Bangalore: Government Press (Vol. II in 4 pts.).
HERTZ, ROBERT
1907 Contribution à une étude sur la représentation collective de la mort. *Année Sociologique* 10: 48–137 (English trans. by Rodney and Claudia Needham, 1962, titled *Death and the Right Hand*. London: Cohen & West.)
HOCKINGS, PAUL EDWARD
1973 John Sullivan of Ootacamund. *Journal of Indian History*, 50th Jubilee no.: 863–71.
1976 Paikara: An Iron Age Burial in South India. *Asian Perspectives* 18: 26–50.
1978 *A Bibliography for the Nilgiri Hills of Southern India*. New Haven: Human Relations Area Files; HRAFLEX Books.
1980a *Ancient Hindu Refugees: Badaga Social History, 1550–1975*. The Hague and New York: Mouton Publishers; New Delhi: Vikas Publishing House Ltd.
1980b *Sex and Disease in a Mountain Community*. New Delhi: Vikas Publishing House; Columbia, Mo.: South Asia Books.
1982 Badaga Kinship Rules in their Socio-Economic Context. *Anthropos* 77: 851–74.
, AND ANTHONY RUPERT WALKER R
1981 Toda Secondary Funeral Ritual. In *Main Currents in Indian Sociology; 5—Religion in Modern India* (Giri Raj Gupta, ed.); 313-42. New Delhi: Vikas Publishing House Pvt. Ltd.; New York: Advent Books.
HOMANS, GEORGE CASPAR
1950 *The Human Group*. New York: Harcourt, Brace.
HOUGH, JAMES
1829 *Letters on the Climate, Inhabitants, Productions, &c. &c. of the Neilgherries, or Blue Mountains of Coimbatoor, South India*. London: John Hatchard & Son.
1860 *The History of Christianity in India from the Commencement of the Christian Era*. Vol. 5. London: R.B. Seeley and W. Burnside.
HULL, EDMUND C.P.
1878 *The European in India; or, Anglo-Indian's Vade-Mecum*. London: C. Kegan Paul & Co.
HUTTON, JOHN HENRY
1926 The Use of Stone in the Naga Hills. *Royal Anthropological Institute, Journal* 56: 71–82.
IBBETSON, DENZIL CHARLES JELF
1916 *Panjab Castes. Being a Reprint of the Chapter on 'The Races, Castes and Tribes of the People' in the Report on the Census of the*

Panjab, Published in 1883 by the Late Sir Denzil Ibbetson, K.C.S.I. Lahore: Superintendent, Government Printing, Punjab.

JAGOR, ANDREAS FEDOR

1876 Die Badagas im Nilgiri-Gebirge. *Verhandlungen der Berliner Gesellschaft für Anthropologie, Ethnologie und Urgeschichte* Jhrg. 1876: 190–204.

KAPP, DIETER BERND

1978a Pālu Kuṟumba Riddles: Specimens of a South Dravidian Tribal Language. *Bulletin of the School of Oriental and African Studies* 41: 512–22.

1978b Childbirth and Name-giving among the Ālu Kuṟumbas of South India. In *Aspects of Tribal Life in South Asia, I: Strategy and Survival, Proceedings of an International Seminar held in Berne 1977* (Rupert R. Moser and Mohan K. Gautam, eds.); 167–80. Berne: University of Berne (*Studia Ethnologica Bernensia* 1/1978).

1978c Die Kindheits-und Jugendriten der Ālu-Kuṟumbas (Südindien). *Zeitschrift für Ethnologie* 103: 279–89.

1980 Die Ordination des Priesters bei dem Ālu Kuṟumbas (Südindien). *Anthropos* 75: 433–46.

1982 The Concept of Yama in the Religion of a South Indian Tribe. *Journal of the American Oriental Society* 102: 517-21.

1985 The Kuṟumbas' Relationship to the 'Megalithic' Cult of the Nilgiri Hills (South India). *Anthropos* 80: 493-534.

KARL, WILLIAM VICTOR

1945 The Religion of the Badagas. Serampore: Serampore College unpublished B.D. dissertation.

KARVE, IRAWATI

1953 *Kinship Organization in India.* Poona: Deccan College Post-Graduate and Research Institute.

KETTI MEDICAL MISSION

1940 *Annual Report, 1940* (K.I. Simon, ed.). Mysore: Wesley Press and Publishing House.

KEYS, WILLIAM

1812 A Topographical Description of the Neelaghery Mountains. In Grigg 1880: xlviii-li.

KING, WILLIAM ROSS

1877 Notice of a Prehistoric Burial Place with Cruciform Monolith, near Mungapet in the Nizam's Dominions. *Asiatic Society (Calcutta), Journal* 3: 179–85.

KROEBER, ALFRED LOUIS

1939 *Cultural and Natural Areas of Native North America.* Berkeley and Los Angeles: University of California Press (University of California Publications in American Archaeology and Ethnology, 38).

LAL, BRAJ BASI

1963 The Only Asian Expedition in Threatened Nubia: Work by an Indian Mission at Afyeh and Tumas. *Illustrated London News* 242: 579–81.

LESHNIK, LAWRENCE SAADIA

1970 A Suggested Dating for the Antiquities of the Nilgiri Plateau, South India. *Acta Praehistorica et Archaeologica* 1: 87–99.

1974 *South Indian 'Megalithic' Burials, The Pandukal Complex.* Wiesbaden: Franz Steiner Verlag GmbH.

LING, CATHARINE FRANCES
1910 *Dawn in Toda Land: A Narrative of Missionary Effort on the Nilgiri Hills, South India.* London: Morgan & Scott Ltd.
1934 *Sunrise on the Nilgiris: The Story of the Todas.* London: The Zenith Press.

LINGUISTIC PROVINCES COMMISSION
1948 *Report of the Linguistic Provinces Commission, 1948.* New Delhi: Government of India Press.

LOGAN, WILLIAM
1951 *A Collection of Treaties, Engagements and Other Papers of Importance Relating to British Affairs in Malabar.* 2nd edition. Madras: Superintendent, Government Press.

LUIZ, A.A.D.
1962 *Tribes of Kerala.* New Delhi: Bharatiya Adimjati Sevak Sangh.

LUSHINGTON, ALFRED WYNDHAM
1902 Hill Forests of North Coimbatore. *Indian Forester* 28: 134–50.

MACLEANE, CHARLES DONALD (ED.)
1893 *Manual of the Administration of the Madras Presidency, in Illustration of the Records of Government & the Yearly Administration Reports. Vol. 3—Glossary, Containing a Classification of Terminology, a Gazetteer and Economic Dictionary of the Province, and Other Information, the Whole Arranged Alphabetically and Indexed.* Madras: Superintendent, Government Press.

MACPHERSON, EVANS
1820 Letter to John Sullivan Esq. Dated June 12, 1820. In Grigg 1880: lv-lx [whence our pagination].

MAJUMDAR, DHIRENDRA NATH
1945 *Races and Cultures of India.* Allahabad: Kitabistan.

MALINOWSKI, BRONISLAW GASPAR
1926 *Myth in Primitive Psychology.* London: Kegan Paul, Trench, Trübner & Co.; New York: W.W. Norton & Company.

MANDELBAUM, DAVID GOODMAN
1941 Culture Change among the Nilgiri Tribes. *American Anthropologist* 43: 19–26. [Reprinted, 1967, in *Beyond the Frontier: Social Process and Cultural Change* (Paul James Bohannan and Frederick T. Plog, eds.); 199–208. New York: The Natural History Press.]
1956 The Kotas in their Social Setting, In *Introduction to the Civilization of India* (Robert Redfield and Milton Borah Singer, eds.); 288–332. Chicago: The College, The University of Chicago.
1960 A Reformer of his People. In *In the Company of Man* (Joseph Bartholomew Casagrande, ed.); 273–308. New York: Harper and Brothers.
1970 *Society in India.* Berkeley, Los Angeles, London: University of California Press; 2 vols.

MARSHALL, WILLIAM ELLIOT
1873 *A Phrenologist amongst the Todas or the Study of a Primitive Tribe in South India. Their History, Character, Customs, Religion, Infanticide, Polyandry, Language with Outlines of the Tuda Grammar...* London: Longmans, Green, and Co.

METZ, JOHANN FRIEDRICH
1856 A Vocabulary of the Dialect Spoken by the Todas of the Nilagiri Mountains. *Madras Journal of Literature and Science* 17: 103–8.

1857a The Toda Vocabulary [contd]. *Madras Journal of Literature and Science* 17: 131–46.

1857b A Vocabulary of the Dialect Spoken by the Todas of the Nilagiri Mountains [contd]. *Madras Journal of Literature and Science* 18: 1–24.

1864 *The Tribes Inhabiting the Neilgherry Hills: Their Social Customs and Religious Rites*. 2nd edition. Mangalore: Basel Mission Press Press.

MISRA, PROMODE KUMAR

1969 The Jenu Kuruba. *Department of Anthropology Bulletin* 18: 181–246.

MISRA, RAJALAKSHMI CHENNAKESWARA RAMANUJA

1971 *Mullukurumbas of Kappala*. Calcutta: Anthropological Survey of India.

1972 Inter-tribal Relations in Erumad. *Eastern Anthropologist* 25: 135–48.

MORGAN, HENRY RHODES

1876 The Hill Ranges of North Coimbatore and Lambton Peak Range. In *The Hill Ranges of Southern India*, Part 5 (John Shortt, ed.); 95–101. Madras: Higginbotham & Co.

MURDOCK, GEORGE PETER, AND SUZANNE F. WILSON

1980 Settlement Patterns and Community Organization: Cross-Cultural Codes 3. In *Cross-Cultural Samples and Codes*(Herbert Barry III and Alice Schlegel, eds.); 75–116. Pittsburgh: University of Pittsburgh Press.

MURTHY, C. KRISHNA

1976 Unique Megalith from Kadiraraya Cheruvu, Andhra Pradesh. *Journal of Indian History* 54: 239–41.

NAGASWAMY, R.

1973 Hero Stones of Tamilnadu. *Journal of Indian History* 51: 265–70.

NAIK, IQBAL ABDUL RAZAK (afterwards WAGLE)

1966 The Culture of the Nilgiri Hills with its Catalogue Collection at the British Museum. London: Ph.D. dissertation, University of London

NAMBIAR, P.K.

1965 *Census of India, 1961, Volume IX, Madras, Part V-c, Todas*. Delhi: Manager of Publications.

NAMBIAR, P.K., AND T.B. BHARATHI

1965 *Census of India, 1961, Volume IX, Madras; Part VI, Village Survey Monographs, 20. Hallimoyar*. Delhi: Manager of Publications.

NOBLE, WILLIAM ALLISTER

1968 Cultural Contrasts and Similarities among Five Ethnic Groups in the Nilgiri District, Madras State, India, 1800–1963. Baton Rouge: Ph.D. dissertation, Louisiana State University.

1976 Nilgiri Dolmens (South India). *Anthropos* 71: 90–128.

, AND LOUISA BOOTH NOBLE

1965 Badaga Funeral Customs. *Anthropos* 60: 262–72.

O'FLAHERTY, WENDY DONIGER

1981 *Siva, the Erotic Ascetic*. Oxford and New York: Oxford University Press.

OLIVIER, GEORGES

1961 *Anthropologie des Tamouls du Sud de l'Inde*. Paris: École Française d'Extrême Orient; Publications hors-série.

OUCHTERLONY, JOHN
1848 A Geographical and Statistical Memoir of a Survey of the Neilgherry Mountains. *Madras Journal of Literature and Science* 15: 1–138 (Reprinted in Shortt 1868).

PANDIT, SHANKAR RAO
1927 Report on the Health of the Toda. Guindy: The King Institute, Guindy (typescript).

'PANTER-DOWNES, MOLLIE' (pseud. of MOLLIE ROBERTSON)
1967 *Ooty Preserved: A Victorian Hill Station in India.* London: Hamish Hamilton; New York: Farrar, Straus and Giroux.

PARSONS, TALCOTT, AND EDWIN A. SHILS
1951 *Toward a General Theory of Action.* Cambridge, Mass.: Harvard University Press.

PERICOTS GARCIA, LUIS
1950 *Los Sepulcros Megaliticos Catalanes y la Cultura Pirenaica.* 2nd edition. Barcelona: Instituto de Estudios Pirenaicos.

PETER, H.R.H. PRINCE OF GREECE AND DENMARK
1951 Possible Sumerian Survivals in Toda Ritual. *Bulletin of the Madras Government Museum* 6, no. 1: 1–24.
1955 The Polyandry of South India: the Todas of the Nilgiris. *Man* 55: 89–93.
1963 *A Study of Polyandry*. The Hague: Mouton Publishers.

PRICE, JOHN FREDERICK
1908 *Ootacamund. A History. Compiled for the Government of Madras.* Madras: Superintendent, Government Press.

PRINSEP, CHARLES CAMPBELL
1885 *Record of Services of the Honourable East India Company's Civil Servants in the Madras Presidency. From 1741 to 1858....* London: Trübner & Co.

RAGHAVAN, MANAYATT DHARMADAN
1929 Jain-Kurumbers: An Account of their Life and Habits. *Man in India* 9: 54–65.

RAJU, R.K.
1972 Todas—the Vanishing Tribe. *Sunday Standard*: January 9: IV.

RAMAN, S.T.
1978 Agriculture in Toda Community. Ootacamund (typescript).

RECLUS, JEAN-JACQUES-ELISÉE
1883 XIV Inde méridionale: Madras, Maïsour, Courg, Cochin, Travancore. In his *Nouvelle Géographie universelle, La terre et Les hommes; VIII L'Inde et L'Indo-Chine....*; 511–78. Paris: Librairie Hachette et Cie.

RENFREW, COLIN
1979 *Before Civilization*. New York: Cambridge University Press.

RHIEM, HANNA
1900 Die Badagąs. *Allgemeine Missions-Zeitschrift* 27: 497-509.

RICE, BENJAMIN LEWIS
1898 *Epigraphia Carnatica, Vol. 4.* Bangalore: Mysore Government Press.

RICHARDS, FREDERICK JOHN
1920 Badaga and Toda Months. *Man* 20: no. 13: 23–5.

RIVERS, WILLIAM HALSE RIVERS
1906 *The Todas*. London: Macmillan and Co.

ROBERTSON, WILLIAM ROWNTRIE
1875 *A Report on the Agricultural Conditions, Capabilities, ana Prospects of the Neilgherry District*. Madras: E. Keys, at the Government Press.

ROOKSBY, RICHARD LIONEL
1961 The Kurumas ot Malabar. London: University of London; Ph.D. dissertation in Anthropology.

ROY, DAVID
1963 The Megalithic Culture of the Khasis. *Anthropos* 58: 520–56.

ROY, J.K., AND SUHAS KUMAR BISWAS
1964 Proteins in the Diets of some Indian Tribes. *Science and Culture* 30: 126–9.

SABERWAL, SATISH
1971 Regions and their Social Structures. *Contributions to Indian Sociology* (n.s.) 5: 82–98.

SAHLINS, MARSHALL DAVID, AND ELMER ROGER SERVICE
1960 *Evolution and Culture*. Ann Arbor: University of Michigan Press.

SAKTHIVEL, S.
1976 *Phonology of Toda with Vocabulary*. Annamalainagar: Annamalai Unversity.
1977 *A Grammar of the Toda Language*. Annamalainagar; Annamalai University.

SAMIKANNU, C. PAUL
1922 Function of Religion among the Badagas. *Madras Christian College Magazine* (n.s.) 2: 26–38.

SAVORY, ISABEL
1900 Ootacamund and Anglo-Indian Life. In *A Sportswoman in India: Personal Adventures and Experiences of Travel in Known and Unknown India*; 325–52. London: Hutchinson & Co.

SAXTON, GEORGE HARPER
1870 A Set of Iron Implements, etc., Found in a Cromlech in the Estate of Major Sweet in the South of the Nilghery Plateau. *Asiatic Society (Calcutta), Proceedings* 1870: 52–4.

SCHEBESTA, PAUL
1927 *Among the Forest Dwarfs of Malaya*. London: Hutchinson.

SCHERMAN, LUCIAN
1942 Von Indiens 'Blauen Bergen' (Nilgiri): Kurumba—Irula—Paniyan. *Journal of the American Oriental Society* 62: 13-35.

SCHMID, BERNHARD
1837 An Essay on the Relationship of Languages and Nations. Of the Dialect of the Todavers, the Aborigines of the Neelgherries. *Madras Journal of Literature and Science* 5: 155–8.

SEN GUPTA, P.N.
1980 Dietaries and Nutrition, A Survey of the Regional Tribes of India. In *Man and His Environment* (Indera Pal Singh and S.C. Tiwari, eds.); 133–75. New Delhi: Concept Publishing Company.

SEWELL, ROBERT
1932 *The Historical Inscriptions of Southern India (Collected till 1923) and Outlines of Political History*. Madras: Madras University.

SHAPIRO, MICHAEL C., AND HAROLD F. SCHIFFMAN
1981 *Language and Society in South Asia*. Delhi: Motilal Banarsidass.

SHARMA, ABHIMANYU
1963 Negrito Problem in India: New Perspectives. In *Anthropology on the March* (L.K. Bala Ratnam, ed.); 89–103. Madras: The Book Centre.

SHORTT, JOHN
1868 *An Account of the Tribes on the Neilgherries, by J. Shortt....and a Geographical and Statistical Memoir of the Neilgherry Mountains, by the Late Colonel Ouchterlony.* Madras: Higginbotham & Co. (Reprinted, 1869, in *Transactions of the Ethnological Society of London* 7: 230–90.)

SINGH, S.P.
1973 Unsolved Riddle of a Dying Race....*Sunday Standard*, August 26: Magazine Section, 34, no. 34:I.

SINHA, SURAJIT
1957 Tribal Cultures of Peninsular India as a Dimension of Little Tradition in the Study of Indian Civilization: A Preliminary Statement. *Man in India* 37: 93–118.

SKEAT, WALTER WILLIAM, AND CHARLES OTTO BLAGDEN
1906 *Pagan Races of the Malay Peninsula*. London: Macmillan and Co.

SLOBODIN, RICHARD
1978 *W.H.R. Rivers.* New York: Columbia University Press.

SMITH, CAROL A.
1976 *Regional Analysis, Vol. 2 Social Systems*. New York: Academic Press.

SRINIVAS, MYSORE NARASIMHACHAR
1966 *Social Change in Modern India.* Berkeley and Los Angeles: University of California Press.

SRINIVASAN, K.R., AND N.R. BANNERJEE
1953 Survey of South Indian Megaliths. *Ancient India* 9: 103–15.

STANFORD, JOHN KEITH
1962 *Ladies in the Sun: The Memsahibs' India, 1790–1860.* London: The Galley Press.

STONE, WINIFRED M.
1925 *Ups and Downs on the Nilgiris.* London: Church of England Zenana Missionary Society.

SUBBIAH, GOVINDARAJULU
1973 A Descriptive Analysis of Kota Language. Annamalainagar: Annamalai University, Ph.D. dissertation in Linguistics.

SUBRAHMANYAM, P.S.
1971 *Dravidian Verb Morphology: A Comparative Study*. Annamalainagar: Annamalai University (Department of Linguistics Publication 24).

SULLIVAN, JOHN
— See Anonymous 1819.

TAYLOR, PHILIP MEADOWS
1865 Description of Cairns [stone circles], Cromlechs [dolmens], and Kistvaens [cists]....in the Dekhan. *Royal Irish Academy, Transactions* 24: 329–62.

THURSTON, EDGAR
1898 Malagasy-Nias-Dravidians. *Madras Government Museum, Bulletin* 2: 119–27.
1906 *Ethnographic Notes in Southern India.* Madras: Government Press.

THURSTON, EDGAR, AND KADAMKI RANGACHARI
1909 *Castes and Tribes of Southern India.* Madras: Superintendent, Government Press; 7 vols.

TIGNOUS, HENRI-PIERRE-JOSEPH-ARTHUR
1911 In the Nilgherries. *Illustrated Catholic Missions* 26: 99–102, 116–19, 154–57.

TUDEN, ARTHUR, AND CATHERINE MARSHALL
1980 Political Organization: Cross-Cultural Codes 4. In *Cross-Cultural Samples and Codes* (Herbert Barry III and Alice Schlegel, eds.); 117-45. Pittsburgh: University of Pittsburgh Press.

TURNBULL, COLIN M.
1965 *Wayward Servants: The Two Worlds of the African Pygmies.* New York: Natural History Press.

TURNER, VICTOR WITTER
1969 *The Ritual Process: Structure and Anti-Structure.* Chicago: Aldine Publishing Company.

UNITED PLANTERS' ASSOCIATION OF SOUTHERN INDIA
1960 *Planting Directory of Southern India.* 16th edition. Coonoor: U.P.A.S.I.
1979 *Planting Directory of Southern India.* 20th edition. Coonoor: U.P.A.S.I.

VIETS, MARINA
1970 Only 879 Left of Toda Tribe Hidden in South India Hills. *New Circle* 15, no. 8: 16–19.

WALHOUSE, MORETON JOHN
1873 On Some Formerly Existing Antiquities on the Nilgiris. *Indian Antiquary* 2: 275–8.

WALKER, ANTHONY RUPERT
1986 *The Toda of South India: A New Look.* Delhi: Hindustan Publishing Corp.

WARD, BENJAMIN SWAIN
1821 Geographical and Statistical Memoir of a Survey of the Neelgherry Mountains in the Province of Coimbatore Made in 1821 under the Superintendence of Captain B.S. Ward, Deputy Surveyor-General. In Grigg 1880: lx-lxxviii.

WATTS, DAVID
1971 *Principles of Biogeography.* New York: McGraw-Hill Book Company.

WEBER, MAX
1946. *From Max Weber* (Hans H. Gerth and C. Wright Mills, trans.) New York, London: Oxford University Press.

WHITEHEAD, HENRY
1921 *The Village Gods of South-India.* 2nd edition. Calcutta: Association Press (Y.M.C.A.)

WILLIAMS, JAMES L.H.
1969 The White Bison Country in the Palni Hills, Madurai District, South India. *Bombay Natural History Society, Journal* 66: 605–8.

WISER, WILLIAM HENRICKS
1936 *The Hindu Jajmani System.* Lucknow: Lucknow Publishing House.

WOOLF, LEONARD
1961 *Growing: An Autobiography of the Years 1904–1911*. London: Hogarth Press; New York: Harcourt, Brace and World.

ZVELEBIL, KAMIL VEITH
1973 *The Irula Language*. Wiesbaden: Otto Harrassowitz (Neuindische Studien, Bd. 2).
1979 *The Irula (Ẹ̄rla) Language*. Wiesbaden: Otto Harrassowitz; (Neuindische Studien, Bd. 6).
1980 A Plea for Nilgiri Areal Studies. *International Journal of Dravidian Linguistics* 9: 1–22.

Biographical Notes

THEODORE (Ted) ADAMS (b. 1944 in Kelso, Washington State), studied French at Linfield College, in Oregon, and then took M.A. at the University of Washington in Linguistics (1970) and a Ph.D. in Anthropology at the same university (1985). His work in India included linguistic study as a Fulbright Scholar in 1970-1, and research on the sociolinguistics of Gudalur Taluk, migration and change, during 1977-86. He has taught at the Universities of Washington and Pennsylvania.

NURIT BIRD-DAVID (b. 1951 in Beersheba, Israel) studied Economics and Mathematics at the Hebrew University, Jerusalem (B.A., 1974), and later specialized in Social Anthropology at Cambridge. Her field research on the Naikens of Gudalur Taluk was conducted in 1978 and 1979. She was awarded the Ph.D. in Social Anthropology by Cambridge University in 1983, and is now a Lecturer in Anthropology at Tel Aviv University.

FRANÇOIS BLASCO (b. 1939 in Meaux, France) studied at the Université Paul Sabatier of Toulouse, where he is now joint director of the Institut de la Carte Internationale du Tapis Végétal. From 1964 till 1974 he was attached to the scientific section of the Institut Français at Pondichery, during which time he made numerous visits to the Nilgiris for an extensive survey of the botanical geography. He was awarded the degree of Docteur es/Sciences d'État in 1971, and now also holds the position of Maître de Recherche with the National Committee for Scientific Research (C.N.R.S., France). He is the author of *Montagnes du Sud de l'Inde, Forêts, Savanes, Écologie* (1971), *Mangroves of India* (1975), and *Tropical Vegetation Mapping* (1982), as well as the main compiler of some important vegetation maps (South India, Cambodia, Indonesia, South America, etc.) and articles.

CHARLES HILTON BROWN (b. 1890 in Elgin, Scotland) was educated at St. Andrew's and London Universities before joining the I.C.S. He served in various administrative positions in South India from 1913 to 1934, and was during 1933-4 Collector of the Nilgiris. He was a prolific poet and novelist, and during 1940-6 worked in the Talks Department at the B.B.C. His written work ranged from serious critical studies, such as *Rudyard Kipling*, to radio talks and the light verse he contributed for *Punch*. His major novels were *Susanna* and *Ostrich Eyes*. He died in 1961, in Nairobi.

JAMES HENRY SPROULL COUSINS (1873-1956) was a prominent poet and playwright of the Irish literary revival. Born in Belfast, he went to Dublin as a young man and associated with 'AE', Yeats, Hyde and others. He wrote prolifically throughout the entire first half of this century. In 1908 he joined the Theosophical Society. In 1915 he and his wife left Liverpool for India, where he spent the rest of his life as an English professor, public lecturer and journalist, except for a year as poetry professor at the City College of New York, and a couple more years in Japan (where he was awarded the D.Litt. degree of Keio University). From 1918 to 1922, and again from 1933 until 1938, he was Principal of the Madanapalle Theosophical College, in Andhra Pradesh. From 1939 until 1943 he lived in the Nilgiris at Kotagiri. His most successful play was

The Racing Lug (1902). His *Collected Poems, 1894-1940*, was published at Madras. He was a careful craftsman whose work was influenced by Irish legend as well as by South Indian theosophy. His poem, 'Longwood Shola', recalls a delightful wood just outside Kotagiri, where he and Margaret E. Cousins often used to picnic.

MURRAY BARNSON EMENEAU (b. 1904 in Lunenburg, Nova Scotia) graduated from Dalhousie University in 1923, then took a B.A. at Oxford (1926; M.A. 1935). He did advanced linguistic work at Yale under Franklin Edgerton and E.H. Sturtevant (Ph.D. 1931), and then under Edward Sapir. Travelling around South India during the years 1935-8, he spent about eight months working with the Todas, as well as studying the Kota, Badaga, Kodagu and Kolami languages. Since 1940 he has been attached to the University of California at Berkeley, where he is now Professor Emeritus of Sanskrit and General Linguistics. He is the author of numerous papers on the languages and ethnography of the Nilgiri tribes, as well as of some of the most useful books published on those subjects: *A Dravidian Etymological Dictionary* (with Thomas Burrow, 1984; *Revised edition*), *Toda Songs* (1971), *Ritual Structure and Language Structure of the Todas* (1974), *Kota Texts* (1944-6), *Dravidian Borrowings from Indo-Aryan* (with Thomas Burrow, 1962), *Dravidian Comparative Phonology* (1970), *Dravidian Linguistics, Ethnology, and Folklore: Collected Papers* (1967), *Language and Linguistic Area* (Collected Papers, 1980), and *Toda Grammar and Texts* (1983).

PAUL EDWARD HOCKINGS (b. 1935 in Hertford, England) is Professor of Anthropology at the University of Illinois in Chicago. He studied Anthropology at the Universities of Sydney, Toronto, Chicago, Stanford and California, receiving the Ph.D. in Anthropology from Berkeley in 1965. He is the author of numerous articles on the Nilgiri Hills as well as three books, *Ancient Hindu Refugees, Badaga Social History (1550-1975)* (1980), *Sex and Disease in a Mountain Community* (1980), and *Counsel from the Ancients, a Study of Badaga Proverbs, Prayers, Omens and Curses* (1988). In addition he has helped produce several films, and has edited *Principles of Visual Anthropology* (1975), *Aezojinruigaku* (with Junichi Ushiyama, 1979), *A Bibliography for the Nilgiri Hills of Southern India* (Revised edition 1978) and *Dimensions of Social Life, Essays in Honor of David G. Mandelbaum* (1987). Since 1962 he has spent a total of three years amongst the Badaga community.

A. WILLIAM JEBADHAS (b. 1944 in Sankanankulam, Tinneveli Dist., T.N.) is a botanist who studied at both Annamalai University and the University of Madras (Ph.D., 1981). His dissertation was entitled *Ethnobotanical Studies on some Hill Tribes of South India*; and he has also published several articles on ethnobotany and sugarcane. He is currently a scientist attached to the Sugarcane Breeding Institute at Coimbatore.

DIETER BERND KAPP (b. 1941 in Heidelberg, Germany) is an indologist holding the degree of D.Phil. from the University of Heidelberg (1971; Habilitation, 1980). He was attached to the South Asia Institute of that University during 1971-82, and did research for over two years among various Kur̠umba groups of the Nilgiris. As well as having written several articles dealing with their languages, beliefs and customs, he is the author of three books: *Das Verbum paraba in seiner Funktion als Simplex und Explikativum in Jāyasīs Padumāvatī* (1972); *Ein Menschenschöpfungsmythos der*

Mundas und seine Parallelen (1977); and *Ālu-Kurumbaru Nāya"—Die Sprache der Ālu-Kurumbas. Grammatik, Texte, Wörterbuch* (1982).

HANS JÜRGEN VON LENGERKE (b. 1945 near Göttingen, Germany) studied Geography and English at the Universities of Hamburg, Bochum and Mainz. After writing a Master's thesis on the Indian Monsoon at Mainz (1969) he did research on the climate of the Nilgiris during 1972-3, and was awarded the degree of Dr Rer. Nat. by the University of Heidelberg in 1976. This study was subsequently published as *The Nilgiris Weather and Climate of a Mountain Area in South India* (1977). Besides the Nilgiri fieldwork, he has worked in Sri Lanka, Bangladesh and South-East Asia, and is now an administrator at the Volkswagen-Stiftung.

DAVID GOODMAN MANDELBAUM (1911-1987) was introduced to Anthropology at Northwestern University through Melville J. Herskovits. He earned his Ph.D. in Anthropology at Yale University in 1936, studying under Edward Sapir, Clark Wissler and Leslie Spier. In 1937 he began research on the Kotas, and for fifty years published articles on them and the other Nilgiri communities, as well as much work on Indian society and more general topics. Since 1947, he has taught at the University of California, Berkeley. His books include *Society in India* (1970), and *Human Fertility in India* (1974). He visited the Nilgiris for research on many occasions.

RAJALAKSHMI MISRA (b. 1930 in Secunderabad, Andhra Pradesh) received her education at the University of Madras (M.Litt. in Anthropology). Since 1959 she has been employed by the Anthropological Survey of India, and is currently Assistant Anthropologist at its Mysore office. She has published a number of papers on the ethnography of South India, as well as several books: *Mullukurumbas of Kappala* (1971), *Jains in an Urban Setting* (1972), *Nomads in the Mysore City* (with P.K. Misra and I. Verghese, 1971), and *Life and Culture of the Mala Ulladan* (with Santibhusan Nandi and I. Verghese, 1971). Her fieldwork with the Mullukurumbas of Gudalur Taluk was conducted mainly in 1965.

WILLIAM ALLISTER NOBLE (b. 1932 in Nagercoil, Tamil Nadu) received most of his primary education in the Nilgiris, before going to the United States. He began field research on five Nilgiri communities during 1962, and has returned to India a number of times since, mostly to work in Kerala and Panjab. In 1968 he was awarded the Ph.D. in Geography and Anthropology by Louisiana State University, and has since published many articles on the ethnography, architecture and human geography of the Nilgiris and Kerala. He is now working on a book about *sati*.

FRANK EUGENE POIRIER (b. 1940 in Paterson, New Jersey) was educated at Paterson State College and then pursued graduate studies in Anthropology at the University of Oregon (Ph.D., 1967). Having made the standard ethological study of the Nilgiri black langur, he has published a large number of papers on this and other species. His fieldwork in the Nilgiris was carried out in 1965 and 1966. More recently he has produced several books on biological anthropology, namely *In Search of Ourselves* (3rd edition 1981), *Fossil Evidence: The Human Evolutionary Journey* (3rd edition 1981), and *Human Evolution: Physical Anthropology and the Archaeological Record* (1972) and (with Suzanne Chevalier-Skolnikoff) *Primate Bio-social Development* (1977). Fieldwork with other primate species has taken him to Puerto Rico, Japan, China, St. Kitts, Micronesia, East Africa and Taiwan. Currently he is Professor of Anthropology at the Ohio State University, in Columbus.

Anthony Rupert Walker (b. 1940 in London) was educated at Tonbridge School, Osmania University, Hyderabad, and St. John's College, Oxford. After completing an initial field study of the Todas in 1962-3, he studied at the Institute of Social Anthropology, Oxford, under E.E. Evans-Pritchard, David Pocock, Edwin Ardener and others, and presented his Toda material (M. Litt., 1965). During the late 1960s he was constantly with the Lahu hill people of northern Thailand, and in 1972 was awarded the D.Phil. by Oxford University for his dissertation on them. He is currently Associate Professor Anthropology at the Ohio State University, in Columbus and has published over 60 articles and three books on the upland peoples of northern Thailand. At the same time he has been returning repeatedly to the Nilgiris, and after 25 years of fairly continuous study of the Toda his book on them, *The Toda of South India: A New Look*, appeared in 1986.

Index

Adam, F. 338
Adams, T. 319–34, 400
administration 11–17
Adondai 306
afterlife 244–7
agriculture 203, 206, 209–10, 236, 282, 284–6, 290, 293–8, 303–4, 317, 339, 372
Aiyappan, A. 305–6
All Karnataka Unification Sangha 324
Amherst, W. 338
Ammanni, L. 335–6
Anglo-Indians 14, 344–6, 350–3
animals hunted 79, 287–8, 291, 362, 374
areal studies 1–19, 138–41
army 338, 343, 352
Australia 347

Baber, T. H. 319
Badaga 2–7, 16–19, 38, 129–31, 145–7, 150, 153–62, 166–74, 176–84, 206–31, 236–41, 253, 278–9, 322–4, 339, 342, 355–6, 363–76
Badaga language 47–50, 133–43
balls 349
barter 252, 258, 262–3, 279, 286–7, 291, 317
Basel Evangelical Mission 355
Bassein 337
Besant, A. 343
Bevan, H. 350–1
bibliography 362, 378–99
Biligiri Rangan Hills 355
bilingualism 136–8
biome 367–70
Birch, DeB. 232
Bird-David, N. H. 249–81, 400
birth 201, 241, 252
Blasco, F. 20–78, 400
Blavatsky, H. P. 337, 343
bonnet macaque 88–91
boundary 323, 333
brahmin 222, 303
Breeks, J. W. 251–2, 256–7, 272, 277, 287–8
brideprice 312
British 2–3, 11–16, 186–7, 207, 306, 334–58, 362, 366, 374–6
bronze 122–3
Brown, C. H. 359–60, 400
Buchanan-Hamilton, F. 250–1, 284, 286
buffalo 189–91, 198–200, 202–4, 211
burial 102–32, 242, 272, 290, 293
Burton, R. F. 343–4, 348–9

calendar 217
Carthage 337
caste society 6–7, 15, 176–84, 306–7, 327, 331–2
cattle 290, 302–3; *see also* buffalo
cemetery 290, 293, 338, 345, 354
census 208, 232, 236, 264–6, 305, 322, 324, 326, 339, 352–3, 362, 365
Channel Islands, 347
Chettiar 300; *see* Wynaad Chetti
Christianity, Christians 173, 188, 204, 308–9, 343, 355–6
chronology 127–9
cinchona 321
circles, stone 104–13, 118–20
cist 114–15
Citragupta 245–7
clan 193–5, 202, 211, 214, 233–4, 307–10
class, social 344–7
Cleghorn, H.F.C. 286
climate 24–43, 218, 319, 342, 367–8
Cockney 344
coffee 259–60, 264, 266, 290, 298–300, 321–2
Coimbatore District 324–5, 336–8, 340
collector 17
communitas 328–30
conservation 99–100
Coonoor 298, 345, 351
council 193–4, 214, 314, 341
Counsins, J.H.S. & M.E. 343–4, 357–8, 400–1
Cyprus 347

dairy 190–2, 197–200, 204
Darjeeling 338
Dawson, H.R., 298
day 215–20, 271
debt 252, 271, 279, 306–7
deforestation *see also* timber deity 204, 220–8, 230–1, 242–7, 272–3, 288 301–2, 306, 315–8
De:na:du 342
Devala 321
dialect 141–2
Diffloth, G.F. 323
divination 301, 315, 317
divorce 311–13, 315

doctor *see also* therapist
dolmen 115–18, 123–30
Dravidian *see also* language
Dubois, J.A. 250–1, 287

ear-piercing 201
East India Co. 320, 335–42
ecology 99–100, 361–77
economic relations 190–2, 196–7, 207–9, 235–8
elephant 265, 288
elevation *see also* climate, topography
Emeneau, M.B. 133–43, 187, 198, 401
endogamy *see also* marriage
English *see also* British
English language *see also* 333, 344, 351
entertainment 348–50
ethnicity 327
Europe 102–3
exogamy *see also* marriage

family 194, 211, 271–2, 308, 317
farming *see also* agriculture
festival 269, 272–3, 300–3
Finicio, J. 2, 186, 188–9, 207, 336
fire 267, 271, 284
Fletcher, F.W.F. 252–3, 257
flora 45–83, 265–6, 286–7, 290–2, 337, 362
Folke, S. 18
food 285–8, 295–8, 303, 311, 372–6
foraging 268–71, 282, 284, 290–1, 304
forest *see also* flora, timber
Francis, W. 252
French 345–7
funeral 202–3, 228–31, 238–242
Fürer-Haimendorf, C. von, 250, 254–5, 292, 366

Gandhi, M. K. 343
Gardner, P. M. 255, 300
geography 1–63
giant squirrel 86
Gir Valley 258–66, 273
gods *see also* deity
gold 259, 320
Gollan 153, 166
Gopallan Nair, C. 249
government 191–2, 237, 263, 290–2, 299, 321, 324, 327–9, 335–42, 345–6, 374–6
grey langur 95–99
Grierson, G. 323
Grigg, H. B. 252, 271–2
Gudalur 18, 264, 319–34
Gulliford, H. 347, 351, 355–6
gun 292, 316

hamlet *see also* village
Handley, M. A. 345
Hanuman monkey 95–9
Harijan 307–327
Harkness, H. 284–8
hartal 329
headman 193, 214, 238, 242, 273, 314–5, 340–1
health *see also* illness, therapist
herding *see also* buffalo
Hette 223–8
Hill Pandaram 270
Hinduism 204, 242–7, 288, 300–4, 306, 317
Hockings, P. E. 206–48, 279–80, 335–77, 401
house 189, 193, 218–20, 267–9, 284, 288, 292, 294, 316, 326–7
Hoysalas 187
Hull, E.C.P. 347–8, 353–4
hunt 310, 316–17, 350, 373–4; *see also* trapping

illness 273–4, 370; *see also* therapist
immigration 320–30, 333–8, 352–3, 364
Indo-German Project 18–19, 191–2
industrialization 340
infanticide 341, 365
Irish 344–6
iron 122–3
Irula 282–304, 363–75
Irula language 138–43

Jebadhas, A. W. 282–304, 401

Kannada language 257, 323–4, 326, 342
Kapp, D. B. 232–48, 254, 277, 401–2
Kāttu Nāyaka *see also* Naiken
Kelso, W. 338
Keys, W. 286–7
Kil-Kotagiri 299–300
Kindersley, N. W. 337
kinship 195, 212–3, 261, 308–13
Kolitippe 241
Koppayur 282, 288–9, 293, 298–300
Kota 2–7, 16, 19, 38, 129, 131, 144–85, 206–9, 238, 339, 341–2, 363–76
Kotagiri 298, 337, 340, 345
Kota language 133–43
Kottayam 306
Kriṣṇa 301–2
Kurichiya 305–7
Kuruba 232–3

Kuṟumba 2–7, 16, 19, 128–31, 145–7, 149–53, 208, 232–55, 266–9, 271–2, 274–80, 303, 305–20, 323, 338, 363–75
Kuṟumba language 133–43, 232–3

land 290, 299, 304, 306–7, 316–17, 320, 327–9, 331, 337–8, 341, 375–6
landscape 3, 17, 20–4, 79–82, 264, 307, 348, 362
language 2, 5, 10, 133–43, 323–6, 330–3, 353–4, 370–1
Lear, E. 344
Lengerke, H. J. von 20–78, 402
Leschenault de la Tour, J.B.L.C.T. 337
library 342
lineage 213, 293, 307–8, 316
Ling, C. F. 13–14
Lingayat 222
Linguistic Provinces Commission 323–4
lion-tailed macaque 84–8
London 335
lowland castes 162–6
Luiz, A. A. D. 305–6

Macartney, G. 335
Macaulay, T. B. 11, 343
Macleane, C. D. 206, 282
magic 149–51
Malabar 325
malaria 320–2, 325, 331, 370
Malayalam language 357, 305, 323–6, 330, 333
Malayalee 325–9
Maleswara 272
Malinowski, B. G. 280
Mandelbaum, D. G. 1–19, 144–85, 366, 402
Mann, H. 298
Mappila 260–3, 306–7, 327
marriage 195, 201–2, 211–3, 252, 255–6, 271–2, 309–14, 347, 350–1, 353–5, 370–1
massacre 150, 240, 338
Mauntadan Chetti 367, 369
Mbuti 279
medicine, folk 84–5
megalith 102–32
memorial ceremony 368, 371
merchant 306, 322, 331–2, 345–7
meteorology *see also* climate
Mettukal 303
Metz, J. F. 305, 355, 361
millet 282
Misra, R. 305–18, 366, 402
missionary 13–14, 168, 173, 186, 188, 343–7, 355–6
modale 273
modernization 317–18, 333–42, 374
monastery, nunnery 364
monkey 79–101
monsoon 24–40, 62
Montessori, M. 343
Morgan, H. R. 234–5, 252, 270, 287
Muḍuga 254, 366
Mudumalai 261, 375
Mullukurumba 254, 305–18, 366
mund 189
music 268–9, 291, 303
music hall 349–50
mythology 223–8, 280

Naiken 249–81, 367–9
Naiken language 257
Nair, Nayar 306, 314, 316
naming 201
Nehru, J. 343
Nelliyalam 259–60, 264–5
New World plants 297–8
Nilgiri langur 91–5
Noble, W. A. 102–32, 242, 282–304, 402
nomadism 265–6, 284–5

O'Flaherty, W. D. 223
Ootacamund, Ooty 203, 214, 264, 322, 325–6, 328, 336–51
Ootacamund Lake 338–9
Ooty Club 350
Ouchterlony, J. 285–7
Ouchterlony Valley 321

Paliyan 255, 300
Pandalur 259, 264, 268
Paniya, 136, 319–20, 323, 326, 367–9, 372–5
paramount chief 214
phratry 214, 233–4
plantation 11–12, 38, 59, 259–65, 270, 290, 298–300, 304, 321, 353–5, 374
Plantation Labour Act 260, 262
planter 343, 346–7, 353–4
plants *see also* flora, timber
poetry 357–60
Poirier, F. E. 79–101, 402–3
police 328–9, 338
politics 328–9
pollution hut 200–1
polyandry 194, 261, 310, 364–5, 370, 377
population 2, 17–18, 145–6, 32, 325–6, 362–5
pottery 119–22
prayer 223, 228–31

prehistory 102–32
priest 234, 239, 282, 291, 300–3
primates 79–101
puberty 272
purity 200

racing 350
racism 351, 354
radiocarbon 127–8
railway 340, 345–6
rainfall *see also* climate, monsoon
Rangawsami Betta 282–3, 290, 300–3
ranking 331–2
Reclus, E. 339
region 1–19
Rivers, W. H. R. 188, 277
road 319, 321, 340
Roman coin 127
rubber 259–62, 264–5

sacred objects 268, 274, 293, 300–2
sanitarium 338
Sanskritization 331
sati (Suttee) 125–7
Savory, I. 350
school 263, 292, 316, 333, 342, 346–7, 356
Scottish 344
sculpture 123–7
season 217–18, 294
settlement pattern 189, 206, 234–6
shaman 273–4
Shortt, J. 286–8
Simla 337
Siva (Shiva) 301–2
slavery 351, 369
slender loris 83
social system 174–84, 192–5, 222–3
soils 43–6, 294
sorcery, 149–51, 234, 237–41, 251, 253, 338
soul 244–7
South Dravidian 132–3
squatter 328–9, 375
Sri Lanka (Ceylon) 260, 325
Srinivas, M. N. 331
suicide 309, 329
Sullivan, H. C. 336, 342
Sullivan, J. 51, 335–43
Sullivan, L. 336
Sullivan, S. J. 335-6
Sullivan's Ghat 340

Tamilian 260, 324–7, 356
Tamil language 63–78, 323, 326, 330–1, 333, 337, 342
tea 290, 298–9, 321–2, 339–40
Telugu language 325–6
temperature *see also* climate
temple 292–3, 304, 314–16
Te:ty 337, 340
therapist 237, 239, 273–4, 278, 310; *see also* medicine, illness
Thurston, E. 299
timber 209, 236, 260-1, 265, 284, 339, 376
time 215–18
Toda 2–11, 16, 19, 28, 39, 128–31, 145–9, 167–70, 186–209, 253, 337, 339–41, 363–77
Toḍa language 133–43, 187
tool 267–9, 285–7, 291, 294, 304
topography 21–4
tourist 203
town 166, 203, 319–34
trader *see also* merchant
trapping 268, 284, 287, 290–1, 304, 310
tree shrew 83, 101
Turnbull, C. 279

Ūrali 254, 317

vegetation *see also* flora, timber
village 189, 193, 206, 211, 234–5, 255–7, 263–7, 273, 284

Wainad *see* Wynaad
Walker, A. R. 186–205, 403
warfare 366
watchman 237–9, 278–9
weather 24–43
wedding *see also* marriage
week 215
Wellington 351
Wellington, Duke of 343
Wenlock Downs 375
West, Lady 337
Whish, J. C. 337
witchcraft *see also* magic, sorcery
world view 220–2, 243
Wynaad 249–81, 305–34, 340, 366
Wynaad Chetti 136, 305–7, 320, 324, 366
Wynaad Gauḍa 259, 261, 264, 366

Yerukala/Korava language 136

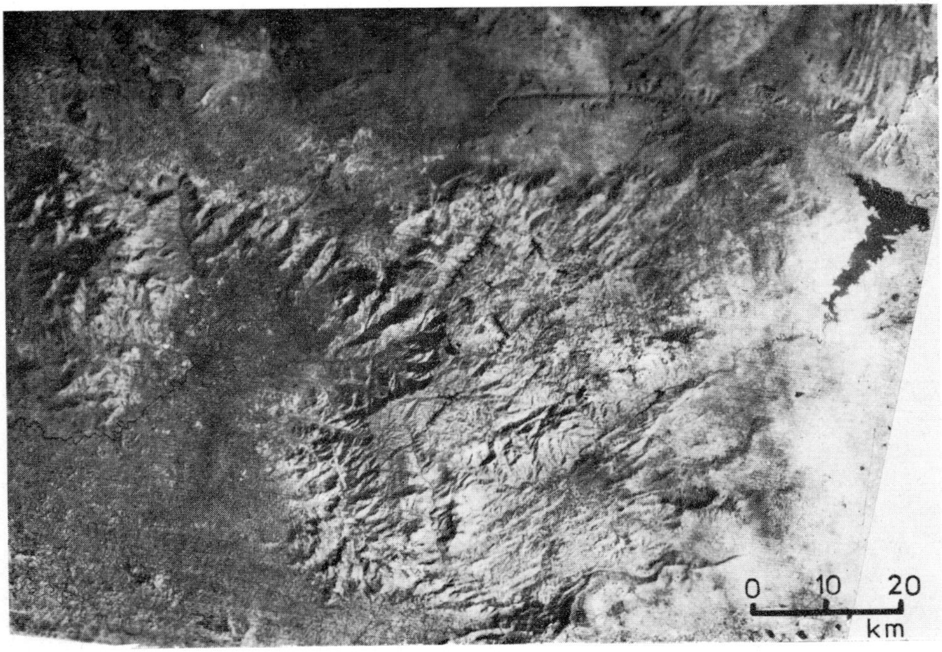

1 The Nilgiris from Space.
Portion of a LANDSAT-1 image (spectral band 7), taken on 23 January 1973, between 9 and 9.30 a.m. centred on the Kundah Range, showing main morphological and hydrographical features including escarpments of Mysore Plateau (Nilgiris-Wynaad) and Western Slopes of the Nilgiris rising above Nilambur Basin (Malabar Plains, southwestern sector), the Moyar 'canyon' running parallel to the Sigur Ghat at a distance of 10–15 km, the major reservoirs on the western Nilgiris Plateau (at low storage level), the Pillur reservoir at the foot of the Coonoor Ghat and the large Bhavanisagar below the (slightly obscured) Eastern Slopes (*Courtesy*: NASA, through EROS Data Centre, Sioux Falls, S.D., U.S.A.; ID Code 81184044505A000).

2 View from the Coonoor Ghat towards Mettuppalaiyam on the Coímbatore Plains. Foreground: Adderley Estate (tea and coffee plantation). Centre: largest contiguous areca palm (*Areca catechu* L.) plantation in South India, and adjacent paddy fields irrigated from the Coonoor River.
(23 March 1973; elevation: 1600 m. This and the following photos were taken by H. J. von Lengerke)

3 Patches of evergreen montane forest ('sholas') in the Upper Bhavani area (3 km NW of Korakundah). Sholas are confined to depression and ravine and surrounded by open grassland. Slope and hilltop to the right are afforested with *Acacia* sp. Note the fire protection line scraped into the savanna below the hilltop.
(9 February 1973; elevation: 2400 m).

4 Vegetation of difficult terrain. Typical distribution of montane forest vegetation in the Upper Bhavani area with rock outcrops even limiting grassland. Failure of *Acacia* plantations due to hard frosts and severe drought on the savanna; 3 km west of Upper Bhavani Dam, on the road to Western Catchment No. 1.
(17 March 1973; elevation: 2320 m).

5 In the Kundah Range. View across the small, westward exposed catchment above the steep Western Slopes towards Mukurti, from the road to W.C. No.2. Isolated *Rhododendron nilagiricum* Zenk. dotting the extensive savannas as 'pioneer' arboreal species of a small shola. The dense mist below the 'edge' of the Nilgiris Plateau is due to up-slope wind before the onset of the SW Monsoon. (25 May 1973; elevation: 2350 m).

6 Upper Bhavani Reservoir. Low water storage level towards the end of the dry season. The maximum storage level is indicated by the barren slopes below the savannas and the *Acacia* plantations in poor state.
(3 March 1973; elevation: 2300 m).

7 Tree fern, *Alsophila latebrosa* Hook., in the shrubby stratum of a shola, on the eastern slope of the Dodabetta Range, indicating perhumid microclimate inside the montane forest.
(4 April 1973; elevation: 2280 m).

8 Physiognomy of a shola tree. Umbrella-shaped crown with gnarled, leafless branches and dense peripheral foliage due to strong winds (SW Monsoon) and intensive solar radiation (NE Monsoon period) on the western Nilgiri Plateau, on the Upper Bhavani Dam—W.C. No. 1 road.
(3 March 1973; elevation: 2360 m).

9 Fire in a shrubbery savanna, 3 km northwest of Korakundah Estate, in the Upper Bhavani area. (17 March 1973; elevation: 2300 m).

10 Early morning hoar-frost. 'Frost basin' of the Wenlock Downs, 9 km northwest of Ootacamund, surrounded by frost-free hillocks carrying large forest plantations (*Eucalyptus globulus* Labill.), from the Ootacamund-Pykara road, at 7 a.m.
(21 January 1973; elevation: 2200 m).

11 The Wenlock Downs. Gently undulating terrain with shola, grassland, introduced (ornamental) tree species along swampy depression, and the recently opened farmland on the Sigur-Pykara watershed. (12 August 1972; elevation: 2200 m).

12 Intensive agriculture. Furrow-irrigated cabbage and (towards forest) potato cultivation on the lower division of Marvahulla (Coffee) Estate, next to the perennial Kukalthoraihalla. Here potato frost damages occurred as low as 1420 m in December 1971.
(17 February 1973; elevation: 1450 m).

13 Features of Megalithic Sites (A-I).

14 Pottery, Terracotta and Bronze Finds (1–38).

15 Carved Orthostats (A-H).

16 Po.*s*: A Traditional Toda Hamlet.

17 Modern Toda Girls in Tamil-style Dress.

18 Toda Woman.

19 Toda Man.

20 A Toda Herdsman become Farmer.

22 Irula Cultivation and Worship (A–H).

21 Irula Sites (A–H).

MAP OF A ... THE N...

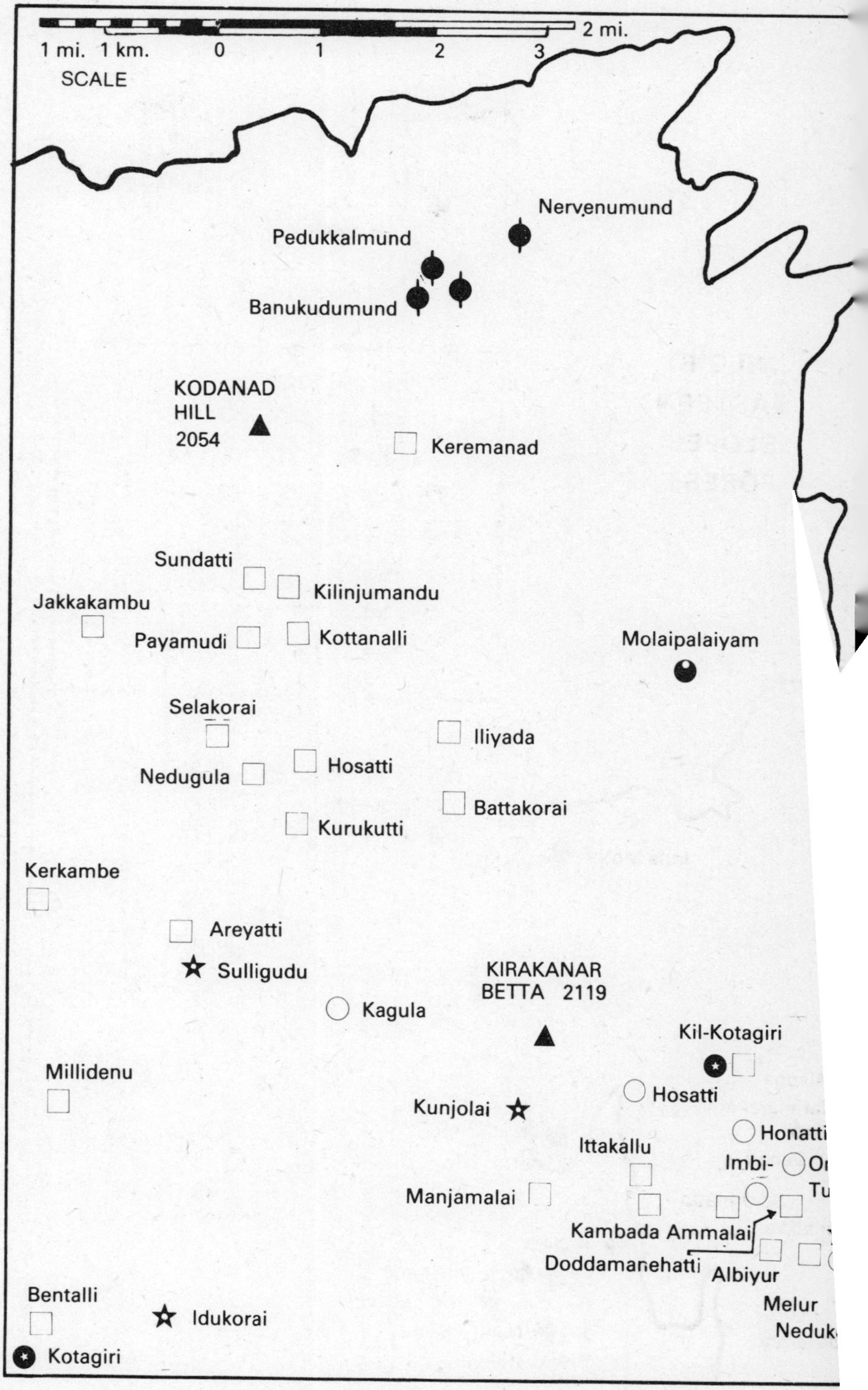

2 mi.
1 mi. 1 km. 0 1 2 3
SCALE
Nervenumund
Pedukkalmund
Banukudumund
KODANAD
HILL
2054
Keremanad
Sundatti
Kilinjumandu
Jakkakambu
Payamudi
Kottanalli
Molaipalaiyam
Selakorai
Iliyada
Hosatti
Nedugula
Battakorai
Kurukutti
Kerkambe
Areyatti
Sulligudu
KIRAKANAR
BETTA 2119
Kagula
Kil-Kotagiri
Millidenu
Hosatti
Kunjolai
Honatti
Ittakallu
Imbi-
Manjamalai
Kambada Ammalai
Doddamanehatti
Albiyur
Melur
Bentalli
Idukorai
Kotagiri